T3-BGM-171

WAR IN THE AGE OF TECHNOLOGY

THE WORLD OF WAR

GENERAL EDITOR
Dennis Showalter

SEEDS OF EMPIRE

The American Revolutionary Conquest of the Iroquois

MAX M. MINTZ

HOW EFFECTIVE IS STRATEGIC BOMBING?

Lessons Learned from World War II to Kosovo

GIAN P. GENTILE

WAR IN THE AGE OF TECHNOLOGY

Myriad Faces of Modern Armed Conflict

EDITED BY GEOFFREY JENSEN AND ANDREW WIEST

EDITED BY
GEOFFREY JENSEN
AND ANDREW WIEST

WAR IN THE AGE OF TECHNOLOGY

Myriad Faces of Modern Armed Conflict

NEW YORK UNIVERSITY PRESS
New York and London

Library of Congress Cataloging-in-Publication Data
War in the age of technology : myriad faces of modern armed conflict /
edited by Geoffrey Jensen and Andrew Wiest.
p. cm. — (The world of war)
Includes index.
ISBN 0-8147-4250-5 (cloth) — ISBN 0-8147-4251-3 (paper)
1. War. 2. Military art and science—History—18th century.
3. Military art and science—History—19th century. 4. Military
art and science—History—20th century. 5. Technology—History—
18th century. 6. Technology—History—19th century. 7. Technology—
History—20th century. I. Jensen, Geoffrey, 1965– II. Wiest, Andrew,
1960– III. Title. IV. Series.
U21.2 .W368 2001
355.02—dc21 00-012556

New York University Press books are printed on acid-free paper,
and their binding materials are chosen for strength and durability.

Manufactured in the United States of America

10 9 8 7 6 5 4 3 2 1

For our parents

CONTENTS

✦ ✦ ✦ ✦ ✦ ✦ ✦ ✦ ✦ ✦ ✦

ACKNOWLEDGMENTS

The authors wish to thank several people and institutions for aiding us in a multitude of ways in the production of this work. Above all, we would like to thank our colleagues in the Department of History at the University of Southern Mississippi. Led by Orazio Ciccarelli and Charles Bolton, the former and present chairs respectively, our colleagues in the history department have recognized the worth of military history and have done much to support and augment our endeavors. With their support we were able to found the Center for the Study of War and Society through which this book was conceived and produced. The support of our colleagues has also made possible a regular faculty exchange with the Royal Military Academy Sandhurst, which was equally important in making this book possible. We would like to thank everyone at Sandhurst for helping make our teaching stints there so productive and enjoyable, but we are especially grateful to Matt Midlane, the late John Pimlott, Duncan Anderson, and Paul Harris for their contributions to the faculty exchange. We would also like to extend our thanks to Tim Hudson, the Dean of International and Continuing Education at USM. His support was crucial in the founding of the exchange with Sandhurst and in our Study Abroad Program in Vietnam—which itself was instrumental to our research on Vietnam and to the chapter on post-traumatic stress disorder in this book. The editors and staff at New York University Press, most notably Niko Pfund and Despina Gimbel, also deserve many thanks for their hard work in making this book a reality. And finally, the authors would like to thank their wives—Laura Morgan and Jill Wiest—for their unfailing support during the work for this project and in our lives.

INTRODUCTION

The Meaning of War in a Technological Age

Geoffrey Jensen

Many military writers view the impact of modern technology on warfare with a mixture of fascination, dismay, and grudging respect, as if the dawn of a new age brought with it an inevitable but nonetheless tragic decline in the very skills and values that once made the waging of war an esteemed art. As William H. McNeill has written, the "technology of modern war, indeed, excludes almost all elements of muscular heroism and simple brute ferocity that once found expression in hand-to-hand combat." The supreme danger in this development, he continues, lies in the unpleasant truth that the "industrialization of war, scarcely more than a century old, has erased the old realities of soldiering without altering ancient, inherited psychic aptitudes for the collective exercise of force."[1]

Such concerns are not new. In fact, even before the atomic horrors of Hiroshima and Nagasaki, the English military theorist J. F. C. Fuller compared military invention to "a Frankenstein monster," warning that it was "destroying man's own work, his own culture, his own civilization, his past, his present and his future."[2] And Fuller—who was anything but a pacifist—had been preceded in this gloomy outlook by Friedrich Engels, who, together with the other founder of modern socialism, Karl Marx, so dramatically forecast the downfall of capitalism. Despite their ostensible enthusiasm for violent revolution, Engels and many socialists actually came to fear the growing destructive power of modern war and the threat it posed to Western civilization. A world war that took advantage of all technological and industrial advances, Engels reasoned, could annihilate everything in its path—including the achievements socialism and the working classes had managed to attain.[3]

Of course, technology of some kind or another has always been an essential ingredient in war, and advances in armaments raised concerns long before Engels. The Spartan king Archidamus, for example, reacted with alarm when he witnessed for the first time a weapon that could shoot darts through the air. "O Hercules," he is said to have exclaimed, "the valour of man is at an end."[4] Needless to say, Archidamus's appraisal was extremely premature, and such later innovations as the crossbow, new kinds of defensive fortifications, and even gunpowder failed to remove individual skill and bravery from their prominent positions in military affairs.

With the Industrial Revolution, however, technology's role in war and society seemed to reach a new level, threatening far more than the valued position held by individual heroism and skill in the "art of war." Humanity's new productive—and thus destructive—capabilities influenced not only military organizations and actions but also their relationships to crucial political, economic, and cultural developments. As the American Civil War soon demonstrated, victory in war could depend as much on each side's ability to harness industrial resources as on battlefield skills. In some cases, at least, technologies of production, weapons, communications, and human and material management seemed to make traditional military abilities and values largely irrelevant.

New technology could also have the effect of alienating soldiers from their comrades and from their own actions on the battlefield. Indeed, the concept of alienation, understood in its broader, psychosocial sense as well as in the strictly materialist terms associated with Marx, easily applies to the character of modern war. If, as Émile Durkheim, Max Weber, and other classical sociologists argued, industrialization brought with it a decline in traditional social norms and community values, then this development must have had an impact on armies. As war entered what one historian calls the "age of systems,"[5] individual soldiers, like factory workers, were more likely to perceive themselves as nothing but cogs in the wheel of a vast, complicated machine, in which their individual actions and heroism were even less important than before. At the same time, new technology and the growing importance of artillery in battle alienated soldiers even further from some of the modes of destruction they employed and their consequences, much as urban workers felt increasingly estranged from the products of their labors. As the invention of new weapons and tactics led to the phenomenon of the "empty battlefield," the notion of soldierly community suffered yet another loss to the onslaught of technology. Whereas classical armies had often worked in single, dense masses conducive to moral cohesion, new levels of firepower forced armies to adopt tactics of dispersion. These changes could create, in the words of one military analyst, a "profound psychological maelstrom."[6]

Curiously enough, at the time leading military theorists paid scant attention to the dramatic changes the new technological and industrial age brought to the art of war, perhaps in part because they could not bear to turn their backs on the sort of military thinking that had dominated the more "heroic" times in which they felt most comfortable. Even Carl von Clausewitz, while certainly aware that armed conflict had changed dramatically during the turmoil of the Napoleonic period he had just experienced, paid scant attention to the potential influences of new technologies on war. And the famous theorist Antoine de Jomini, who at the time was actually far more influential than Clausewitz, maintained that age-old strategic and tactical precepts remained essentially unchanged. Because the epoch of the French Revolution and Napoleon had represented a historical aberration, Jomini believed, the prosecutors of future wars had little need to consider the social or military ramifications of mass armies and popular nationalism.

In fact, thinkers whose primary interests lay in global and national economics grasped the deep significance of the new technologies on armies and war before most military writers did. Marx and Engels shared a strong interest in military history and affairs, and technology figured prominently in their many writings on the subject.[7] For Marx, military force and its uses easily fit into the basic outline of economic history. "Nowhere is the relationship between factors of production and the structure of society more clearly illustrated than in the history of the army," he wrote.[8]

Yet it was Engels above all who would happily combine his lifelong fascination with armies and war with the Marxist ideas he so fervently espoused, even as he also came to fear the destructive power of military technology. In the newspaper articles he wrote on the Crimean and American Civil Wars, for example, he repeatedly stressed weapons advances and industrial potential as crucial elements in helping achieve victory on the battlefield. And when he went beyond these journalistic endeavors to pen some of his more doctrinaire economic writings, he made it clearer still that he felt war should be analyzed primarily in terms of technological developments. "Generals of genius" have not radically altered warfare, he wrote. Instead, "the invention of better weapons and changes in the human material" lie at the bottom of military revolution. Thus, even in such theoretical tracts as his famous critique of the rival economist Eugen Karl Dühring, Engels deemed new developments such as breech-loading rifles, heavier guns, and four-inch-thick armor plating on warships worthy of discussion. Admittedly, he stressed that the inventions in themselves—and the tactical developments they fostered—were not so much military as economic in character. Thus the first uses of gunpowder and firearms did not simply represent "an act of force"; instead, these developments were "a step forward in industry, that is, an economic advance."[9]

Yet Engels did not stop there. He realized that the material changes in military affairs were closely linked to the societal aspects of armies and war, which in turn coincided as much with the Industrial Revolution as with the political and military turmoil Europe had experienced after 1789. Clausewitz had shrewdly grasped how the political unrest of the French Revolution and Napoleonic period had brought war to a near "absolute" state, thus leaving a permanent imprint on military art and science. But Engels went a step further, deeming technology and the economy just as important to future military leaders as the political transformations that had ushered in the era of mass armies and nationalism. Technology had not just altered the structure of armies themselves, adding to them such components as "a new and entirely industrial sub-section, the corps of engineers"; it had also accompanied the rise of universal male conscription and the modern army's concurrent transformation into a dominating and ultimately uncontrollable force. "The army has become the main purpose of the state, and an end in itself," Engels maintained, while "the peoples are only there in addition in order to provide and feed the soldiers." But, he reminded readers, "this militarism also carries in itself the seed of its own destruction," since arms races inevitably lead states to financial disaster. At the same time, the increasing militarization of society was "in the long run making the whole people familiar with the use of arms" and thus more liable to stage a successful revolution.

Engels then added a second lesson of history to his essay, thereby bringing together all the military, socioeconomic, and technological elements that comprise the investigative framework scholars now call "War and Society." In his eyes,

> the whole organisation and method of fighting of armies, and along with these victory or defeat, proves to be dependent on material, that is, economic conditions; on the human material, and the armaments material, and therefore on the quality and quantity of the population and on technological development.

Of course, Engels was hardly the first person to emphasize the historical relationship between war, technology, and society. Not only was it a logical corollary to socialist thought, but—as he himself noted—military officers themselves openly acknowledged the growing impact of material and economic factors on the outcome of wars. It is not surprising, then, that the relationship between war and technology also played an important role in subsequent writings by Engels—and later by Rosa Luxemburg and V. I. Lenin—on imperialism, armaments production, and other aspects of world capitalism. It would underlie, moreover, the subsequent rise in appreciation for the operational level of war, to which Soviet military theorists would devote so much thought. They realized that an age of capitalist technology and industry requires a new conceptual category, "operations," which belongs somewhere in the traditional division between strategy and tactics. As Niall Barr shows in his contribution to this anthology, Soviet thinkers actually derived some of their ideas on operational art from studies of the final, militarily "successful" period of World War I—history's first truly industrial war.

The impact of technology on war in general was a major theme of interwar Soviet military thinker Aleksandr A. Svechin's opus *Strategy*. Svechin, who was the first person to employ the term *operational art* as a conceptual tool to bridge the gap between tactics and strategy, defined strategy itself as something akin to what Western theorists now call "grand strategy." It thus encompassed war in its economic, political, and geographical dimensions, which meant that technology and industry were crucial to its success.[10] Although he warned against trying to use technology to overcome weaknesses in morale, he repeatedly stressed its importance for war in general. But rather than portraying the opposition between "moral" and "material" forces in war as a crude dialectical contest, his writings suggest that he perceived a more complicated relationship between the two. "Modern military technology," he wrote, can serve as "a source for the consolidation of order, and a certain amount of morale is generated in the labor it mechanizes."[11] Hence even Svechin, who clearly felt technology had crucial consequences for military actions on the battlefield, did not neglect its more imprecise but equally important influence on the moral dimensions of warfare, which figured prominently throughout his military writings.

Some Soviet strategists subsequently moved still farther away from purely materialist interpretations of technology in war, emphasizing even more strongly the power of Clausewitzian morale over military equipment and technological developments. This paradox is not really surprising, since, as the historian Bernard Semmel points out, even the Soviet military thinkers' leading ideological ancestor, Engels, had revealed a basic contradiction in his writings on armies and war. In some of his most substantial, earlier works on military history, the younger Engels—whose keen interest in military affairs led friends to begin calling him "the General"—had paid remarkably little attention to the impact of economics and class conflict on war. His first writings on military history were overwhelmingly traditional in analysis, straightforward, and largely devoid of the sort of economic criticism now associated with "orthodox" Marxism and his own later works, emphasizing instead the role played by the traditionally cherished military attributes of leadership, tactical skills, organization, and morale in the winning of wars.[12]

In the twentieth century, moreover, some of the most forceful arguments for an economic-determinist view of war came from military theorists from the capitalist West, rather than from the military doctrines of the Soviet Union. As Semmel writes, a perfect example of such an apparent contradiction was the non-Marxist military historian and theorist J. F. C. Fuller, better remembered today as a prolific and often highly insightful writer who produced some of the earliest tracts on armored warfare, before falling into disrepute during the 1930s because of his links to British Fascists. In fact, Fuller's writings became a target for some Soviet military writers, who—apparently without irony—charged him with placing too much emphasis on material factors over morale.[13] His book *Armaments and History,* written during the final part of World War II, was the first major study in the English language of the historical relationship between technology and war by a preeminent military scholar. Although Fuller made mistakes and was rather disingenuous and self-serving at times in his book, at the base of his sometimes-flawed account is an intelligent and sweeping analysis of the impact of war and technology on human history.

Like so many historians before him, Fuller viewed the discovery of gunpowder as a crucial historical juncture, and he devoted much attention to its subsequent impact on political, religious, socioeconomic, and imperial developments. In his eyes, gunpowder "revolutionized not only the method of fighting, but also the way of living, and in consequence civilization itself." Eventually, he wrote, military demands fostered mass production of new armaments, "which, in turn, stimulated the growth of armies and the advance of Capitalism." In his eyes, Marx and Engels were thus "the apostles of the maturing wardom [*sic*] which was to engulf Christendom by unchaining mass man."[14] Fuller also quoted Jan Bloch, the brilliant turn-of-the-century analyst of the relationship between technology, industry, economics, and war, who had all-too-accurately forecast the trench stalemate and slaughter of World War I. "The soldier by natural evolution has so perfected the mechanism of slaughter," Bloch had written, "that he has practically secured his own extinction."[15] Fuller then took Bloch's prediction to its logical extension, fearing that the advent of the atomic bomb might fit horribly well into the historical destiny of war and technology. After all, "every great scientific discovery has either been turned to profit-making or to war, and profit-making is a breeding ground of war because in itself it is so largely war in the making."[16]

Like many contemporaries, Fuller confirmed the importance of technological advances in general on military art, stressing the oft-lamented decline in the worth of military "valor" that they fostered. But he also seemed to grasp that technology in war implied more than just advances in weaponry, and that a broader understanding of the term—to include technologies of human and economic systems and management—could serve historians well. A decline in the importance of military valor corresponded not just with the use of new weapons, Fuller argued, but with improvements in "military organization" and thus the rise of the role of human intelligence and "cunning" in war.[17]

The most fascinating aspect of Fuller's book is its attention to the moral dimension of armaments in history. He wrote his book shortly before the atomic horrors of Hiroshima and Nagasaki, subsequently adding a preface and additional chapter for the British edition that appeared in 1946. Not surprising for the time, the general theme of morality pervaded the entire work. He wrote that "when we think it out, it is the inventive genius of man which has obliterated his sense of moral values. From the javelin and the arrow to the super-fortress and the rocket-bomb, the very power to destroy, first slowly and then at terrific speed, has intoxicated man."[18] Significantly, he argued that the historical strengthening of the importance of human intelligence over traditional warrior skills corresponded to a decline in the moral influences that had once served to help limit war's destructive fury. He therefore believed the invention of gunpowder to have marked humanity's passage into the "technological epoch of war, the hidden impulse of which is the elimination of the human element both physically and morally, intellect alone remaining." Such reasoning led him to conclude pessimistically that war, "a prime mover in technological civilization," had come to replace religion in fulfilling the modern world's spiritual needs. Thus, in the present age of "wardom," men and women may tolerate the curtailment of liberties and comforts imposed by states geared for war—"so long as their animal instincts are first aroused." Totalitarian states, he believed, had already made this fact painfully clear.[19]

Many of Fuller's views were echoed in another thought-provoking work, the main body of which was also written shortly before Hiroshima, by Fuller's self-described "disciple" and a very famous military historian and theorist in his own right: B. H. Liddell Hart.[20] Largely in agreement with Fuller's observations on the relationship between technology and war, Liddell Hart also examined the meaning of modern war for the force that had just proved itself so powerful: nationalism. He wrote:

> The advent of "automatic warfare" should make plain the absurdity of warfare as a means of deciding nations' claims to superiority. It blows away romantic vapourings about the heroic virtues of war, utilized by aggressive and ambitious leaders to generate a military spirit among their peoples. They can no longer claim that war is any test of a people's fitness, or even its national strength. Science has

> undermined the foundations of nationalism, at the very time when the spirit of nationalism is most rampant.[21]

Much as Fuller related technological war to a general abasement of moral values, Liddell Hart linked the new, horrible level that the fighting of World War I had reached to a "decline of civilized behaviour." In fact, Liddell Hart's portrayal of war seems to suggest that he viewed it as a form of "play," which in a technological world of so much destructive potential desperately required curtailment through "manners." He concluded his book with a hopeful plea for civilized conduct, writing that "only manners in the deeper sense—of mutual restraint for mutual security—can control the risk that outbursts of temper over political and social issues may lead to mutual destruction in the atomic age."[22]

Quite clearly, some observers have long realized that a full understanding of the role of war and technology in history requires analysis of its military, social, political, and cultural aspects. It comes as no surprise, then, that two recent historical works on the subject—by William H. McNeill and Martin van Creveld—venture well beyond technical military history and employ broad and fruitful investigative approaches.[23]

This book, however, goes even further, with a very diverse mixture of topics that sometimes share little with one another except the context of an age when technology seems to reign supreme. The selection of chapters sheds light on the wide variety of approaches available to students of modern war in its totality, and the book's strength is meant to lie above all in this diversity. The ever-rising belief in the centrality of technology in war is a crucial perception that touches—in one way or another—on all aspects of modern military history, and it is for this reason that we chose to highlight the theme of technology. But we are also well aware that technology can only serve as an imperfect source of thematic unity, and that beyond the basic, age-old elements of violence and organized armed force, little really unifies a military age of total, limited, conventional, and nonconventional conflicts waged by various combinations of small and professional forces, mass and "amateur" armies, and even guerrilla bands.

We trace the origins of the modern age of technology in war to around the time of the French Revolution, when political turmoil in the West roughly coincided with the beginning of a dramatic spread of new technologies and industry. All war—like human history in general—may have been "technological" in a sense, but it was in the immediate wake of this period that technology became widely perceived as a dominant feature of life in general, with important consequences for the prevailing ideas of war and peace.[24] As Engels wrote, history has been profoundly shaped by the interaction between developments in technology, politics, economics, and society, and one cannot understand modern war without taking all these elements into account.

This interaction thus figures prominently in the first section of the book, which is devoted to general overviews of several basic problems of war in the age of technology. In the first chapter, Paddy Griffith surveys the link between technological innovation and military methods on the battlefield, shedding light on the evolution and frequently changing direction of what does not always appear to be a rational cause-effect relationship. Focusing on a subsequent historical period, Warren Chin then offers another interpretation of the same issue, examining as well the meaning of technological change for strategy, operational art, and post-1945 military affairs in general. Finally, Stephen Badsey's chapter analyzes one of the most fascinating but poorly understood aspects of modern military affairs: the profound influence of communications and the media on the conduct of war. Of course, military analysts, media critics, and even the popular press itself have devoted much attention in recent years to such buzzwords as "media war" and the "CNN effect." Moreover, "smart" weapons and the latest so-called revolution in military affairs have clearly changed the public's expectations of armies and war. But, as Badsey demonstrates, the influence of communications technologies on "the people"—which together with government and the military comprise Clausewitz's famous trinity of war—is far from new.

The next section of the book focuses on the epitome of truly modern, industrial-technological conflict: World Wars I and II. The Great War shocked the world with its ferocity and length, appearing to set the standard for total war. The death toll reached 9 million, as men and women from all over the world participated in what some commentators came to describe as Europe's suicide and the horrible, final climax of the Industrial Revolution. But, as G. D. Sheffield's chapter illustrates, in spite of the war's many undeniably novel aspects, the importance of such a "human factor" as soldiers' morale in battle continued to be crucial—regardless of how the new technologies otherwise changed the character of war. The British, it seems, stumbled on the idea of small-unit cohesion as a key unifying force of modern mass armies, which may explain why the morale of the British army stood reasonably firm in comparison to its German, Russian, and French counterparts. Sheffield's study is complemented by the chapter that follows, in which Chris McCarthy examines how soldiers on the Western Front spent their time while not in combat. Along with an unprecedented level of industrial-technological activity came alienation and boredom, making the life of a soldier even more like that of the factory workers Marx and Engels described. Army leaders thus devoted at least some of their endeavors to trying to control their men's activities off of the battlefield—whether in music halls, brothels, or elsewhere. Robert McClain then examines some oft-forgotten participants in the Great War, the British Empire's Indian Corps, revealing how total war can have social, cultural, national, and racial consequences for armies and the states they serve. When defensive firepower and fortifications helped turn the conflict into a longer, bloodier affair than military planners had expected, they reluctantly turned to Asia for more troops to help offset losses. According to McClain, the Indian Corps that was then sent to fight in Europe served a crucial—if neglected—military role in the first part of the war, and its history reveals much about other aspects of British and Indian history as well.

The last two chapters on World War I, by Brian Bond and Niall Barr, focus on the latter half of the conflict. Bond examines the historiographical life of what many consider one of the war's most exemplary battles and a tragic symbol of mass, pointless death at the hands of a military high command unable to grasp the new, deadly power of weaponry. He suggests that the popular view of the battle of the Somme as the epitome of unremitting slaughter and pointless sacrifice may have more to do with perceptions of Vietnam than with contemporary views and events surrounding the battle. Barr focuses on the last year of the war, when the British employed tactics, strategy, and even operational methods that demonstrated a good understanding of new technologies. Their learning process, he suggests, in turn helped lead the way to German blitzkrieg and Soviet operational art.

The second section of the book ends with two studies of World War II. In the first, Eric Bobo explores the links between technological innovation and politics, examining British scientific endeavors in the fields of rocketry and radar. Neil McMillen's study, in contrast, centers on the experiences of African Americans in the U.S. armed forces, revealing some parallels—and many differences—with the situation that Indians fighting for Britain faced in the previous world war. Relying primarily on oral sources, McMillen contends that African American soldiers—considered unfit for more technologically advanced tasks—did *not* consider themselves to be fighting for justice at home as well as abroad during the Second World War.

The last section of the book concentrates on the cold war and conflicts thereafter. The product of a joint effort by a military historian, a clinical psychologist, and a professor of social work with extensive experience in veterans affairs, the chapter by Andrew Wiest, Leslie Root, and Raymond Scurfield examines post-traumatic stress disorder (PTSD) from a historical and a medical perspective. As the marked contrast in the authors' approaches to the problem demonstrates, attempts to treat PTSD stand to benefit from increased historical awareness, while historians in turn can learn much about the realities of the wars they write about by examining current ways of dealing with its profound psychological and social effects.

Sean McKnight then looks at the history of the Iraqi armed forces, which failed so miserably in the face of the "high-tech" weapons of the United States military and its allies in the Gulf War. Analyzing the ability of the reputedly third-world power Iraq to prosecute industrial war in an age of high technology, this chapter validates Clausewitz's emphasis on the role of society and government in the prosecution of modern war. According to McKnight, social weaknesses in and outside the Iraqi armed forces have prevented the country from reaching its military potential, and U.S. military planners should not assume that their technological expertise and tools will always allow them to prevail against powers the size of Iraq. Finally, Michael Orr examines how, since the cold war, the Russian armed forces have dealt with the problems of postindustrial warfare. Focusing on the recent Chechen wars, he argues that a uniquely Russian approach to modern war—inherited from the Soviet Union—has combined with a failure to appreciate the continued importance of the human element in military affairs to produce disastrous consequences. This chapter and the one by McKnight offer views of recent military history, technology, and the future of war that at times conflict with those expressed by Chin much earlier in this book, thereby reminding us how far we remain from arriving at a general consensus on the character and fate of contemporary armed force and conflict.

It would be absurd to predict with too much certainty how future wars might be waged and the magnitude of their military and social impact. If the many, varied topics covered in this anthology have any common lesson, it is that modern technology has made armies, wars, and their effects increasingly complicated and unpredictable—regardless of any promises inventors may have made to the contrary. Indeed, huge investments in defense systems and technologies may fuel unexpected but profound social and political transformations. But more familiarity with the myriad faces of war in modern history will at least, we hope, help us confront whatever does come, at the same time strengthening our endeavors for peace whenever possible.

NOTES

1. William H. McNeill, *The Pursuit of Power: Technology, Armed Force, and Society since A.D. 1000* (Chicago: University of Chicago Press, 1982), viii.

2. J. F. C. Fuller, *Armament and History: A Study of the Influence of Armament on History from the Dawn of Classical Warfare to the Second World War* (London: Eyre and Spottiswoode, 1946), x. Fuller completed the main body of this book in March 1945. After the atomic bombs were dropped on Japan in August of that year, he added a preface and final chapter for the book's publication in Britain in 1946 (ibid., v).

3. Sigmund Neumann and Mark von Hagen, "Engels and Marx on Revolution, War, and the Army in Society," in Peter Paret, ed., *Makers of Modern Strategy from Machiavelli to the Nuclear Age* (Oxford: Oxford University Press, 1986), 262.

4. Cited in B. H. Liddell Hart, *The Revolution in Warfare* (London: Faber and Faber, 1946), 30.

5. Martin van Creveld, *Technology and War from 2000 B.C. to the Present* (London: Brassey's, 1991).

6. James J. Schneider, *The Structure of Strategic Revolution: Total War and the Roots of the Soviet Warfare State* (Novato, Calif.: Presidio, 1994), 19.

7. For an overview see Neumann and von Hagen, "Engels," 262–80.

8. Quoted in W. H. Chaloner and W. O. Henderson's introduction to Friedrich Engels, *Engels as Military Critic* (Manchester: Manchester University Press, 1959), xix.

9. This and the following quotations by Engels are from Friedrich Engels, "The Force Theory", in Bernard Semmel, ed., *Marxism and the Science of War* (Oxford: Oxford University Press, 1981), 49–57. (Originally appeared in Friedrich Engels, *Anti-Dühring; Herr Eugen Dühring's Revolution in Science* [1878], part II, chap. III.)

10. Recent discussions of Svechin's military thought and its impact include Jacob W. Kipp's introductory essay "General-Major A. A. Svechin

and Modern Warfare: Military History and Theory," in Aleksandr A. Svechin, *Strategy,* ed. Kent D. Lee (Minneapolis: East View Publications, 1992), 23–56; and Schneider, *Structure,* 129–61. I am grateful to Tim Bean of the Royal Military Academy, Sandhurst, for pointing me toward Svechin's writings on technology and war.

11. Svechin, *Strategy,* 178.

12. Semmel, ed., *Marxism,* 8–12.

13. Semmel, ed., *Marxism,* 34. In some ways, Winston Churchill's views on the relationship between technology and war preceded those of Fuller, as Eric Bobo's chapter in this book demonstrates. On Fuller, see Brian Holden Reid, *J.F.C. Fuller, Military Thinker* (London: Macmillan, 1987); A. J. Trythall, *'Boney' Fuller, the Intellectual General 1878–1966* (London: Cassell, 1977); and Azar Gat, *Fascist and Liberal Visions of War: Fuller, Liddell Hart, Douhet, and Other Modernists* (Oxford: Clarendon, 1998), 13–42. J. P. Harris, *Men, Ideas and Tanks: British Military Thought and Armoured Forces, 1903–1939* (Manchester: Manchester University Press, 1995), includes much critical analysis of Fuller's actual contribution to British military ideas on armored warfare.

14. Fuller, *Armament,* 93, 102–3, 124.

15. Bloch is quoted in Fuller, *Armament,* 126. On Bloch (also known by his Polish name, I. S. Bliokh), see Schneider, *Structure,* 86–92.

16. Fuller, *Armament,* 187.

17. Fuller, *Armament,* 60, 36.

18. Fuller, *Armament,* x.

19. Fuller, *Armament,* 83, 164, 175.

20. Liddell Hart, *Revolution.* Like Fuller, he added an epilogue after the dropping of the atomic bombs on Japan. He is quoted as a self-described disciple of Fuller in Harris, *Men,* 207.

21. Liddell Hart, *Revolution,* 33.

22. Liddell Hart, *Revolution,* 60, 93. I use the term play here in the manner of the late Dutch cultural historian Johann Huizinger, following the example Martin van Creveld sets in his chapter on "Make-Believe War" in *Technology,* 285–96.

23. McNeill, *Pursuit;* van Creveld, *Technology.*

24. Maurice Pearton, *Diplomacy, War and Technology since 1830* (Lawrence: University Press of Kansas, 1984). McNeill, *Pursuit,* 185–222, also stresses the dual impact of the French and Industrial Revolutions on war.

PART I

✦ ✦ ✦ ✦ ✦ ✦ ✦ ✦ ✦ ✦ ✦

TECHNOLOGY AND THE MILITARY

On and Off the Battlefield

CHAPTER 1

Infantry Armament and the Perception of Tactical Need, 1789–1918

Paddy Griffith

It has often been said that the need for a new weapon is identified by one set of people; then the weapon is designed and built by a completely different set of people; and finally, it is used in battle by a third group yet again, who will have had nothing to do with the deliberations of the first two groups.[1] Typically the first group will be politicians; the second group will be engineers; while only the third will be the fighting soldiers whose lives actually depend on just how well their tactical needs match up (or don't) with the weapon that is finally procured.[2] There are many opportunities within this fragmented structure for misunderstandings and poor analyses of real tactical needs, and hence for major wrong turns in procurement. These may involve the creation of costly white elephants that are useless to the frontline troops, or conversely, if there is excessive cost cutting and skimping, there may be an assumption that the frontline troops do not need to replace obsolete weapons with anything new at all.

Many cautionary tales may be told of the disasters that can occur. The whole history of the submarine from the American Revolution to the start of the First World War is a case of "aspiration unsupported by technical feasibility," accompanied, unfortunately, by a small yet continuing list of fatalities among those who volunteered for the service.[3] There was both a political and a tactical need for this weapon, but the engineers simply could not build one that worked. The same could be said of airpower, which represented an even more politically glamorous achievement, widely trumpeted when the first military balloon supposedly "helped" in the French victory at Fleurus in 1794. In practice, the aerial dimension proved incapable of delivering any meaningful tactical benefit before August 1914, when the early airplanes of the British Expeditionary Force proved to be rather better than cavalry at unmasking the German advance through Belgium and Picardy. Airpower never looked back from that moment onward, although it is difficult for us in the year 2000 to remember that this decisive turning point occurred some seventeen years later than the halfway point in the whole history of combat flying.

"Political aspirations" may interfere with weapons procurement in many other ways, for example, when the Congreve rocket was forced on an unwilling Duke of Wellington in 1813–15, with little reference to the true tactical need, merely by virtue of the enthusiastic inventor's personal influence with the Board of Ordnance. At Waterloo, this weapon proved distinctly two-edged in nature, causing as much consternation among friendly forces as among the enemy.[4] It would finally come of age on the battlefield, as a reliable instrument of massed terror bombardment, only after vastly more sophisticated fuses, explosives, and propellant fuels had become available during the Second World War.[5]

Then again, it was politics that distorted British procurement priorities through much of 1941, when the prime minister became fixated with the belief that if only a few fast convoys could carry a critical mass of tanks to Egypt via Malta (as opposed to the usual much longer route via Capetown), then the whole war in North Africa could be won almost instantly. Churchill believed that the tactical need in the desert was simply for more tanks, and so he strained all the resources of the Royal Navy, of the Royal Air Force (RAF), and especially of the tank-producing factories to provide what he saw as the necessary wherewithal. He failed to understand that the tanks were effectively quite useless, once they arrived at Alexandria, until trained crews had been found to man them, and until they had been laboriously fitted with dust filters and other essential modifications for local conditions of climate and terrain. Then, when the tanks were finally committed to action, it turned out they were incapable of delivering any useful results until the whole "bull-at-a-gate" tactical culture of the Eighth Army had itself been radically reformed—which would not happen until well into 1942.[6] Hence it was not merely Churchill who misunderstood the tactical needs of the army—it was the army itself. If it had studied the problem with a little less precipitation but in a little more depth, it would have understood that it needed not more tanks but better ones—together with better antitank guns—and a vital year of research and development might have been gained. As it was, the "definitive" British tank of the Second World War—the Centurion—entered the field only at about the same time as the Red Army was entering Berlin.

Also too late into the firing line was the "definitive" British infantry weapon of the same era: the innovative EM-2 "Bullpup" assault rifle (with a short 7 mm round). This was developed in the late 1940s as a direct result of combat experience, but it had to be abandoned almost immediately, due to the United States' doctrinaire political commitment to standardization on the long (and much heavier) 7.62 mm NATO round.[7] The story gained credence that the real reason for abandonment had been the perceived unsuitability of the squat EM-2 for drill outside Buckingham Palace; but this was doubtless a piece of official disinformation that attempted to capitalize on the common belief that the truth is normally stranger than fiction.

In an ideal world, the procurement process should surely always start with a careful analysis of the tactical need and only then lead on to decisions of financing, of engineering design, and of higher politics. In eighteenth-century Britain, this was well understood by the navy—by far the most technical service—which always tried to maintain as tight a supervision as possible over the building of any new warship. Not only would specialized naval architects be employed directly by the Admiralty—rather than by the contracting dockyards—but the man who was destined to captain a new ship would normally be intimately involved in every stage of its building. Some famous altercations with the dockyard authorities were the inevitable result, and the navy did not always win them. Nevertheless, the Admiralty at least understood that every detail of a ship of the line, which represented a major strategic unit, needed careful inspection by the men who had to use it and live in it, rather than merely by land-based civilian accountants or entrepreneurs.[8]

In the case of the army, however, matters looked considerably less technical, and the need for close supervision of manufacturing never seemed to be quite so pressing. Everyone knew what a "greatcoat" was supposed to be and how a ragged soldier was supposed to patch it up when it wore thin. Even if the cloth was not of the highest quality, it could still be of some use to its wearer and could always be supplemented by additional material obtained from civilian sources. Alternative sources of supply might admittedly be less easy to find in the case of muskets and ammunition; but even then there was no real perception of urgency. The scale of the strategic importance of any individual weapon was vastly less great than that of a capital ship. Thus, if any one musket were found to be defective, it might represent a loss in combat efficiency of only one man out of one hundred thousand (or 0.001 percent of the total army, as opposed to a capital ship's value of perhaps 1 or 2 percent of the total navy); hence it was scarcely a matter of concern to the Army Board. In any case, a considerable body of professional opinion believed that it was only the front rank that could deliver effective fire, and so, in a three-deep line, only one musket in three needed to be in working order. The final volley, when it came, was seen as only one out of three elements of shock action, the other two being a rousing cheer and a menacing charge with cold steel. If properly delivered in rapid succession, this trilogy was the best way to break an enemy's cohesion and chase him away. Nor is it difficult to find opinions that were still more extreme than this. The French Revolution produced many highly placed politicians and generals who genuinely believed that pikes and bayonets could win battles on their own, with no need for musketry at all. For them, presumably, it was not a matter of concern if even 100 percent of all the muskets were technologically useless![9]

In Napoleonic times the army, unlike the navy, saw its main tactical asset as its drill, discipline, and cohesion, rather than any aspect of its technology. Nor can we really say that it was wrong to take this view. Time and again it was shown that badly trained militias were all but useless in open battle, however good their armament, whereas well-trained troops could win victories even with second-rate weapons. As Napoleon pointed out, an undisciplined Mamluk warrior could defeat a French cavalryman every time in an individual duel of one man against another; but the odds became even if a troop fought against a troop, while the French would win every time when a squadron faced a squadron.[10] This was the same Napoleon who, despite their undoubtedly superior accuracy, saw no need for rifled muskets in his skirmish line. This decision has been attributed to the muskets' expense and slow rate of fire, but it is just as likely that he rated cohesion and teamwork far higher than individual marksmanship. It was already well known that accuracy of firing in a real battle always fell massively below accuracy in calm and deliberate range tests, often with just 1 percent of the hits expected in the latter situation being recorded in the former.[11]

This is a truly alarming statistic for anyone who believes in the effectiveness of individual Napoleonic musketry, and it shows why most casualties were inflicted either by artillery or by massed infantry firing unaimed volleys only when the enemy was very close indeed. It also seriously undermines the traditional myth of the rifleman, whether we look at the American War of Independence, the "hunters of Kentucky" in the War of 1812, or the British rifle regiments of the Peninsular War. In all these cases, some highly effective skirmishing was undoubtedly achieved, in the sense of harassing enemy formations and inflicting attritional losses on them; but that is far from asserting that a high rate of hits was achieved. If that had actually happened, the skirmishers would quickly have become the decisive and dominant arm, rather than the indecisive auxiliary that they actually proved to be. This was especially true at long range, where the effects of their marksmanship were often negligible. Hence, at the battle of New Orleans in 1815, the American riflemen appear to have held their fire until a range of just "fifty feet."[12]

The universal adoption of strong units of skirmishers during the years around 1800 was accompanied by a misplaced faith not only in the accuracy of their shooting but also in their native cunning and individual sense of initiative. These were believed to dispense with the need for long training in formal drill, so that once the right (free and democratically enthusiastic) marksmen had been selected, they could be let loose directly onto the enemy. Alas for such hopes, however, it turned out that skirmishing actually required a higher, rather than a lower, standard of formal training. The poorly trained soldier could make many more fatal mistakes (or might simply desert) if he were left alone to his own devices than if he remained in closed ranks under the watchful eyes of his sergeant.[13] Hence the introduction of a light-infantry skirmish line actually intensified the tactical need for veteran soldiers, rather than diluting it.

In these circumstances, "the perception of tactical need" was for good men rather than for good weapons. Commanders constantly demanded veteran troops who were experienced, well drilled, and fully indoctrinated with the army's view of the world, although opinions differed widely over just how much time in the ranks constituted the minimum requirement. Everyone agreed that soldiers were worthless (and especially prone to desertion) during their first two or three days under arms; but in his desperate cannon feeding in later years, Napoleon seems to have relied increasingly on "training on the march" or a crash program of "encadrement," by which the raw conscript became socialized into the army during the few months needed to move him from his regimental depot to the seat of war. By 1814, as the seat of war moved remorselessly nearer to Paris, entire divisions were even being thrown into battle after their men had been "soldiers" for just two short weeks.[14] More conventionally, however, it was felt that at least a year or two of army life were required to make a man "fit for service"—and in some cases, ten or even twenty years were considered preferable.[15] According to this analysis, therefore, the main technologies needed to win battles had nothing to do with guns or bombs but everything to do with such things as pipe clay, boot polish, and the sergeant's pace stick (or even his cat-o'-nine-tails).

This perception of tactical need seems to have been disturbed—though hardly overthrown—by two extraneous impulses. The first came from the many hopeful light-infantry officers who kept making claims to have invented new training methods that might at last solve the frustratingly broad gap in marksmanship standards between the firing range and the battlefield. As we have seen, however, this was a vain hope, doomed to disappointment. The basis on which such claims were made may be dismissed as pseudo-scientific rather than accepted as a genuine new technology.

The second source of questioning came from inventors. These were often civilians who possessed no tactical perception or even the remotest clue about what a battle entailed but who did possess some highly tuned skills in gun making (normally for sporting purposes), in physics or chemistry, or—in some cases—in simple self-advertisement unsupported by any science whatsoever. Such people would merely notice that the world seemed to enjoy going to war and would ask themselves the question "How can I sell my particular skills to the people who are fighting these wars?" Their answers often turned out to be spectacularly unrealistic. (For example, this author would dearly like to know more about the airplane and the tank that were offered, by two different inventors, to Marshal Soult in 1831.)[16] Nevertheless, the farther the Industrial Revolution progressed, the more it became clear that a few well-defined advances in weaponry were in fact possible, practicable, and—perhaps most important of all—realistically procurable. Out of a confused babble of hopeful speculations there gradually emerged a consensus about just where "military science" ought to be concentrating its attention, and in particular, the debate increasingly centered on the rifled musket.

During the 1830s and 1840s inventors perfected muzzle-loading percussion muskets that, unlike the weapons of Napoleonic times, could be rifled without any loss of speed in loading. There was only a slight increase in the rate of fire but a truly spectacular increase in the range at which accurate fire was theoretically possible. An expert marksman could now hope to hit his mark at one thousand yards' range rather than one hundred yards, at least in the practice butts.[17] A number of manufacturers waxed lyrical about the "revolution in warfare" that this might be expected to create, including some inflated claims to the effect that infantry now outranged artillery and could shoot down any cavalry long before it could arrive at effective range for saber or lance. Light-infantry officers embraced this dogma with delight, since it seemed to offer a final breakthrough in the problems with marksmanship that had previously seemed so intractable. Thus, when the French Chasseurs à Pied were created at Vincennes in 1838–40 as specialists in the new technology, they were quick to claim not only that they constituted a completely new arm of the service but even that they had made all other arms (and, indeed, all existing generals) obsolete overnight. It was a vertiginous moment that brought together many different strands of innovative thinking and set half the agenda for everything that was to follow.[18]

According to the Chasseurs' half of the picture, there was a long-standing tactical need to make radical improvements in the marksmanship of infantry, especially at long ranges. Because the technology existed, it should now immediately be applied throughout the whole army, so that no enemy could henceforth continue to live within a zone of around one thousand yards from a French infantry firing line. This stirring critique was echoed in many other countries, and by the 1850s and 1860s, most armies were indeed able to rearm their infantry with modern rifle muskets. In Britain, the .577 Enfield was quickly adopted and enjoyed great success in the

Crimean War, where it asserted a distinct superiority over the smoothbore muskets of the Russians. In the United States, the Ordnance Board opted heavily for the .58 Springfield version of the same weapon, which implied that the American infantry was all but completely committed to at least the outlines of the tactical analysis laid down by the Chasseurs à Pied. General Hardee's translation of the Chasseur drills in 1855, as official U.S. doctrine, only reinforced this decision, which remained effective from the 1850s until after the Civil War.[19] In its defense, we can easily see that the Ordnance Board was influenced by many telling considerations of price, robustness, and ease of engineering; but with hindsight, we can also see that the rifle musket was really a blind alley. It should not have been adopted so completely, so unquestioningly, and with such a deleterious effect on subsequent government production of more advanced long arms.[20]

Even before the end of the Crimean War, it was starting to be suspected that the Chasseurs à Pied were something of a fanatical sect, on the outer fringes of the true tactical debate. Persuasive battlefield evidence began to accumulate that suggested there was not, in fact, any real requirement for accuracy at long range, and even if there were, that the rifle musket could not provide it to the levels demanded. In the Italian war of 1859, the French defeated Austrian rifle muskets not by long-range fire but by aggressive skirmishing and rapid bayonet assaults at close range. The Chasseurs were used as elite shock troops rather than as marksmen, and overall, the tactics relied on small columns operating in relatively close country with short fields of fire. They would not have seemed out of place in Bonaparte's Italian campaign of 1796–97.[21]

In the American Civil War, equally, there seems to be little convincing evidence that officers ever taught their men to fire at ranges much greater than one hundred yards. Several factors contributed to this, notably the heavily wooded terrain in which many of the battles were fought. On many occasions we read of regiments having to clear the underbrush ahead of them, to enable fire to be delivered as far afield as fifty or eighty yards. Nor did many regiments have available the ammunition necessary for sustained courses of target training. Of still greater significance in technical terms, however, was the fact that these low-velocity black-powder weapons still had a very short "point-blank" range, at which the bullet traveled out of the muzzle in a straight line before it started to "droop" under the influence of gravity. In these circumstances, it was essential to know the exact range to the target, beyond about one hundred yards, so that sights could be adjusted and the men instructed to "lay off" for the plunging of the shot. This, in turn, normally meant that range markers would have to be laid out on the ground, at intervals of fifty or one hundred yards, and that trained noncommissioned officers (NCOs) would then shout out the necessary adjustments as the enemy approached. Normally this was found to be cumbersome and overtechnical for the massed militia armies available in America during the 1860s, and this author has been unable to identify a single occasion on which it was even attempted. The conclusion is that officers did not usually ask their men to fire at ranges much beyond one hundred yards; in other words, they were perfectly content to leave some 90 percent of their rifles' theoretical capabilities unused.

The development of the rifle musket was a logical technological advance over the flintlock smoothbore of the Napoleonic era, but it does not seem to have changed tactics in any way. It did represent a relatively small improvement in the battlefield performance of the earlier weapon, but it did not find a new solution to any properly perceived tactical need. It did not improve the standards of marksmanship or even of skill in skirmishing. It certainly did not, as is sometimes claimed, make either the artillery or the cavalry obsolete; nor did it really change either the scale of casualties or the way in which the infantry fought. In the American Civil War, just as in the Napoleonic wars, the infantry often advanced in massed formation and exchanged fire at close range. Losses were naturally high when the two opposing lines were only a few yards apart, although we must remember that in "America's bloodiest day," at Antietam in 1862, the total casualties were still only around twenty-two thousand, or around one-third of those suffered at Borodino in 1812 or at Waterloo in 1815. Equally, when Major General Emury Upton's new American drill appeared in 1867, encapsulating all the experience of the Civil War, it still looked very similar to Hardee's old French drill of 1855; there was a slight lightening of the line but no great revolution in tactical methods.[22] Indeed, it cannot be stressed too often that the percussion rifle musket changed nothing very much except the rhetoric used by a small minority of gun enthusiasts and rifle manufacturers. The vast majority of officers and men continued to use it in exactly the same way that their predecessors had used the smoothbore flintlock—en masse and at close range. They thought of the new weapon as merely a small, natural evolution of the old one, not as a revolutionary conceptual breakthrough.

What really changed tactics in the 1860s was not the muzzle-loading rifle musket, theoretically accurate at long range, but the advent of breech-loading rifles that could and did deliver a significantly higher rate of fire at close range. Where these were used in significant numbers, as in the Union's dismounted cavalry actions of 1864–65 with repeating (magazine-loaded) Henry and Spencer carbines or in the Prussian "needle-gun" battles of 1866 with single-shot weapons, they could provide great tactical advantages.[23] In cavalry actions against other cavalry, the Colt revolver was found to offer an equivalent advantage by its high rate of fire at very close range. What was suddenly discovered in the 1860s was that long range scarcely mattered in practice, whereas a high volume of fire did. A force could normally adjust itself to a protracted attritional drizzle of shots from distant positions and could keep open the option of controlled retreat before its casualties mounted to seriously threatening levels. By contrast, if the same force were suddenly confronted by a fast-firing enemy at close range, it might well suffer an instant, catastrophic, and decisive collapse in morale. Such an attack represented a modernized, technological version of the Frederican and Napoleonic mixture for shock action, which had classically consisted of "volley–cheer–bayonet." With more modern weapons, a workable equivalent could now be delivered by fewer men, who might be less closely packed together and who might need less "military spirit" and cohesion than their Napoleonic predecessors in order to pull off a comparable trick.

This "high volume–close range" vision of fire tactics represented the opposite half of the debate that had been joined by the French Chasseurs à Pied around 1840. Whereas they had drawn their pedigree from the rhetoric of earlier skirmish tactics, the school of decisive shock action looked back to the older and harsher tradition of close-order, strongly disciplined fighting that had often been mocked as "outdated" by the Chasseurs themselves. To modernists and self-appointed "scientists," the idea of forcing soldiers into closely regulated lines, depriving them of the use of cover, and then pushing them close into the enemy's fire was not just counterinstinctual—it smacked of positively medieval oppression. Yet the "big idea" that lay behind the shock-action school was that it was only by getting close and risking something that you might possibly win everything. Paradoxically, it turned out that the more advanced generation of firearms that superseded the rifle musket owed its tactical effectiveness to the generation of tactical thinking that had actually preceded the rifle musket.

In terms of the processes of procurement, however, this was not a finding that had emerged as the result of any scientific analysis of tactical need. In the American case, at least, the newest and most experimental breechloaders had been sold to the government by relatively small private manufacturers when they found themselves excluded from the mainstream rifle-musket market by the massive output of that weapon from official arsenals. These private manufacturers looked around for a commercial edge and found it in novel and untried technologies. Then they used all the guiles of the salesman to persuade the government to buy the product. If endorsements from famous military men formed a part of the package, they were driven by the urging of the manufacturer rather than by any perception of tactical need originating in the army.[24] Indeed, the whole process is best summed up by the two "summit meetings" that took place between Christopher Spencer, the civilian producer of repeating carbines, and Abraham Lincoln, the civilian politician. It was not a military occasion when the president personally fired and admired the revolutionary new weapon.[25]

In Prussia the breechloader, in the form of the Dreyse needle rifle, was already being adopted by 1840, and by the 1860s the whole of the infantry carried it; but once again, it had been adopted merely because it was available and apparently "better," not because of any perceived tactical need. On the contrary, the tacticians worried that it lacked sufficient range to compete with the Austrian rifle muskets in 1866, and then in 1870 with the more advanced (because second-generation) French Chassepot breechloaders. Such worries proved groundless in the case of the Austrians, since the firefights still took place at relatively close range and the Prussians had an obviously superior technology. Apart from their high rate of fire, they could reload in the prone position, using cover. With the French the picture was rather more complicated, since by that time the infantry battle was being set in the context of a whole new generation of artillery and—at long last—a combat-worthy machine gun.[26]

The French had adopted the Chassepot and the machine gun because the technology existed and because the enemy seemed to be stealing a lead. The emperor was personally interested in weaponry and was supported by an ingenious artillery establishment, whose responsibility it was to arm the infantry. The tactical assumptions that they made between them, however, seemed to be at odds with the traditional close-range shock action expected by the infantry officers themselves. A completely new set of tactics was invented by which new weapons were supposed to create a blanket of fire extending well over one thousand yards forward from the infantry line. It was an idea that harked back to the Chasseur thinking of the 1830s, although this time the fire was to be unaimed and based on high volume rather than on the precision of individual shots. At the longer ranges, the infantry was encouraged to use "coffee-mill" *(moulin à café)* fire from the hip, where the firing hand was closest to the cartridge pouch. Four or five times as many rounds per minute could be fired as had been possible with muzzle-loading rifle muskets, to which were added roughly 150 rounds per minute from each machine gun supplementing the line.[27] All this added up to a genuine novelty in tactics, and it proved very damaging when properly applied, notably against the Prussian Guards at St. Privat. Nevertheless, we can still say it perpetuated the same fallacy as did the earlier Chasseurs, since the potential shock value of both the rifles and the machine guns was dissipated at long range. The Prussians were not stopped dead in their tracks at one thousand yards, but were able to absorb their casualties until they went to ground at around four hundred yards from the French. An indecisive firefight then ensued, which either remained static—representing a defensive French victory of a sort—or was broken by the intervention of superior Prussian artillery. Thus the French actually squandered the capability that was given to them by their new weapons, of making a truly devastating fire ambush at close range. This was surely because the tacticians were starting with the new weapons and trying to stretch their long-range potential to the limit, rather than analyzing the battlefield from the infantry's proper perspective. They were adopting a theoretical artillerist's model of the battlefield, rather than one that was based on experience.[28]

The British somewhat blindly followed roughly the same path, although they were mercifully excused the necessity of fighting against a major modern army until 1914. They also placed more stress on individual marksmanship than the French had done and made a sustained attempt to give all their soldiers, throughout the empire, plenty of target practice at long ranges. In the 1879 Zulu War, they duly opened aimed fire with their Martini Henry breechloaders at around one thousand yards, only to find that at Isandlwana they had disastrously run out of ammunition by the time the Zulus arrived at close range.[29] The enemy could be kept at bay as well as the Prussians had been at St. Privat, for as long as ammunition was available; but after that there was no defense. Thus the British tacticians paid the price for forgetting the disciplined close-range volley and countercharge that had won so many battles for the duke of Wellington.

After Isandlwana the British even made a virtue of carefully aimed long-range fire, at a slow tempo, since it could be portrayed as saving ammunition, at least when contrasted with the rapid, unaimed blanket fire used by the French. When the magazine rifle was adopted in the 1890s, moreover, the British also specified a magazine cutoff, which was at least an early example of feedback from battlefield experience getting to the arms manufacturers. The cutoff meant that the rounds stored in the magazine could be kept in reserve at long range while hand-fed single rounds were being fired slowly, whereas at close range the whole magazine could be emptied all at once, in a "mad minute" of high-volume fire.[30]

The more the British studied these problems and compared their experiences in various colonial campaigns, the more they began to move their emphasis away from careful long-range shooting and toward frenzied, rapid fire at close range. This was the tactic that succeeded magnificently at Mons and Le Cateau in 1914, and it actually represented a rejection of some sixty years of ill-informed procurement philosophy. Nevertheless, the issue was fudged, and even the mad minute of Mons was supposed to be aimed fire, though in reality it was not. The problem was that the theory of the mad minute had originated in the school of musketry at Hythe, where the whole ethos of the establishment was always built on marksmanship.[31] Hence the idea of taking careful aim has remained a shibboleth in the British army throughout the twentieth century, even though most other armies have tended to neglect it.

The First World War normally saw infantry delivering fire at relatively close range, perhaps up to three hundred yards, even though they now had rifles that had been procured on the basis that they would be "accurate" at ranges up to fifteen hundred yards or more. In practice, the main burden of long-range infantry fire now passed to the belt-fed machine gun, which was capable of sustained indirect-fire barrages at ranges up to around two thousand yards. A heated debate persisted over whether or not this was actually the best use for machine guns, since it kept them as more of an artillery weapon than an infantry one and took them out of the arena of close-range shock tactics. In the British service, at least, there was a strong sense that military tactics were being determined by technical capabilities rather than by the needs of real combat, and there was also a suspicion of devious political motives among the Machine Gun Corps, which wanted to assert its independence from the infantry.[32] However that may be, this doctrinal confrontation at least led to a demand from the infantry for plenty of machine guns of its own, for use at close range. Although the result was not, as they had hoped, a multiplication of Vickers guns under infantry control, they did get a whole new breed of light automatics that could be used in shock action. If a mad minute delivered from magazine rifles was good, then it would be better still if it could be supplemented by a Lewis gun in each platoon.[33] Thus, for the first time, the infantry began to define its requirements in terms of what it needed in combat, rather than simply choosing from what happened to be available in the arms manufacturers' catalogs.

The traumatic shock of trench warfare ensured that a similar reversal of earlier procurement practice would occur in many other fields. Trenches were designed to give protection from direct fire, whether from rifles or machine guns, so new weapons were needed that could reach into them by some form of indirect fire. The hand grenade offers a dramatic case in point, since the frontline troops were actually making their own, to meet their perceived tactical needs, a few weeks after the mobile operations of 1914 had become static. There were many different local designs, and many accidents, before the arms procurement machinery in Britain finally arrived at a reliable pattern in 1916. The story of trench mortars is very similar, with many improvised designs used during 1915 until the incomparable Stokes mortar was finally adopted the following year.[34] It was found, in fact, that the pressure for arms procurement was at last coming from the bottom up, rather than from the top down.

A part of this change may be explained by the sheer novelty of the problems that the frontline soldiers faced. They had no option but to look for new weapons and tactics, especially since failure to do so would inevitably multiply the already horrifically long casualty lists. The very structure of trench warfare also contributed, since it was a prolonged and static experience in which there was plenty of time for innovative officers to imagine new solutions and then disseminate them through the various training camps and tactical schools that quickly sprang up behind the front. Equally, the great advances in technology since 1870 meant that civilian inventors trying to sell their ideas to the army now had many more workable gadgets or combinations of gadgets to offer, in many diverse disciplines. This gave them a correspondingly greater ability to tailor their products to the customer's precise requirements.

The length of the war also gave time for the industrial infrastructure at home to reorganize itself and begin to listen to the demands coming back from the front. The army's own experimental establishments, at Hythe and elsewhere, could now at last test their peacetime speculations against the hard experience of real modern combat. The whole of the arms industry was radically expanded to meet a sudden new level of demand that few had imagined in their worst nightmares. There was a general acceptance of the need for change, and this, in turn, was exploited by ambitious politicians such as Lloyd George and Churchill, who alleged that they alone could provide the necessary innovative solutions. These leaders did some good in helping along the existing processes of arms procurement, although they also did enormous long-term damage to the reputation of the army. By suggesting that only they, as civilians and outsiders, could truly understand the tactical requirements of the front line, they effectively denigrated the great effort of innovation that the front line itself was already making.

Despite the interference of self-interested politicians, the requirements of the front line were, by and large, successfully conveyed to the arms industry during 1915–18 in a way that was new for the army, if not for the navy. Doubtless many mistakes were made along the way, and as much by the military authorities as by civilians. Many of the civilian inventors who came forward were politely told that their ideas were inappropriate—including Frederick Stokes himself, before his dogged perseverance eventually got his mortar accepted. Conversely, many of the inventions that were seized on by the army turned out to be tactically useless. The Barratt forcing jack (or "pipe pusher") makes a spectacular example. It was a machine employed on the line of communication to bore holes horizontally through the ground to take water pipes, telephone cables, or, in mining, explosive charges. In 1916, it was seen as potentially capable of meeting the perceived tactical needs of the front line, and so, during the first half of the battle of the Somme, it was used on a number of occasions to push explosives or gas under or into the enemy's trenches.[35] If it had worked as well as hoped, it would have been every bit as useful as the tank; but alas, it proved to be extremely unreliable and sometimes lethal to its operators. It did have some successes but rather more failures.

If the pipe pusher was an existing machine that was instantly enrolled to frontline use as the result of a perceived tactical need, the tank enjoyed a much more elaborate and colorful procurement history. It had been a fond aspiration of military theorists since at least the time of Leonardo da Vinci, and even as far back as the mobile armored siege towers of Roman times. With the advent of the internal combustion engine, it was promoted into the realms of what may be called "attainable science fiction," although the practical difficulties of producing a working prototype remained daunting. In the event, its development had to await a combination of the war crisis, some highly individual personal initiatives (notably from Churchill), and perceived tactical requirements for an assault weapon that could outface machine guns. Thus, in common with the 1840s' rifle musket, the tank relied in part on a preexisting mythology; but unlike the rifle musket, it was designed to meet a real and pressing need, rather than merely a theoretical one. Even so, when it was first used during the second half of the Somme battle, many soldiers believed it would be no more successful than the pipe pusher. Despite ecstatic reports in the propagandist press, its operations on September 15, 1916, were distinctly disappointing and posed as many new problems as they solved.[36] These continued unresolved for over a year, with undistinguished performances at Arras and Messines and complete disasters at Bullecourt and Ypres. It was only at Cambrai in November 1917 that the tank started to fulfill some of its original promise, and even then, its many limitations remained obvious to all who experienced it firsthand. It was never the war winner that its propagandists claimed, but by the end of 1917, it had at least grown to the status of a useful auxiliary to the infantry. It was always expensive in manpower, treasure, and scarce industrial capacity: it was unreliable, cumbersome, slow-moving, vulnerable to many types of fire, and notoriously short in range. It could not in itself effect a break-in, let alone a breakout, but could only assist the efforts of the other arms, particularly by crushing wire. That it managed to survive its first year of setbacks was perhaps due more to its unique mythological and symbolic aura than to anything it actually achieved on the battlefield. One may certainly speculate that if the development of pipe pushers had received even one-tenth as much support from politicians and Douglas Haig's high command, then they, too, could have risen to equivalent status with tanks and might have taken an equally honorable place in the victory offensive of autumn 1918.

The tank was a very specialized luxury in the Great War, from which much official support was promptly removed after the peace. If its second generation was destined to enjoy a greater glory in the 1940s, that was only after it had evolved into a very different machine to fulfill very different roles. It is nevertheless worth remembering that the original development of the tank, however incomplete and disappointing in its results, had arisen from a wider revolution in arms procurement practices during the Great War, whereby the real needs of the infantry had at last begun to be analyzed and attended to by those who controlled financial and industrial policy. In later years, this precedent would be enlarged and institutionalized, especially with the rise of a science of "operational research"; but it was the shocking novelty of trench warfare in the winter of 1914–15 that initially sparked it off.

In Napoleonic times, the infantry had been expected to fight mainly by its cohesion and discipline, supplemented by whatever provision of muskets the government could offer, which was often inadequate. As the Industrial Revolution gathered momentum after Waterloo, there was a corresponding growth in speculation about weapons that might revolutionize the battlefield; but most of this came either from outside the army or from self-interested light-infantry officers who normally lacked direct combat experience. The theoretical tactics that became widely accepted in the era of the rifle-musket, from about 1840 to 1865, were essentially those of the Chasseurs à Pied, which relied on an ideal of marksmanship at long range. In the event, such marksmanship singularly failed to materialize, and in practice, the Chasseurs' tactics were quickly overtaken by a new generation of rifles. These were breechloaders, with the key quality of delivering a much higher rate of fire, albeit at considerably shorter range, at least before 1870. Even then, however, the armies tended to persist with the unrealistic Chasseur ideal of accurate marksmanship at long range rather than shock action at close range. It took half a century of trial and error to disabuse them of the myth of long-range accuracy, and it was really only during the Great War that events finally made the switch unavoidable. During the trench fighting on the Western Front, it was the infantry's direct experience of combat that forced them to design—and, in some cases, even build with their own hands—a whole new set of weapons, scarcely imagined by prewar arms manufacturers. The seriousness of the crisis at long last enabled the infantry to impose their requirements on politicians and industrialists alike, and so, for the first time, it began to be the front line that determined what was produced in the factories, rather than vice versa. There were admittedly still many wrong turns and failures of communication between the two sides, but at least the essential start had been made.

NOTES

1. This chapter is an attempt to draw together the thoughts about procurement that have arisen during my writings about several different periods in the history of infantry tactics, especially in *Battle Tactics of the Civil*

War, 2d ed. (Swindon: Crowood; London and New Haven: Yale University Press, 1989); *Military Thought in the French Army 1815–51* (Manchester: Manchester University Press, 1989); *Forward into Battle,* 2d ed. (Swindon: Crowood; Novato, Calif: Presidio, 1990); "British Armored Warfare in the Western Desert 1940–43," in *Armoured Warfare,* ed. J. P. Harris and F. H. Toase (London: Batsford, 1990), 70–87; *Battle Tactics on the Western Front 1916–18* (London and New Haven: Yale University Press, 1994); *The Art of War of Revolutionary France* (London: Greenhill Books, 1998).

2. I wish to thank Jim Wallman of Megagame Makers for this insight.

3. Bernard Brodie, *From Crossbow to H-Bomb* (New York: Dell, 1962), 116–18, 164–67.

4. William Congreve, *The Details of the Rocket System* (London, 1814); John Naylor, *Waterloo* (1960; reprint London: Pan, 1968), 32–33.

5. The Katushyas and Nebelwerfers of World War II had their immediate precedents in the fiendish Livens projector of the Great War (see my *Battle Tactics on the Western Front,* 119).

6. See, for example, my article on "British Armored Warfare in the Western Desert," in Harris and Toase, ed., *Armoured Warfare.*

7. Jac Weller, *Weapons and Tactics, Hastings to Berlin* (London: Nicholas Vane, 1966), 108–9. There would be many hollow laughs around Enfield Lock when the Americans rushed to embrace the M-16 assault rifle in 1966, even though it fired a distinctly nonstandard short 5.56 mm round.

8. Brian Lavery, *Nelson's Navy: The Ships, Men and Organisation 1793–1815* (London: Conway Maritime Press, 1989; rev. ed., 1990), 58, 99, 112.

9. See, e.g., my *Art of War of Revolutionary France,* 215–17 and passim; John A Lynn, *The Bayonets of the Republic: Motivation and Tactics in the Army of Revolutionary France, 1791–94* (Urbana: University of Illinois Press, 1984; Boulder, Colo.: Westview Press, reprint, 1996), 186–91.

10. See, e.g., my article "'Keep Step and They Cannot Hurt Us': The Value of Drill in the Peninsular War," in Ian Fletcher, ed., *The Peninsular War: Aspects of the Struggle for the Iberian Peninsula* (Staplehurst, Kent: Spellmount, 1998), 163–72.

11. Brent Nosworthy, *With Musket, Cannon and Sword: Battle Tactics of Napoleon and His Enemies* (New York: Sarpedon, 1996), 202–9.

12. See my discussion in *Forward into Battle,* 47.

13. Compare the Prussian "democratic" movement, described in Peter

Paret, *Yorck and the Era of Prussian Reform, 1807–1815* (Princeton: Princeton University Press, 1966), with some of the liberal values attributed to the British light-infantry movement (e.g., in J. F. C. Fuller, *Sir John Moore's System of Training* [London, 1924] and corrections in D. Gates, *The British Light Infantry Arm, c. 1790–1815* [London: Batsford, 1987]). In fact, these systems both relied on an elite of especially well trained men, rather than simply on inspired patriots. The case of the French was slightly different, since they had indeed won notoriety for their disorganized but massed skirmishing during the early 1790s, at a moment when they had been able to field very little apart from "inspired patriots." Yet their most authoritative and influential book on the subject (C. Duhesme, *Essai historique sur l'infanterie légère* [1814; 3rd ed., Paris, 1864]) was largely made up of a concentrated attack on untrained skirmishers, in favor of very highly trained ones.

14. Alan Forrest, *Conscripts and Deserters: The Army and French Society during the Revolution and Empire* (first published as *Deserteurs et insoumis,* Paris, 1988; English trans., Oxford and New York: Oxford University Press, 1989).

15. J. A. Houlding, *Fit for Service: The Training of the British Army 1715–1795* (Oxford: Clarendon Press, 1981); and compare the post-Waterloo French debate in my *Military Thought in the French Army.*

16. See my *Military Thought in the French Army,* 77.

17. Jac Weller, "Shooting Confederate Infantry Arms," in *American Rifleman,* (April, May, June 1954); and contemporary discussion in Hans Busk, *The Rifle and How to Use It* (ca. 1850; 4th ed., 1859; reprint, Richmond, Surrey, 1971).

18. See my *Military Thought in the French Army,* 107, 115–17, 125–30.

19. See my *Battle Tactics of the Civil War,* 26, 76–90, 99–102. Fascinating, if uneven, general coverage of the many weapons is available in William B. Edwards, *Civil War Guns* (Harrisburg Pa.: Stackpole, 1962).

20. The best analysis of procurement in the Civil War is Carl L. Davis, *Arming the Union: Small Arms in the Union Army* (Port Washington, N.Y.: Kennikat Press, 1973).

21. Discussion of the combat psychology in Ardant du Picq, *Battle Studies,* trans. J. N. Greely and R. C. Cotton (New York: Macmillan, 1921), especially appendixes on the Crimean and Italian campaigns, 263–71. For the formations used, see Steven T. Ross, *From Flintlock to*

Rifle: Infantry Tactics, 1740–1866 (Cranbury, N.J.: Associated University Presses, 1979), 165–6.6

22. See my *Battle Tactics of the Civil War,* 19, 103–4, and passim.

23. Ibid., 184–88, 191–92.

24. Edwards, *Civil War Guns,* 149–50.

25. Ibid., 159–63.

26. Wilhelm Duke of Württemberg, *The System of Attack of the Prussian Infantry in the Campaign of 1870–71,* trans. C. W. Robinson (Aldershot, 1871).

27. Ibid., 8–10.

28. Michael E. Howard, *The Franco Prussian War* (1960; reprint, London: Fontana, 1967), 36, repeats the long-standing criticism that the machine guns should have been treated as infantry weapons for use at short range, rather than as long-range artillery. For an infantry view that was influential from 1828 onward, see T. R. Bugeaud, *Aperçus sur quelques détails de la Guerre,* 24th ed. (Paris, 1873).

29. Donald R. Morris, *The Washing of the Spears* (1966; reprint, London: Sphere, 1968), 368–90.

30. In much the same spirit, the British would not allow an "automatic" mode to be fitted to their new self-loading rifle in the 1950s: Weller, *Weapons and Tactics,* 109.

31. Shelford Bidwell and Dominick Graham, *Firepower: British Army Weapons and Theories of War, 1904–45* (London: Allen and Unwin, 1982), 26–32ff.

32. See my *Battle Tactics on the Western Front,* 120–34. The remainder of my coverage of the Great War is based on the other sections of this book.

33. In later years, the platoon or squad automatic would be supplemented by short-range "sub" machine guns that could be carried by every man. By the late 1940s, these in turn were given greater range, until they became true "assault rifles."

34. See my *Battle Tactics on the Western Front,* 112–18.

35. Ibid., 118–19.

36. J. P. Harris, *Men, Ideas and Tanks* (Manchester: Manchester University Press, 1995), 65–67.

CHAPTER 2

Technology, Industry, and War, 1945–1991

Warren Chin

The Industrial Revolution changed all aspects of how states fought wars. It may well represent one of only three revolutions to have affected the domain of war over the last fourteen thousand years,[1] and it was so important that it became one of the fundamental pillars of military power in the nineteenth and twentieth centuries.[2] The first and most important aspect of this revolution was that it completely changed the material conditions of war. In the past, the scarcity and cost of resources required to wage war were a principal limitation on the character, intensity, and duration of such conflicts. In contrast, from the second half of the nineteenth century, the creation of a mass-consumer, mass-production society and the transfer of these techniques to the realm of armaments production resulted in the erosion of this limitation. States now possessed almost limitless means with which to wage war. More important, in the context of the rivalry between the great European powers in the period leading up to World War I, the desperate need to secure victory in war resulted in an imbalance between the means used to fight war and the grand strategic goals for which the war was fought.[3] As a result, a new phenomenon was born: total war.[4] Total war was a condition peculiar to the bloody wars of the twentieth century, and its character was very much shaped by the power of modern manufacturing.

Such was the importance of industry in war that it affected even our conception of strategy and tactics. As Martin van Creveld explains, whereas in the preindustrial world strategy was purely concerned with the concentration of force at the decisive point, as a result of the Industrial Revolution, it acquired an added dimension. By the late nineteenth century, it was no longer simply concerned with waging military operations: instead, it became necessary to coordinate all areas of activity within the state: political, economic, and industrial. Thus was born the term *grand strategy.*[5]

The new material conditions also changed the conduct of war in that it became less decisive, and as has already been said, industrialized war demonstrated a propensity to escalate toward totality. In the preindustrial era, armies were small and tended to march and fight as a single, concentrated mass. Because of the inability of principalities and kingdoms to replace their losses, defeat in a few battles usually made it imperative for the losing side to negotiate an end to the war. In the era of mass-industrialized warfare, however, the enormous size of armies and the ability to replace losses quickly served to make it almost impossible to achieve a decisive victory in a single battle or campaign. Victory depended on a willingness to fight a protracted and often attritional warfare, and it went to the side best able to meet to material demands of this new style of war.[6]

World War II represented the apotheosis of total war and demonstrated in the clearest possible terms that, in the era of mass-industrialized warfare, quantity possessed a quality all its own. Thus one of the strongest images we have of this war is that of a highly skilled German military machine being overwhelmed by the material superiority of their militarily less skilled opponents. The best illustration of this view is Germany's fight against the Russians on the Eastern Front (1941–45) where, it was believed, superior numbers allowed the Soviets to prevail over the numerically inferior but more skilful German army.[7] Although this is a simplistic view, there is little doubt that the economic strength of the Allied powers and the efficient mobilization of their economies constituted a significant reason for why they prevailed.[8] Indeed, it is estimated that the United States alone raised over one hundred divisions during the war and supplied aid to both Britain and the Soviet Union equivalent to the creation of a further two thousand divisions.[9] In 1944 the Germans produced 17,800 tanks and 39,800 aircraft. This output was dwarfed by the Allies, who constructed over 51,000 tanks and 167,654 fighter aircraft in the same year.[10]

The Industrial Revolution also changed the material conditions of war by establishing the idea of military invention as a permanent and systematic feature of modern war. Of course, the invention of new weapons was not an activity peculiar to the Industrial Revolution—history is littered with examples of such efforts. What made the era of mass-industrialized warfare different from preceding ages was that military research began to be conducted on a sustained basis and sought to take advantage of the rapid technological change being experienced in industry and society at that time.[11] This practice dates back to the 1850s, when industrial magnates began to make a systematic effort to transfer the skills and technological inventions of the Industrial Revolution to the realm of the military. It was at this point that a new form of arms race was born: the qualitative arms race, a competition based not on numbers but on the performance of weapons.[12]

Increasingly, the invention of new weapons was believed to be the key to victory. How obsessive this compulsion became can be seen by examining the views of the military historian J. F. C. Fuller, who argued that weapons had the potential to account for ninety-nine percent of victory in war.[13] This was not an isolated or unusual view. In the 1880s, the British state began to make increasing demands on private industry to provide better armor and guns for the Royal Navy.[14] As a result, it was not long before the state began to push industry to work beyond the state of the art and to devise new and better weapons.[15] The result was the creation of what have been described as "command technologies." These were not products devised by industry using its own resources to fulfill a clearly perceived need in the market. Instead, the government and the armed forces increasingly pushed for the development of more effective weaponry. The defense industries were not able or willing to finance such financially hazardous projects on their own, and so it was that the state began to finance private companies on a large scale to conduct ambitious defense research.[16] In the United Kingdom, this tendency was most pronounced in the fields of naval construction and aircraft production.[17] The relationship between the state and industry became increasingly close as a result of the demands of World War I. The application of state-sponsored scientific research resulted in the refinement or development of such systems as tanks, aircraft, machine guns, radio communications, and poison gas.[18] Financial constraints in the interwar period inevitably dampened the pace of qualitative improvements in weaponry. State-sponsored military research remained prominent, however, and considerable emphasis was placed on the preservation of defense research and development facilities within the manufacturing center. An excellent illustration of this support was the aid given by the British government to the aircraft industry during the interwar period.[19]

The importance of defense research increased during World War II. At the outset of war, the scale and intensity of this activity were generally underestimated. Thus the British Admiralty allowed for a wartime expansion of forty scientists and technicians in its research and development facilities. By the end of the war, however, the Admiralty was employing more than forty thousand such people.[20] This process also proved to be expensive. For example, in the United States, military research increased from $13 million in 1939 to $1.5 billion in 1944.[21]

In the stress of war, however, the ambitiousness of defense research was balanced against the need to get equipment into service as a quickly as possible. As a study of British military research during World War II demonstrates, time pressures and the need to ensure that new equipment could be mass-produced quickly and efficiently imposed constraints on the depth of research undertaken by the Allies. They realized that in mass-industrialized war, quantity mattered as much as quality, and that their technical superiority over the Germans had to be expressed as much in terms of mass production as of invention.[22]

Although this was a war characterized by mass armies sustained by mass production, it nevertheless saw the establishment of an important precedent in the field of defense research. During the war, a significant shift occurred in the emphasis of defense research, in that it no longer just tapped into the existing realm of knowledge in science and engineering.[23] Thus the war was not significant just because of the quantity of material produced but also because of the technological quality of some of the equipment developed. This was to have profound consequences for the conduct of military research after the war. The most obvious illustration of this trend was the creation of the atomic bomb.[24] Yet nuclear physics and the related engineering were not the only spheres of the unknown that were explored. The development of other military technologies indicated the application of basic research to even the field of conventional weapons.[25]

The precedent for ambitious and highly specialized research was firmly established in the midst of World War II. This development was to mature, however, and come to fruition in the postwar era and the decades of relative peace that followed. One of the most significant aspects of the cold war was the unprecedented scale and pace of technological change. In the past, military invention was confined primarily to periods of war. Once these were over, the process of innovation either slowed or came to a complete stop. In the midst of the cold war, the situation was quite different, and military research and development continued at a heightened state in a period of peace.[26] Thus, according to one view in 1960, the world was experiencing a technological revolution in the art of war approximately every five years.[27] Evidence of the increasing importance of scientific innovation in defense during the cold war can be seen by examining the exponential growth of the defense research-and-development R and D) budget. After all, heavy military research expenditures were essentially a post–World War II phenomenon. Before 1940, total R and D expenditure by both industry and government in the United States was about $1 billion a year. By the mid-1980s, this figure had increased to $100 billion a year.[28]

Equally important was the fact that the proportion of money spent on research and development and on production also changed during the cold war. Just after 1945, around 5 percent of the defense procurement budget was committed to military research and development and 95 percent to production. By the mid-970s, the ratio was fifty-fifty.[29]

Such was the dominance of this trend that defense research became the vanguard of scientific advance. From the 1940s until the 1970s, military research and development led rather than followed the ideas of industry in a number of critical product sectors, such as electronics and aerospace. For example, World War II and the cold war really provided the impetus for the development of the computer. Although there is no doubt that the computer would have been developed without the financial and material support of the military, this process of development would not have happened as quickly if such support had been absent. Thus it has been estimated that the technology to make computers was in place twelve to fifteen years before its normal course of development because the U.S. Army intervened and invested in its development. The most significant problem that restricted the extent to which the computer could be used was its size. Continued investment by the military resulted first in the development of the transistor in 1949, which achieved a considerable reduction in the size of computers and their energy requirements. A decade later, the first integrated circuit was developed. This permitted the miniaturization of electronics and made it possible to develop computerized weapons, ranging from strategic ballistic missiles to radar, communications, and precision-guided weaponry. In effect, investment in defense research provided the basic foundation of what was a quantum jump in technology in the 1970s, leading to the information revolution.[30]

An important question to ask is why states placed such a premium on defense research during the cold war. One of the significant characteristics of the international system during this period is that while a great potential existed for war between the great powers, based on an intense ideological, political, and economic competition, no such conflict ever erupted. The absence of such conflict is often attributed to the presence of nuclear weapons.[31] The most important impact of these weapons was that they made war between the superpowers irrational under all but the most extreme circumstances. As the U.S. strategist Bernard Brodie explained in 1946, conventional military wisdom in the form of the principles of war simply did not apply in the nuclear world. Surprise and an overwhelming concentration of force were not likely to guarantee victory. As long as the state being attacked possessed even a fraction of its own atomic arsenal to retaliate, it would, in all probability, inflict a high price on the attacker's population.[32] Moreover, the sheer destructiveness of the atomic bomb and the fact that the detonation of even a small number of bombs could devastate the urban and industrial heartland of a nation made an effective defense against this form of attack almost impossible.[33] As a result, whereas in the past the main purpose of the military establishment was to win wars, with the advent of the atomic age its primary purpose was to prevent them; it had no other useful function.[34]

But why did this condition result in an unprecedented period of technological innovation in military research? After all, once both sides possessed nuclear weapons, it seemed entirely logical to accept the existence of a military stalemate and instead rely on a large nuclear arsenal to deter attack, rather than investing in more innovation. The problem was that the West sought to deter potential Soviet aggression through the deployment of technically sophisticated weapons. As such, the deterrent was sensitive to technological change. This required a continuous process of investment in defense research and development. Failure to do so risked the collapse of the deterrent, as the Soviets continued to make qualitative improvements in their weaponry.[35] Moreover, unlike wartime procurement, where the pressure to get equipment into service imposed a limit on defense research, no such barriers existed during the cold war. Consequently, there was a great temptation to explore what was scientifically desirable rather than necessary.[36]

The West's dependence on technology as the basis of its deterrent strategy can in part be explained in terms of the influence of the

United States' military—the dominant power in the North Atlantic Treaty Organization (NATO) coalition—which was recognized as having a strong predilection for technological solutions to its security problems. But it is also apparent that other, more practical considerations played a role in the West's deterrent strategy. Of primary importance was the unwillingness of Western governments to pay the economic cost to compete with the Soviet Union in terms of the size of its armed forces. Technology presented a less costly solution to the problem of how to deter aggression and acted as a significant force multiplier. A perfect illustration of such thinking can be seen in the American policy of "massive retaliation" and its British equivalent in the 1950s.[37]

The deterrent was not just sensitive to technological innovation on the part of the Soviet Union; it also proved vulnerable to changes in the political and military strategies adopted by both the Soviets and the Chinese Communists. The effectiveness of the Western deterrent depended on the willingness of the United States to fight a total war if either it or its allies were attacked by the Soviet Union. While no one questioned that, in the event of a direct attack on the United States, it would respond with all means at its disposal, it was less certain that the strategy of massive retaliation would automatically be implemented in the event of an attack against one of the its allies. It was in these peripheral areas that the Communist bloc began to launch a series of political crises and limited wars. As the famed theorist of limited war Robert Osgood pointed out, this allowed the Communist bloc to achieve its aim without having to risk total nuclear war. No one of these crises justified a general nuclear war, but when combined, they represented a significant threat to Western interests.[38]

The response of the United States was to develop its own limited war capability, and the emergence of this new strategy of limited war represented an unprecedented development in the history of warfare. Although there was nothing intrinsically new about the practice of limited war, one of the principal reasons for such limitation was the lack of resources to fight any other type of war. In the case of the United States, in contrast, the means at its disposal to fight a war were almost limitless. Limited war, as practiced by the United States, was based on the application of massive self-restraint.[39] Its aim was to permit the use of force in such a way that conflict did not escalate into a general nuclear war. In essence, it was designed to ensure that a balance between the goals of war and its cost were maintained. For limited war to work, it was important that the belligerents agreed on the rules of the game and were willing to operate within agreed parameters. In addition, this was a war of negotiation; the purpose of military power was not to secure military victory but to assist the bargaining process. Its function was to pressure and coerce the opponent to accept the preservation of the status quo. Hence the advent of nuclear weapons resulted in the suppression of the almost natural process that drove war toward totality.[40]

Although the new limited-war doctrine went some way to address the weaknesses of the deterrent, the strategy created as many problems as it solved. The most fundamental of these was that it represented a direct challenge to conventional military wisdom. The aim of such a strategy was not necessarily to win the war, at least not militarily. Even if the enemy's center of gravity was known, fear that attacking such a target might cause the war to escalate was enough to ensure that it was not targeted. Similarly, the strategic offensive was also rejected. Most important, it was believed that in order for the government to retain control of policy, no effort could be made to mobilize the support of the people, because they might pressure government to expand the war.[41] In practice, this meant the United States was unwilling to mobilize the resources needed to react to the pervasive threat of wars in the third world, and that operations were conducted under stringent political controls.

To overcome these problems, the United States military relied on the application of science and technology to develop the means that would allow it to prevail in these new wars. A good example of how these restrictions operated and how technology was used as a solution to overcome them can be seen in the intervention of the United

States in Vietnam, where the limitations imposed as a result of an adherence to the strategy of limited war simply did not make military sense. The United States did not mobilize the full potential of its available manpower, even though the military requested the call-up of reserves.[42] On the ground, military operations were confined to South Vietnam when it might have been more effective to have either invaded North Vietnam or fought the war in the border areas of Laos and Cambodia, where there was a greater chance of finding and stopping the North Vietnamese army before it infiltrated South Vietnam.[43] Similarly, important constraints were imposed on the use of airpower in the bombing campaign against North Vietnam.[44]

Yet, at the time, there was no doubt that the United States, once committed to this conflict, would prevail, and any disadvantages imposed on U.S. forces by the requirements of fighting a limited war were, to a large extent, addressed through the application of technology, which acted as a force multiplier. Through this approach, the United States was able to demonstrate that it possessed the means and the capability to meet the challenge of perceived Communist expansion in the third world. Thus, through air mobility, an infantry unit's area of control increased threefold. This meant it was possible to defend South Vietnam with fewer troops and that no general call-up of the reserves was necessary. The importance of the approach in terms of allowing fewer troops to cover more ground can be seen in the battle of the Ia Drang in 1965.[45] The First Air Cavalry Division fought on a battlefield that stretched over an area of fifteen hundred square miles. Technology proved to be of vital importance because the dense jungle and mountainous terrain that dominated the country prohibited movement on the ground. The problem of mobility was compounded because of the destruction of the transport infrastructure in South Vietnam in the preceding French-Indo-Chinese War.[46]

Significant efforts were also made to capitalize on the use of firepower. Thus close air support to infantry on the ground was provided by such exotic systems as B-52 strategic bombers, each armed with thirty tons of bombs. Then there were converted C-130s, each armed with a 105 mm cannon and 20 mm Gatling guns. Of particular importance was the increased use of air-launched guided weapons and smart bombs designed to destroy armored and other vehicles, enemy radar, and missile batteries. The classic example cited to illustrate the effectiveness of these weapons was the destruction of the Thanh Hoa Bridge. By 1972, 871 sorties had been flown against the bridge without success. Eleven aircraft had also been lost in these attacks with dumb bombs. In the end, the bridge was destroyed by four laser-guided Paveway bombs.[47]

Confronted by an inability to stop the enemy from infiltrating South Vietnam, and then unable to find them once in the country, U.S. forces once again relied on technology to provide a solution. The answer was the creation of an electronic battlefield, in which sensors able to detect the movement of Communist forces hiding under the triple-canopy jungle were scattered in the border regions. This information was then processed by computer, and air strikes were launched against the suspected targets. In fact, the track record of this method was rather patchy, and its development can actually be compared with that of the tank in World War I. It had great potential—which was subsequently demonstrated during the Gulf War—but it suffered from being in its first experimental generation.[48] Over $3 billion was spent on the development of such a system to enhance the effectiveness of aerial interdiction of North Vietnamese forces moving down the Ho Chi Minh Trail. Unfortunately, the system proved susceptible to relatively cheap, low-tech countermeasures.[49] So heavily dominated was this war by technology that it has been described as "techno-war" and "capital-intensive" war.[50] Wherever possible, machines, not men, fought this war; dollars were cheaper and easier to replace than the lives of men. As the war progressed, the political need to control casualties served to increase the importance of technology.

In the aftermath of the Vietnam War the drive for greater innovation in defense technology did not lessen. Soviet endeavours to improve the technological superiority of both its conventional and nuclear arsenal in the 1960s and 1970s placed NATO at an increasing disadvantage. It was feared that, with this new capability, the Soviets would be able to launch a blitzkrieg style attack with its conventional forces against Western Europe whilst using its nuclear arsenal to negate the possibility of a nuclear strike by the United States. Given the quantitative superiority of Soviet conventional forces and the erosion of the West's technological superiority, a debate ensued on how to address this new threat. The resulting debate focused on what new technologies could be used and how best they could be used in the construction of a defense of Western Europe. Of particular importance was the belief that technological developments in the field of conventional weaponry offered a potential solution to the threat of Soviet blitzkrieg and that it might even be possible to defeat such an attack without having to resort to nuclear weapons.[51]

This debate was particularly active within the U.S. Army. Its defeat in the Vietnam War resulted in a reassessment of all aspects of its training, doctrine and equipment. The outcome of this debate was significant because it brought about a revolution in war fighting and signalled the birth of a new era in war. The Vietnam War played a pivotal role in this evolution. Not only did it result in a new generation of military equipment, but it also provided key operational concepts such as air mobility.[52]

The significance of this revolution was not fully appreciated at the time. The first effort to develop a new doctrine based on current technological developments was completed in 1976. Known as "Active Defense," it relied heavily on the recent experience of the Yom Kippur/Ramadan War to interpret what effect modern technology would have on the future battlefield. What was particularly noticeable in this war was the scale of destruction suffered by both the victor and the vanquished. After three weeks of fighting, Arab and Israeli losses in tanks and artillery exceeded the total U.S. tank and armored forces losses in Europe during World War II.[53]

Active Defense focused on how to maximize the effectiveness of modern firepower. As such, it concentrated on developing new techniques for the smallest formations in battle to employ. The lethal nature of weapons was now such that unless new techniques were adopted by the lowest tactical formations, failure would result at the higher levels of war. The Yom Kippur War also confirmed that new weaponry enhanced the power of the defense. It was best used in static, well-concealed positions. The accuracy of the firepower produced was expected to achieve a kill ratio of five or six to one in favor of the defense.[54]

The publication of this doctrine sparked much criticism. It was said that too great an emphasis was placed on the power of the defense at the expense of the offense, and that it ignored the psychological dimensions of war. The manual also focused too much on fighting a war in Europe.[55] Perhaps, however, its greatest weakness was its failure to recognize that the most important actions in war happened in the theater of operations as a whole, rather than on the battlefield. Thus one of the most ardent critics of this doctrine, William Lind, argued that Active Defense played into the hands of the Soviets because it failed to recognize that perceived main axes of attack might be feints or that the Soviets would encourage the Americans to concentrate at one particular point and hold the force with an attack while launching a massive offensive on the depleted sector of the front.[56] Active Defense envisaged the use technology in an almost reactionary fashion. It represented a continuation of the tradition of mass-industrialized warfare, in which forces were deployed on linear front from where they attempted use superior firepower to effect maximum attrition among the enemy. While the doctrine provided a clear idea of how to organize the battalion, and to a lesser extent the division, it offered no advice on how the Corps battle was expected to develop and provided no guidance on how the corps commander was to deal with the problem of the Soviets' follow-on echelons. In essence, Active Defense failed to recognize the importance of operational art.[57]

Concern over the efficacy of Active Defense resulted in the implementation of a series of war games. The results were extremely worrying because they showed that even when all the available new weaponry was factored in, NATO still lost the battle. The problem was that while NATO's firepower could defeat the first Soviet echelon, the initial engagements also resulted in the exhaustion of NATO forces, and consequently, they were unable to deal with Soviet reserves, which could then attack where they pleased. The doctrine assumed that the attacker required a six-to-one superiority to overcome the modern defense. This analysis of combat power, however, did not take into account the intangible elements of war that have often contributed to victory: initiative, maneuver, the achievement of surprise.[58]

As military critics have noted, a true military revolution requires more than just the incorporation of a new weapon into an existing force. Equally important is the development of a doctrine and organization capable of using technology in an unexpected and innovative way. Such action leads to a new mode of operation in war.[59] Dissatisfaction with Active Defense resulted in the emergence of a new doctrine in 1982. Called AirLand Battle, this doctrine was designed to address the perceived weaknesses of Active Defense, providing the means and the techniques required to defeat Soviet blitzkrieg.

AirLand Battle rejected the simple equation of force ratios to determine the relative combat power of two forces. Instead, it attempted to incorporate both tangible and intangible factors in an effort to show how to fight outnumbered and win.[60] Unlike Active Defense, AirLand Battle stressed the importance of coordinating action in the entire theater of operations rather than just on the battlefield. The most significant aspect of AirLand Battle was that it recognized the importance of being able to conduct deep attacks behind the forward edge of battle. The aim of these deep strikes was to stop the Soviets from massing their reserve forces on selected axes of advance and to create windows of opportunity to counterattack and defeat the enemy in detail.[61]

Although the ideas of deep battle and deep operations were hardly new, what made this doctrine revolutionary was that it relied on the development of new conventional technologies to create what the Soviets called the reconnaissance strike complex. Of particular importance was the development of VISTA (very intelligent surveillance and target acquisition) technologies. These systems were able to see into the operational depth of the enemy's positions. Improvements in data processing and the creation of secure high-speed data links allowed information to be provided almost instantaneously, and from this it was possible to create a real-time picture of enemy activity. The enemy's command and control, transport infrastructure, logistic facilities, and force concentrations could then be attacked in long-range precision strikes by conventional missiles, airpower, and special forces.[62]

The development of these more powerful information systems and improvements in both conventional munitions and their delivery systems in the 1980s marked the beginning of what the Soviet Union termed a military technical revolution.[63] The renowned "futurists" the Tofflers also see this time frame as marking the start of the oft-cited "revolution in military affairs": a military revolution based on the more widely based information revolution that was having a marked effect on society and the economy. In essence, it heralded the end of the era of mass-industrialized warfare.[64] This new doctrine and the development of the technology to achieve it were designed to achieve the goal of victory in war.[65]

The clearest evidence that this goal could be achieved and that the era of mass-industrialized warfare was at an end may be seen in the United States–led coalition's war against Iraq in 1990–91. What is interesting about this war is that the Iraqis at least attempted to fight using traditional methods of war, based on the deployment of a force of between 380,000 and 500,000 men in Kuwait. These were organized in a linear defense that ran the entire length of the Saudi-Iraqi border. The tactical zone of defense was arranged in depth, and the entire system was supported by local armored reserves and, further back, an operational mobile reserve in the form of the Republican Guard. With this defense, the Iraqis hoped to break the coalition's attack against a wall of firepower. In essence, their aim was to fight a battle of attrition using direct-fire weapons in the forward edge of battle.[66]

Given this formidable defense, it was expected that efforts to remove Iraq from Kuwait would entail massive casualties: possibly 170 aircraft out of a fleet of 1,700 aircraft and anywhere from between 30,000 to 100,000 soldiers out of a total force of 795,000. Contrary to expectation, however, the coalition's actual losses were only thirty-three aircraft and 240 soldiers killed in the land war. In contrast, the Iraqis suffered heavily in this war: at least twenty-five thousand killed, more than sixty thousand wounded, and at least eighty-six thousand prisoners captured.[67] This represented an unbelievably low loss rate for the coalition: one fatality per three thousand soldiers. This loss rate was only one-tenth of losses suffered by the Israelis in either the Six-Day War or their invasion of Lebanon in 1982. It was also less than one-twentieth of the loss rate the Germans experienced in their wars against Poland and France in 1939–40.[68]

The significance of the Gulf War outcome lay not simply in the fact that the coalition inflicted such a bloody defeat on the Iraqis; equally important was the speed with which they achieved this outcome. The air campaign lasted for thirty-nine days and was followed by a ground war that only four days. According to one school of thought, this victory represents a return to the era of decisive battle. In effect, technological innovation has provided the means to overcome the resistance of mass-industrialized warfare.

How did technology achieve the neutralization of such a formidable system? What made the Gulf War radically different from its predecessors was the application of an "electronic fire operation." In essence, this represented the use of electronic surveillance, electronic warfare, stealth fighters, conventional cruise missiles, and smart bombs to disrupt, disorganize, and destroy the political leadership of Iraq, its command and control, the transport infrastructure, and the Iraqi armed forces. It was because of the electronic fire strike that the ground war was won with such ease.[69] Through such equipment, the killing zone was elevated from the tactical forward edge of battle to the theater of operations as a whole. As such, it became possible to target and destroy the Iraqi army's systemic organization and effectively unhinge its defense and thus destroy its cohesiveness and its ability to coordinate military action. Long-range conventional strikes so weakened the Iraqi defense that tactical action became entirely dependent on paralysing strikes launched against strategic and operational targets throughout the enemy's territory.

Of particular importance in fulfilling this mission was the use of airpower and space power. Thus, in the view of the Soviets, the Gulf War was revolutionary in its conduct because, whereas in the past victory depended on ground forces routing the enemy's army and then destroying their political and economic infrastructure, what was most important here was the use of airpower to launch a strategic offensive against Iraq. This action, rather than the land war, decided the outcome of the coalition's campaign.[70] As Christopher Bellamy points out, the success of the coalition's land campaign did not depend, as in previous war, on the classical form of offensive action, which entailed tactical breakthrough at the forward edge of battle. In the case of the Gulf War, disruption of the enemy's operational depth was achieved through long-range conventional strikes in conjunction with the double envelopment of Iraqi forces through the air and on the ground. The tactical breakthrough battle was determined by these operational attacks.[71]

In this context, it is clear that while the outcome of the Gulf War conformed to the ideal of the strategically decisive battle, it emerged from more than a single tactical action. Success was due to the ability of the coalition to attack the Iraqi army through a series of related actions that took place at the strategic and operational levels of war. The neutralization of the enemy on such a scale was achieved because of the increased range and accuracy of conventional weaponry and through the use of modern surveillance systems, which allowed the coalition to see what the Iraqis were doing in real time throughout the Kuwaiti theater of operations. The use of such technology and the development of a doctrine and military organization capable of translating it represented the highest evolution in operational art.[72]

Ironically, the technological capability to deliver such a stunning victory also has made the instigator of this style of war more vulnerable to defeat because of the high cost of such forces and the difficulty of replacing them if lost. It is interesting to note that while defense spending generally increased in real terms throughout the cold war, in contrast, the size of armed forces shrunk. This reflected the increasing unit cost of each new generation of equipment compared to its predecessor.[73] On average, new equipment cost two to three times as much as the kit it replaced. The result of this trend was a phenomenon known as structural disarmament.[74] To compensate for this reduction in terms of physical mass, greater efforts were made to improve the combat power of new equipment being developed. Such was the drift down this path that there was genuine concern that, by sacrificing quantity for quality, the armed forces would lack the critical mass needed to fight a future war. Thus, in 1955, the U. S. Department of Defense spent $7 billion (in 1982 dollars) to procure fourteen hundred aircraft; by 1982, it was spending $14 billion a year for approximately two hundred aircraft.[75] One analyst predicted that, based on contemporary cost trends, in 2054 the U.S. defense budget would buy a single fighter![76]

A direct consequence of striving for greater performance from new equipment was an increase in the technical complexity of modern weapons. So complicated did weapons become that Mary Kaldor referred to this phenomenon as the emergence of "Baroque technology." In her view, such systems were ineffective because they cost too much, were unreliable, were vulnerable to attack from cheaper weapons, and the improvements made in each new generation did not result in a significant improvement in performance when compared to its predecessor.[77]

The strategic consequence of this development was the death of mass production in modern war, as industrial production nearly reverted to consisting of a highly skilled cottage industry characterized by low productivity. The production rates of many new weapons and smart munitions cannot be greatly increased in a few months or even a year. Their complexity and their heavy dependence on microelectronics create a fundamentally different fabrication task than was faced in mobilizing industry to produce thousands of ships, aircraft, and tanks during World War II.[78] A good illustration of the problems created in manufacturing such weapons was provided by Northrop's experience when developing a guidance system for the MX missile in the early 1980s. The unit was about the size of a basketball, but contained more than nineteen thousand parts.[79]

Neither is it at all clear that significant increases in production can be achieved even after two to three years of mobilization. In the event of war, it is likely that once existing equipment stocks have been used up, the belligerents will have to rely on less sophisticated kits that can be mass-produced quickly. It is not clear, however, that modern industry will be able to fall back on that kind of production without major changes in plant and machinery.[80]

The fragility of military power and the vulnerability of the state to the delivery of a knockout blow has been reinforced because the use of such equipment requires a considerable degree of skill on the part of the user. Hence the soldier requires a lengthy period of training before becoming proficient in the use of such weapons. The most practical solution to this problem has been to rely on a professional armed forces. This is very costly, however, and if heavy casualties were incurred, it would be extremely difficult to replace such losses in the short term.

In the context of the cold war, it was assumed that a conflagration between the superpowers would either be stopped quickly or escalate into a possible nuclear war. Under these circumstances, the such weaknesses described above were irrelevant. But in our post–Cold War world these weaknesses are more problematic. Today, the cost of modern forces and the risks associated with their use are such that governments will be extremely reluctant to commit their forces unless they are sure of success. In this sense, technology has perhaps taken us back to a style of war similar to that of the cabinet wars of the eighteenth century. Like their counterparts in that historical epoch of limited wars, today's armed forces represent an expensive asset that, if used and lost, will not be easy to replace. Unfortunately, like their eighteenth-century counterparts, the high-tech professional military establishments of the West face a threat in the form of a political revolution that could expose their critical weaknesses and thus overwhelm them. In the past, the threat came in the form of the French Revolution. Today it is the collapse of the nation-state, the rise of alternative sources of organized violence within these states, and the emergence of new threats to our security. As Martin van Creveld writes, when confronted by this revolution in security affairs, the present armed forces are obsolete and, in fact, represent everything one must avoid if military power is to be effective.[81]

NOTES

1. A. Toffler and H. Toffler, *War and Anti-War: Survival at the Dawn of the Twenty-first Century* (London: Little, Brown and Company, 1994), 19–20.
2. P. Kennedy, *The Rise and Fall of the Great Powers* (London: Macmillan Papermac, 1989).
3. B. Brodie, *War and Politics* (London: Cassell, 1977), 1–28.
4. E. Ludendorff, *My War Memories 1914–1918* (London: Hutchinson, 1919), 328–400.
5. M. van Creveld, *Technology and War* (London: Collier, Macmillan Publishers, 1989), 164.
6. See H. S. Orenstein, ed., *The Evolution of Soviet Operational Art: The Documentary Basis, Vol. 1: Operational Art, 1927–64* (London: Frank Cass, 1995).
7. See A. Clark, *Barbarossa: The Russian German Conflict 1941–45* (London: Collier, Macmillan Publishers, 1989). For a more balanced assessment of the war on the Eastern Front see D. Glanz, *When Titans Clashed: How the Red Army Stopped Hitler* (Lawrence: University Press of Kansas, 1995).
8. R. Overy, *Why the Allies Won* (London: Jonathan Cape, 1995), 134–79.
9. H. P. Wilmott, *The Great Crusade* (London: Pimlico, 1989), 293–94.
10. Kennedy, *Rise and Fall*, 455.
11. A. Echevarria, "Tomorrow's Army: The Challenge of Non-Linear Change," *Parameters* (Autumn 1998): 85–97.
12. van Creveld, *Technology and War*, 224. W. McNeill, *The Pursuit of*

Power: Technology, Armed Force and Society since A.D. *1000* (Oxford: Basil Blackwell, 1982), 223, 13.

13. J. F. C. Fuller, *Armament and History: The Influence of Armaments on History from the Dawn of Classical Warfare to the End of the Second World War* (1945; reprint, New York: Da Capo Press, 1998), 31.

14. McNeill, *Pursuit of Power,* 278–79.

15. M. Pearton, *The Knowledgeable State: Diplomacy, War and Technology since 1830* (London: Burnett Books, 1982), 209–10.

16. McNeill, *Pursuit of Power,* 280–87.

17. Pearton, *Knowledgeable State,* 172.

18. G. Hartcup, *The Challenge of War: Scientific and Engineering Contributions to World War Two* (New Abbot: David Charles, 1970), 21.

19. K. Hayward, *The British Aircraft Industry* (Manchester: Manchester University Press, 1989), 12.

20. Hartcup, *Challenge of War,* 27.

21. C. Hable Gray, *Postmodern War: The New Politics of Conflict* (London: Routledge, 1997), 129.

22. Hartcup, *Challenge of War,* 24.

23. S. Zuckerman, *Scientists at War: The Impact of Science on Military and Civilian Affairs* (London: Hamilton Publishers, 1966), 28–29.

24. Gray, *Postmodern War,* 128–59.

25. Pearton, *Knowledgeable State,* 246.

26. J. Garnett, "Technology and Strategy," in J. Baylis, K. Booth, J. Garnett, and P. Williams, *Contemporary Strategy* (London: Holmes and Meier, 1987), 91–103.

27. H. Kahn, "The Arms Race and Its Hazards," *Daedalus,* 89 (Fall 1960): 765.

28. J. S. Gansler, *Affording Defense* (Cambridge, Mass.: MIT Press, 1991), 215.

29. Ibid.

30. Gray, *Postmodern War,* 144–46.

31. For example, K. Holsti, *Peace and War: Armed Conflicts and International Order 1648–1989* (Cambridge: Cambridge University Press, 1991), 285–86.

32. B. Brodie, *The Absolute Weapon: Atomic Power and World Order* (New York: Harcourt, Brace and Company, 1946), 73.

33. Ibid., 25–30.

34. Ibid., 76.

35. B. Buzan, *An Introduction to Strategic Studies in Military Technology* (London: Macmillan, 1987), 216.

36. G. Hartcup, *The Silent Revolution: The Development of Conventional Weapons 1945–85* (London: Brassey's, 1993), xviii.

37. B. Brodie, *Strategy in the Missile Age* (Princeton, N.J.: Princeton University Press, 1959), 248–251. For a detailed analysis of the Britain's policy on nuclear weapons, see M. Navias, *Nuclear Weapons and British Strategic Planning: 1955–58* (Oxford: Clarendon Press, 1991).

38. R. Osgood, *Limited War: The Challenge to American Security* (Chicago: Chicago University Press, 1957), 5.

39. B. Brodie, *Strategies in the Missle Age,* 310–311.

40. See T. Schelling, *Arms and Influence* (New Haven, Conn.: Yale University Press, 1966); and M. Halperin, *Limited War in the Nuclear Age* (New York: John Wiley and Sons, 1963).

41. R. Osgood, *Limited War Revisited* (Boulder, Colo.: Westview Press, 1979), 1–11.

42. H. Summers, *On Strategy in Vietnam* (Novato, Calif.: Presidio Press, 1982), 32–41.

43. See C. Browser, "Strategic Reassessment in Vietnam," *Naval War College Review* (Spring 1991): 20–51.

44. See M. Clodfelter, *The Limits of Air Power: The American Bombing of North Vietnam* (New York: Free Press, 1989).

45. For a detailed account of the first use of air mobile forces on a large scale, see Lt. Gen. H. G. Moore and J. Galloway, *For We Were Soldiers Once and Young* (Shrewesbury: Air Life Publishing, 1994).

46. Lt. Gen. P. B. Phillip, *Vietnam at War: The History 1946–75* (London: Sidgwick Jackson, 1988), 37–42.

47. J. Dunnigan, *Digital Soldiers: The Evolution of High Tech Weaponry and Tomorrow's Brave New Battlefield* (New York: St. Martins Press, 1996), 129.

48. P. Griffiths, *Forward into Battle: Fighting Tactics from Waterloo to Vietnam* (Chichester: Anthony Bird, 1981), 111.

49. Lt. Gen. Hay, *Tactical and Material Innovations in Vietnam* (Washington, D.C.: Department of the Army, 1973), 6–11.

50. W. Gibson, *Perfect War: Techno War in Vietnam* (Boston: Atlantic Monthly Press, 1986).

51. See C. Bertram, *New Conventional Weapons and East West Security* (London: Macmillan, 1979); J. J. Mearsheimer, *Conventional Deterrence* (Ithaca, N.Y.: Cornell University Press, 1983); and J. J. Mearsheimer, "Precision Guided Munitions and Conventional Deterrence," *Survival* (March–April 1979), 68–76.

52. S. Badsey, "The Doctrines of the Coalition Forces," in J. Pimlott and S. Badsey, eds., *The Gulf War Assessed* (London: Arms and Armour Press, 1992), 57–80.

53. P. Herbert, *Deciding What Has to Be Done: Gen. W. E. Depuy and the 1976 Edition of FM 100–5*, Leavenworth Papers no. 16 (Leavenworth, Kans.: Combat Studies Institute, 1988), 30.

54. Ibid., 55.

55. Ibid., 67.

56. W. Lind, "FM 100–5 Operations: Some Doctrinal Questions," *Military Review* 3 (March 1977): 54–65.

57. Gen. D. Starry, "The Extended Battlefield," *Military Review* (March 1981): 32–50.

58. Ibid., 54–58.

59. E. A. Cohen and T. A. Keaney, *Revolution in Warfare? Air Power in the Persian Gulf* (Annapolis, Md.: Naval Institute Press, 1995), 188–212.

60. See *U.S. Army Field Manual 100–5: Blue Print for AirLand Battle* (McLean, Va.: Brassey's, 1991).

61. Starry, "Extended Battlefield," 54.

62. C. Campbell, *Airland Battle 2000* (London: Hamlyn Publishers, 1986), 67–110.

63. See M. Fitzgerald, *The New Revolution in Russian Military Affairs*, (London: Royal United Services Institute, 1994), 1.

64. Toffler and Toffler, *War and Anti-War.*

65. Starry, "Extended Battlefield," 32.

66. For a more detailed analysis of the Iraqi army, see Sean McKnight's chapter in this volume.

67. See A. H. Cordesman and A. H. Wagner, *Lessons of Modern War, vol. 4: The Gulf War* (Oxford: Westview Press, 1996), chap. 3, "Forces Engaged"; chap. 6, "Shaping Coalition Air Power and the Air Campaign"; chap. 7, "Offensive Air Power"; and chap. 8, "AirLand Battle."

68. S. Biddle, "Victory Misunderstood: What the Gulf War Tells Us about the Future of Conflict," *International Security* (Autumn 1996): 142.

69. Fitzgerald, *New Revolution,* 12.

70. Ibid., 18.

71. C. Bellamy, *Expert Witness: A Defense Correspondent's Gulf War* (London: Brassey's, 1993).

72. S. Naveh, *In Pursuit of Military Excellence: The Evolution of Operational Theory* (London: Frank Cass, 1997), 323–31.

73. For more information on the causes of cost escalation of military equipment see D. L. I. Kirpatrick, "The Rising Unit Cost of Defense Equipment: The Reasons and the Results," *Defense and Peace Economics* 6 (1995) 263–88.

74. D. Smith, *The Defense of the Realm in the 1980s* (London: Croom Helm, 1980).

75. J. S. Gansler, *Affording Defense* (Cambridge, Mass.: MIT Press, 1991), 7.

76. See N. Augustine, *Augustine's Laws* (New York: Viking Press, 1986).

77. M. Kaldor, *The Baroque Arsenal* (London: Andre Deustch, 1982).

78. Lt. Gen. W. E. Odom, *America's Military Revolution: Structure and Strategy after the Cold War* (Washington, D.C.: American University Press, 1993), 146.

79. K. Mayer, *The Political Economy of Defense Contracting* (New Haven, Conn.: Yale University Press, 1991), 54.

80. Odom, *America's Military Revolution,* 146.

81. M. van Creveld, *The Transformation of War* (Toronto: Free Press, 1991).

CHAPTER 3

✦ ✦ ✦ ✦ ✦ ✦ ✦ ✦ ✦ ✦ ✦

The Impact of Communications and the Media on the Art of War since 1815

Stephen Badsey

Introduction

The very start of the twenty-first century has witnessed a preoccupation in the developed world with both communications technology and the mass news media. This has included a belief (often expressed in terms of very deep conviction) that both communications and the media have played a critical role in the wars of the 1990s and will do so again in the future. In light of this preoccupation, it is understandable that historians have begun to trace the impact of communications and of the news media on warfare back through past conflicts. Most of this research is new, emerging in a comparatively underdeveloped field. What is presented in this chapter therefore represents a synthesis of historical work in progress, rather than a definitive account.

A critical finding of this research has been that, in historical practice, military communications have developed intertwined with civilian communications, and not separately; and both have developed in a wide social and political context. In some cases the development has been principally military, and the effects on wider society have been correspondingly delayed. In others, military communications and the art of war have been changed by what have been essentially civilian developments. A recent example of the first case is the Internet, which was conceived in the 1960s as a "post-apocalypse command grid,"[1] a method of maintaining strategic military communications in the event of a surprise nuclear attack. An example of the second case is television, an essentially civilian development of the 1930s that has had a large impact on military affairs, including communications. Sometimes the two developments have been virtually indistinguishable, in cases such as the civil and military evolution of radio communications.

While studies of military communications have tended to ignore or marginalize the issue of civilian war reporting, much research in the field of military-media relations has focused chiefly on the role of civilian war correspondents, whose writings and memoirs are often a valuable historical source. It is also quite common for general histories of wars to quote from contemporary newspapers or other media sources as authoritative evidence, rather than reflecting on the circumstances that led to their production. But the role of the war correspondent in campaign has been only the most visible facet of a more complicated interrelationship that reveals much about the nature of industrialized society as well as about industrialized warfare. Part of the value of studying any government's behavior toward the mass media at war, and also the response of the mass media, is that historians have found such behavior to represent an extreme form of peacetime practices, rather than an exceptional case.[2] Any historical assessment of the role of the mass communications media in war and of the military-media relationship since about 1815 must therefore take into account a wide range of factors. Among these factors are: how far the country had embraced concepts of civil society, and the role of the media within them; the media within that society as an institution, and its structures; what the nature of warfare was, and was believed to be at the time; and the physical means available both for military and for media communications. Above all, there must be an understanding of how the relationship between government, armed forces, and "public opinion"—however that difficult term is defined—worked, or was believed to work. A simple description of who the war correspondents were and how they reported the war completes such a picture, rather than defining it.

Research into communications, media, and warfare not only has revealed complex interrelationships but has provided valuable new insights into the nature of how warfare has developed. It is well understood that technological developments do not necessarily impact on human society to cause change immediately but may be taken up or delayed in their impact through a wide variety of social, cultural, or practical factors. This has turned out to be as true of developments in communications and the media as in military technology.[3] It has also been the case that some—if by no means all—influential military writers since 1815 have included an understanding of changes in communications as part of their theories of warfare. Some successful commanders also have included an understanding and exploitation of the mass media within their own methods and philosophies of war. Generally, it is coming to be understood that the role of communications and the media has, on a number of occasions, been central and fundamental, rather than peripheral, to the art of war.

Transportation and Information: The Double Revolution

Although early industrialization played some part in the French Revolutionary and Napoleonic wars, it is the period since 1815 that has seen its major effects on warfare, usually considered in two main forms. One of these has been the vastly increased impact of firepower, in all its varieties, on the conduct of battle, which has shown little sign of slowing or diminishing over two centuries. Indeed, entire books have been written, and theories and models of warfare developed that have considered virtually nothing but this aspect of the military art.[4] It is also fair to say that, throughout the industrialized period, most practitioners of warfare have thought about their profession largely in terms of firepower and how to direct it.

Away from the immediate battlefield, the other main impact of industrialization that has attracted attention has been on the wider conduct of military strategy. In particular, the social changes and population growth in Europe and North America during the nineteenth century have been recognized as the basis for the mass armies and navies that remained a feature of industrialized countries until almost the end of the twentieth century. Developments in physical communications or transport systems have long been understood as an important aspect of this second area of impact. Both contemporary military writers and subsequent historians have attached considerable significance to the development of railroads in Europe and the United States in the nineteenth century, as helping determine the manner in which industrialized wars were fought. The British mid-twentieth century military theorist J. F. C. Fuller went so far as to argue, "It was George Stephenson more so than Napoleon or Clausewitz who was the father of the nation-in-arms," through Stephenson's building of the world's first true railroad, the Stockton-Darlington Railway of 1825 in northern England. By 1840, Great Britain had five hundred miles of railroad line, increasing to thirteen hundred miles within a decade.[5] Historians have analyzed the military input into the building of the European railroads that followed, particularly those of Austria, France, and Germany, together with the interrelationship between the use of railroads for strategic deployment and the rise of the Prussian General Staff under Field Marshal Graf Helmuth von Moltke (the Elder).[6] By 1914, Germany had forty thousand miles of track, while Belgium, with over five thousand miles of track, had the densest railroad network in the world.

In the 1960s some historians took the issue of this interrelationship between railroad communications and warfare much further, arguing that the railroad timetables developed for mobilization by most European powers in 1914, particularly the controversial German "Schleiffen Plan," deprived its leaders of choice. In this hypothesis, the First World War was "imposed on the statesmen of Europe by railway timetables."[7] Whether or not this is true (and subsequent research suggests it is not), railroads began to have a significant impact on military strategy in the middle of the nineteenth century, with the Franco-Austrian War (1859), the American Civil War (1861–65), and the Austro-Prussian War (1866). They continued to do so well into the Second World War, perhaps most famously with the German "Operation Barbarossa" offensive against the Soviet Union in 1941.[8] In terms of strategic land movement, there remains no substitute for railroads in industrialized warfare. European railroads even played a significant role in the deployment of American and British forces from Germany to Saudi Arabia for the 1991 Gulf War, and it is difficult to point to a single major land campaign since 1859 in which railroads have played no part at all.

Whereas the impact of railroads on military strategy began in the early nineteenth century, they have had only occasional impact on battlefield tactics.[9] The true impact on the battlefield itself of changes in transport systems was not seen until the beginning of the twentieth century, with the development of the practical all-terrain internal-combustion engine vehicle, particularly in its armored form as the tank, first seen in 1916. It may be argued that the tank and accompanying armored vehicles, in transforming the size and scale of the battlefield, led directly to a new area of twentieth century military theory, the "operational level" lying between conventional battlefield tactics and wider strategy. Although a valid attempt has been made to find its origins in the concept of "grand tactics," proposed by the Swiss theorist Antoine Henri de Jomini in 1838, the true development of the operational level of war began only in the 1920s in the new Soviet Union, with the school of military thinkers under M. M. Tukhachevskii. Their ideas, given a first clear expression in 1929 by V. K. Triandafillov, proposed a "revolution in military affairs," the product of industrialization, the motorization and mechanization of armies (including airpower and sea power), and concomitant political and social changes.[10] The existence of the tank and of large-scale mechanized warfare also gave new importance to the use of roads in warfare, as well as railroads (although military roads have existed since the time of the Romans, if not before).

As the British science writer Arthur C. Clarke has pointed out, this transformation in communications has actually been a double one: a greatly increased ability to move physical objects (including people) at speeds unknown before 1815 has developed in parallel with the ability to move information over even greater distances, achieving in practice simultaneous communication.[11] Experiments in the development of the steam engine as the basis for the railroad (and steamship) took place from the middle of the eighteenth century, alongside the earliest experiments in using electricity for sending information. The first practical telegraphs were actually mechanical semaphore systems, based on ideas and technology that had existed for centuries but first built in the Revolutionary and Napoleonic wars: the French Chappé chain of 1794 and the British Admiralty chain linking London with Plymouth by 1806.[12] The practical electric telegraph connected through cables did not appear until the 1840s. The difference to both peaceful commerce and warfare was made not so much by the railroad as by the railroad *timetable*: the ability to transmit information at speeds much faster than the trains themselves (which, by the end of the nineteenth century had achieved speeds of over 100 mile per hour), in order to control the rail network. By 1900 the entire industrial world was connected by underwater and overland telegraph cables, largely commercially owned.[13]

The idea that increases in the advance of communications technology might produce perfect communications, whether for the military or for the news media, has been a persistent hope of industrial-age warfare, not least from its inventors and manufacturers. Surprisingly, the concept has not met with much favor from experienced practitioners of war, including Moltke "the Elder," who reflected on the first impact of industrialized military communications in a famous (if usually paraphrased) passage:

> But the most unfortunate of all supreme commanders is the one who is under close supervision, who has to give an account of his plans and intentions every hour of every day. This supervision may

> be exercised through a delegate of the highest authority at his headquarters or a telegraph wire attached to his back. In such a case all independence, rapid decision, and audacious risk, without which no war can be conducted, ceases.[14]

Other aphorisms by Moltke that have passed without much reflection into twentieth-century military folklore, "mistakes in the original deployment cannot be corrected" and "no plan survives contact with the enemy,"[15] derived from such thinking, and from the specific levels of industrial development of railroads and telegraph communications in Europe of the 1860s and 1870s. A study undertaken in 1985 under the Israeli historian Martin van Creveld showed that from Napoleon onward, and virtually regardless of the level of technology available, commanders and their headquarters functioned best in war if they were trained and structured to minimize formal orders and accept with equanimity occasional communications gaps and breakdowns. Despite the skepticism of many military practitioners, what van Creveld has described as "the quest for certainty" in military communications has continued through the development of aviation and radio in the early twentieth century, finding its most recent manifestations in claims for military surveillance technology of the computer age.[16]

The Mass Media and Civil Society

The rise in importance of communications for moving material as well as information under the effects of nineteenth-century industrialization was accompanied by a corresponding rise in the importance of the mass communications media, both in commercial terms, as businesses, and in constitutional terms, as part of the concept of wider civil society. Throughout history, government production of official communiqués and sponsored media, as well as the habit of commanders publishing their own dispatches and justifications for their actions, has been quite common, predating the first independent mass media and war correspondents by several centuries. The world's first newspaper—so described—probably appeared in Central Europe in 1609, and established daily journals first appeared in London about a century later. But the concept of a mass communications media as part of civil society evolved only slowly and patchily in Europe of the nineteenth century. It has been reasonably argued, "In both Britain and France before 1800, the press only touched the lives of a minority of people. The French Revolution may have led to the creation of a press on a scale never before seen in western Europe, even in Britain, yet it left vast areas of society unaffected."[17] Levels of urbanization, literacy, and lower-class spending power were insufficient to maintain a mass market for newspapers, and little effort was made to create one. Broadly, mass civil society developed later and less rapidly on the continent of Europe than in Great Britain, which was unusual in having no central government censorship system for the press. Although probably the world's first news agency was founded in Paris in 1832, it was these British liberal traditions and the development of the telegraph that led to the much more significant establishment of Reuters in London in 1851.

The first important military writer to express an interest in the emergence of the new mass-industrialized society was the Prussian major general Carl von Clausewitz, in his seminal but posthumous work, *On War,* of 1832. Although subject to fashion ever since, even in his home country of Prussia/Germany, Clausewitz has probably been more influential on military thought as a whole than any other single theorist.[18] Unsurprising given the period during which he wrote, Clausewitz has little to say about the press, except for a passage acknowledging its growing impact even on his native Prussia:

> The movement of the enemy's columns into battle can be ascertained only by actual observation—the point at which he plans to cross a river by the few preparations he makes, which become apparent a short time in advance; but the direction from which he threatens our country will be usually announced in the press before a single shot is fired. The greater the scale of preparations, the smaller the chance of achieving a surprise.[19]

Even so, one of Clausewitz's most enduring concepts from *On War* has been what he called his "remarkable trinity," a three-sided relationship between leader (or government), people, and armed forces that he argued was fundamental to understanding any country at war. Viewing the media as part of the essential communications by which that relationship is maintained fits well not only with eighteenth-century notions of the role of the media in wider civil society but also with later twentieth-century ideas about the mass media and their importance.

By conventional agreement, the first major war to be reported in a systematic manner to newspapers at home, and in which newspaper reporting played a significant part in the war's conduct, was the Crimean War of 1854–56, with the Anglo-Irishman William Howard Russell, reporting for *The Times* of London, being the first to be described as a "war correspondent" (a title that he actively disliked).[20] The new technology of telegraph and steamship also played a significant role in getting news reports back from the Crimea to London and other capitals. Since the voting public of Great Britain at the time represented only 7 percent of the adult population (men and women), however, the impact of Russell and his fellows should be judged as falling more on government and political and social elite circles.

A much more important case is the role of reporting in the American Civil War. Both sides encouraged reporting from overseas, particularly from Great Britain and France, and the impressions of visiting reporters and attached officers form one of the more useful historical sources for the war, as well as sparking a debate on how well nineteenth-century Europe learned its lessons.[21] But, although propaganda played its part in the war, and the survival strategy of the Southern Confederacy in particular depended heavily on achieving European support and recognition, there is little evidence of attempts by either the Lincoln or the Davis government to appeal to the people of Europe through their mass media, or even to consider such an option. This appears to have been partly due in the Confederate case to overoptimism.

Russell as a reporter described numerous conversations with Southerners who "seemed to consider the British Empire as a sort of appendage to their cotton kingdom."[22] Partly this perception was due to the relative underdevelopment in any European country—including Great Britain—of the concept of mass society and the mass newspaper press that went with it. It was only in the 1860s that France pioneered "the really large-circulation newspaper, thus breaking a trail which the rest of the continent was later to follow."[23] Typical was *Le Petit Journal,* founded in Paris in 1863, which had achieved a circulation of 5 million two years later.

In these terms the United States, both during the Civil War and for some years after, had the most developed and complex relationship between government, press, and people in wartime of any country in the world. Universal manhood suffrage was joined with a high basic literacy rate, at least in the Northern cities. The constitutional separation of executive from legislature, the protection of the press under the First Amendment, and the sheer size of the United States all contributed to an exceptional role for the mass news media in American society and politics and so, by extension, in war. By comparison, it has been estimated that "in Prussia the circulation of newspapers in the 1860s in proportion to population was only a sixth as great as England, and that of the 19 million inhabitants of the state there were not more than 1.5 million with a reasonable understanding of public affairs derived from the press."[24] In Russia of the same period, even the concepts of public opinion and civil society were only just coming into existence.

Nevertheless, the ability of the press both to report the Civil War and to influence its conduct remained severely limited by technological and institutional constraints. There was no national press; one of the largest-circulation newspapers, the weekly edition of the *New York Tribune,* had sales of only 220,000 in 1862 (although, like all newspapers, it was read by many more people than bought it). Rather than employing specialist war correspondents, newspapers often relied heavily on "clipping" stories from the news agencies or from each other, with results that were frequently wildly inaccurate.

The Confederate fortress town of Vicksburg was reported in the Northern press as captured at least three times before it finally fell in July 1863. Confederate soldiers were even worse off, and their diaries and letters "suggest that they suffered less anxiety through ignorance than from 'knowing' things which had in fact never happened."[25] American newspaper reporting of the frontier wars that followed, and newspaper commentary on those wars, was equally haphazard, with accounts of the Fetterman Massacre of 1866 or Little Big Horn ten years later sometimes bearing almost no relation to reality.[26]

It was one of the oddities of the Crimean War that, in response to newspaper criticism of its conduct, Prince Albert (husband to Queen Victoria) encouraged the use of new media technology to reveal the nature of the war, through the photographer Roger Fenton. Although thereafter war photographs developed a small niche for themselves in the form of exhibitions, they fit best into the continuing domination of newspapers over the mass news media market in most countries, taking their place alongside the earlier kinds of war illustration but not starting to replace them until the very end of the nineteenth century.[27] Cine-cameras also first began to record wars from about the Spanish-American War of 1898 onward. But, although the cinema newsreel grew increasingly in importance, reaching a high point in the 1930s, it never seriously challenged the domination of the newspapers. Research has confirmed that although typically cinema audiences enjoyed newsreels, they saw the cinema principally as a source of entertainment rather than news.[28] Certainly, even as early as the First World War, some newsreel films played a major part in mass perceptions of warfare, notably the remarkable British production *Battle of the Somme* of 1916.[29] But the major impact of visual images on the military-media relationship came with further technological developments later in the twentieth century, most obviously that of television.

Although some of the same technology that went into making lenses for cameras contributed to the new gun sights that were to transform artillery between 1880 and 1914, photography and film became a central part of warfare only by combining with the emerging transport technology of powered flight. Coupled with other technologies and techniques, this combination was to bring about the major changes in artillery that have recently been acknowledged by historians as one of the critical developments of the art of war during the First World War.[30] The other major communications technology development of the late nineteenth century had, like the railroad, an immediate impact on strategy but took some decades to have any significant impact on the conduct of battles. This was the development of the telephone in the 1870s, followed by wireless telegraphy at the turn of the century, with Marconi sending his first signal across the Atlantic in 1901. This, in turn, called into existence a new branch of warfare, "signals intelligence," which played a significant role in both world wars. In the First World War, as well as having considerable value on the battlefield, signals intelligence was responsible for British code breakers obtaining the "Zimmermann Telegram," an episode that helped precipitate the United States' entry into the war in 1917.[31] In the Second World War, the contribution of signals intelligence, particularly through the Allied breaking of the German Enigma codes, was of such importance that it is still being researched and evaluated.[32]

The introduction of these new technologies and many others both coincided with and assisted in a transformation of society throughout Europe from the 1870s onward, including unprecedented emigration to the United States and the colonial empires, and in the rise of "technocrats" in government in every major country. The phenomenon the creation of a mass popular press, often in imitation of the American "New Journalism" style of the 1880s,as part of a widening of the political nation in each country, often accompanied by an extension of the franchise. From having four daily newspapers in 1878, Paris had seventy of all kinds by 1914, while, by 1896, London had the first mass circulation newspaper deliberately aimed at a working-class market, the *Daily Mail*.[33] Most political leaders at the turn of the new century were conscious that "public opinion" had taken on a new meaning.

Colonial War, Communication, and the Mass Media

In addition to this social and political transformation, the era 1871–1914 saw relatively few wars between industrialized nations, and none at all between the great powers (despite repeated war scares), but also considerable expansion by the United States and the major colonial empires, particularly that of Great Britain. Notoriously, the British imperial expansion was the result not of central policy but of local political leaders creating wars despite the indifference or even opposition of the government in London. They succeeded in doing so in part by exploiting those gaps and delays in the telegraph system that still existed before 1900, as well as exerting influence over the provincial and sometimes the metropolitan press.

Perhaps the earliest case of this happening, according to recent interpretation, involved the governing authorities in New Zealand in the Northern War of 1845–46, who managed to present a limited defeat against the indigenous Maoris as a triumphant victory, with "the final and complete subjugation of the rebels," by the use of misleading dispatches back to London and careful influence over the local newspapers, so paving the way for further attempts at conquest.[34] Certainly, the Zulu War of 1879 was precipitated by the governor general of Cape Colony, Sir Henry Bartle Frere, depending on the slow speed of communications with London.[35] Most notorious of all, his later successor Sir Alfred Milner precipitated the South African or Anglo-Boer War (1899–1902) by actions that included planted stories in the London and the Cape press to create war fever.[36] In a famous essay written in 1940, the British writer George Orwell noted why such imperial—not to say imperious—behavior in fighting colonial wars had vanished along with its practitioners:

> The thing that had killed them was the telegraph. In a narrowing world, more and more governed from Whitehall, there was every year less room for individual initiative. . . . By 1920 nearly every inch of the colonial empire was in the grip of Whitehall. Well-meaning, over-civilised men, in dark suits and black felt hats, with neatly-rolled umbrellas

> crooked over the left forearm, were imposing their constipated view of life on Malaya and Nigeria, Mombassa and Mandalay. The one-time empire builders were reduced to the status of clerks, buried deeper and deeper under mounds of paper and red tape.[37]

Orwell might have added that developments in mass media also brought the military methods used to expand and control the British and other empires under much closer scrutiny by a much larger voting public. This new and often unfavorable publicity played an important part in the response to such events as the Amritsar Massacre of 1919 and in subsequent opposition to British rule in India and elsewhere.[38]

The same set of circumstances also produced, in the same late-nineteenth-century period, a new relationship between senior officers on campaign and the news media, based largely on the lack of opportunity in all armies for promotion by formal means on merit. Generally, and regardless of the degree of democracy in the countries for which they pledged to fight, senior officers were governed far more by their own personal attitudes toward the press than by any regulations or official guidance, and they were usually deeply hostile. In the United States, William T. Sherman was notoriously intolerant of the press, both during the Civil War and later. The manner in which his armies famously "vanished" into the interior of the Confederacy in November 1864 was due in large part to cut telegraph wires and Sherman's blunt refusal to allow reporters to accompany his march.[39] Only a few years later, in 1871, Colonel Sir Garnett Wolseley, the leading British army reformer, wrote in his wide-selling *Pocket Book* of advice to officers:

> The English general of the present day is in the most unfortunate position . . . [of] being surrounded by newspaper correspondents, who, pandering to the public craze for "news" render concealment most difficult. However, transport and telegraph will always be in the general's hands, so he can lay an embargo on the mails whenever he wants it, without its being known for a long time; or he can, by spreading false news among the gentlemen of the press, use it as a medium by which to deceive an enemy.[40]

Despite this, Wolseley rose to become commander in chief in 1895, in part through his knowledge of how to use the press to promote his own image on campaign. Such behavior was often frowned upon, but it was also an essential part of the campaigning style of some notable commanders. When "General" (actually Lieutenant Colonel) George Armstrong Custer was killed at Little Big Horn in June 1876, there is a strong possibility that he was trying to achieve a great victory close to the one hundredth anniversary of U.S. Independence on 4 July.[41] But the greatest success in creating a self-image through the press in war was Lord (Frederick) Roberts of the Indian army, commander in chief of British forces in South Africa in 1900, who managed to convince Great Britain that the Anglo-Boer War was won and was recalled as its hero and victor (taking almost all the British war correspondents with him), when in fact the war still had considerably more than a year to run.[42]

This short-lived and very personalized relationship between senior officers and the press was the product of the small-scale nature of nineteenth-century colonial warfare, known to the British as "small wars." The combination of wars fought at a distance, fragile communications, an expanding market for mass media at home, greater government sensitivity to public opinion, and the need for publicity perceived by some senior officers all gave the war correspondents importance, so that within a few years this era would be mythologized as the "golden age" of war reporting.

Mass Armies, Mass Media, and National Ideology

The requirements and problems of European-style *grande guerre* (or industrialized warfare) raised altogether different issues to those of small wars. At the time when Clausewitz wrote, the ideas of mass warfare drawing on the traditions of "people's war" that he appeared to advocate were by no means popular with the governments of Europe. Antiliberal or reactionary European political leaders of the age of Metternich sought, if possible, to remove the memory of people's war, with its associations of patriotic fervor and mass conscription in the French Revolution. This was particularly true of Prussia, where the army was firmly "deliberalized" between 1815 and 1848.[43] But particularly after the events of 1848 showed that the forces created by the revolution were not going to dissipate so easily, politicians of the age of Bismarck sought instead to channel such behavior into a new militarism that would provide such mass conscript armies, but under state control. In Prussia especially this was done partly through press and propaganda campaigns directed increasingly at the middle and lower classes. Bismarck himself helped found a newspaper (the *Kreuzzeitung*) and was accomplished both as a propagandist and at exploiting journalistic and telegraph service contacts.[44]

After the Franco-Prussian War, virtually every European country adopted some form of Prussian-style short service conscription. The problem that this provided for military institutions and their theorists up to the First World War was how to balance the need for mass armies with the fear of mass rebellion. In France and Germany, some revolutionaries favored conscription (to the bafflement of their British and American colleagues), since among other advantages it taught military skills to potentially revolutionary workers. Hence the adoption also of fierce "Prussian" discipline and military style in most armies, and also the increasing emphasis on military nationalism that was a feature of the last decades before the First World War, stretching even to calls for peacetime conscription in Great Britain. In the United States, the same issues produced in the early 1900s the "Uptonian pessimism" debate, an argument that mass armies could neither exist nor fight effectively without such concomitant militarized traits, and that the United States could become an effective military power only by abandoning its democratic and citizen-soldier traditions.[45] It was the experience of the First World War that convinced the United States that democratic citizen armed forces, motivated by an independent mass media in cooperation with the state, could win major wars.

An important part of this Prussian style, which was to endure until the end of the Second World War, was the brutal repression of popular uprisings and other manifestations of people's war by occupying forces. Moltke the Elder, writing in about 1880 of his difficulties in the Franco-Prussian War with French *franc-tireur* irregulars, produced this passage of self-justification:

> An armed mass of people is far from being an army. Leading such a mass into battle is pure barbarism. . . . Their gruesome work had to be answered by bloody coercion. Because of this, our conduct of the war finally assumed a harshness that we deplored but which we could not avoid. The *Franctireurs* were the terror of all the villages; they brought on their own destruction.[46]

These ideas led directly to the brutal German policy toward Belgium in the first months of the First World War known as *Schrecklichkeit* (translated into English at the time as "frightfulness"), a belief that only outright terror could suppress popular guerrilla uprisings. An important part of *Schrecklichkeit* was that, to be effective as a weapon of popular control, German reprisals required publicity, including the use of the international mass media. Journalists such as the American Gerald Morgan were invited to witness such brutal episodes as the sack of Louvain.[47] As one British historian of the First World War has observed, from the propagandist's viewpoint, "the Germans were in many ways the perfect enemy. Their conduct throughout the war seemed almost designed to offend British liberal sensibilities and to galvanize public opinion in support of the war effort."[48] This apparently bizarre behavior came not from a failure to appreciate the importance of the mass media but from the very depths of German war-fighting doctrine and methods, and from their belief as to how the media should be involved to aid them. In 1937, the Imperial Japanese Army used similar arguments in inviting the world's press (including, this time, newsreel cameras) to witness the "Rape of Nanking" as part of their occupation of China.

This emphasis on national ideology and the militarized "nation in arms" reached its high point in the First and Second World Wars, and the role of propaganda and the media in these wars continues to be widely studied. Indeed, one historian has written that although these were wars of firepower and industrial strength, "This is not to underestimate the importance of propaganda, for the Second World War more than any previous conflict since perhaps the seventeenth century, was a battle for men's minds. Entire peoples had to be convinced that the war was worth fighting and that ultimate victory was assured."[49] In almost all countries, the main focus of mass media was on maintaining public morale and support for the war at home. In Great Britain and the United States and in other democratic countries, this was achieved in both wars by a fusion of government strategies with the media industry's skills, in what American cinema owners described in 1917 as "practical patriotism."[50]

One particular feature of the First World War recently noticed by historians was its role in the creation of a national identity for Australia, through the promulgation of Australian military achievements by the mass media at home in a very deliberate manner. United only in 1901, when it became a commonwealth, Australia was also possibly the most democratic country in the war, with universal suffrage including women (but not Aborigines); and the need to create such an identity from war may reflect this unusual combination of a wide social base for civil society with underdeveloped political institutions. Taking a process common to many warring countries to its greatest extent, the Australians appointed their own official war correspondent to accompany their forces overseas, Dr. C. E. W. Bean. Later transformed into the Australian official historian, Bean played a critical role both during the war and for decades afterward in creating and sustaining the "ANZAC myth" (from "Australian and New Zealand Army Corps").[51]

With public morale seen as an institutionalized government responsibility, control of the press became, in most major countries, much less a personal matter for individual generals and much more part of the state apparatus, even before 1914. The role of the mass media was so important that even the most reserved and autocratic of First World War commanders, such as the British Sir Douglas

Haig, found it necessary to associate with news reporters as part of the art of generalship.[52] By the twentieth century, the mass media were not something that any industrial state at war could ignore or regard with indifference, and increasingly as the century progressed, neither could senior military figures in the conduct of their battles.

One misconception in the art of warfare to emerge from the First World War was a deep belief in the power of the mass media, harnessed through state propaganda, to undermine the morale of the enemy population and armed forces on a large scale. This was closely associated with the fanciful German "stab-in-the-back" claim that their home society had collapsed in 1918 and not their armies. The German magazine *Simplicissimus* provided an appropriate cartoon of the British Lord Northcliffe, owner of *The Times* and the *Daily Mail* and also director of Propaganda in Enemy Countries, being welcomed into hell by the devil with the words "Welcome Great Master! From you we shall learn the true science of lying!"[53] The broadening of the franchise across Europe; the success of Fascist demagoguery, starting with Mussolini in Italy in the 1920s; even the growth of popular interest in psychology, all combined to produce, by 1939, the vague belief that people en masse were vulnerable to forms of mind control induced through the media.

Technical abilities to transmit speech radio had been advanced in the course of the First World War, first by the British through use of "continuous-wave" sets, supplementing field telephones and wireless telegraphy.[54] This helped produce the postwar boom in domestic radio sets, which became the main addition to the mass news media in the 1920s apart from the growth in cinema newsreels. The difference between government attitudes toward print media and toward radio (and later television) was significant. Whereas, by the twentieth century, newspapers independent of direct government control were considered part of democratic culture, a degree of government control over radio was regarded as mandatory, particularly in time of war. In the United States, where government regulation was weak, the Radio Corporation of America was founded in 1919; in the first quarter of 1921 alone, 219 commercial radio stations were registered, and by 1922 there were fifteen thousand retail outlets selling radio sets.[55] A special place in the history of both communications and warfare belongs to the British Broadcasting Corporation (BBC), established in 1926, which shortly afterward began experimenting with broadcasts around the world and in 1932 established its global "Empire Service."[56]

Like the cinema, radio was seen as a source of entertainment more than a source of news, but by the 1930s it was undoubtedly the dominant popular entertainment and communications medium and had begun seriously to challenge newspapers as the dominant news medium. Of 32 million family homes in the United States in 1938, 27.5 million had radios, a higher proportion than had telephones, automobiles, plumbing, or electricity or that took newspapers or magazines. In the Second World War, radio came into its own for news purposes, including through presidential and prime-ministerial broadcasts. But already a startling episode had seemed to confirm all fears of the vulnerability of mass society to the power of radio. Orson Welles's Mercury Theater radio play adaptation of H. G. Wells's *War of the Worlds* on 30 October 1938, transmitted from New York as a factual account of a Martian invasion of New Jersey and heard by an estimated audience of 6 million, briefly caused mass panic.[57] The Soviet military theorist and exponent of the operational level of war, V. K. Triandafillov, included a special section in his writings on the importance of propaganda and media manipulation. The role of press propaganda remained central to Soviet views of warfare throughout the existence of the Soviet Union, with "propaganda companies" interacting in battle in the Second World War in a manner similar to that of their German equivalent but largely unknown to the western Allies.[58] The United States and Great Britain did use institutionalized and uniformed propaganda units (seldom, if ever, given that title) to carry out specific tasks such as psychological warfare.[59] But each country also maintained a largely separate structure for their war reporters, who were at least notionally independent of direct government control, even if often required to wear military uniform.

Senior commanders of the Second World War, much more than their First World War counterparts, also acknowledged the importance of the mass media in maintaining morale as part of their war-fighting process. The most forthright view was that of the United States' General Dwight D. Eisenhower, supreme Allied commander in Europe 1944–45: "Public opinion wins wars."[60] In November 1944, Eisenhower even called for a major Allied propaganda and media offensive to lower German military and civilian morale, in order to reduce the prospect of hard fighting as the Allies entered Germany. Eisenhower's own conclusion from this controversial episode was that "the expenditure of men and money in wielding the spoken and written word was an important contributory factor in undermining the enemy will to resist."[61]

Given the particular place of the mass media in United States political culture, it was expected that American generals would emphasize their connections with the press and its value to the war effort. One of the most controversial cases was the control of both official communications and the press that allowed General Douglas MacArthur to mitigate his defeat against the Japanese at Bataan in the Philippines in 1942, and his own responsibility for it. MacArthur told his press officer, Carlos Romulko, "The sword may rest, but the pen never does." The stream of over 140 official communiqués in seventy-eight days from MacArthur's headquarters—from which he seldom, if ever, ventured, earning himself the nickname "Dugout Doug" from his troops—was coupled with his careful cultivation of journalists, including those from *Time* and *Life* magazines. By little more than skillful use of the media for his own ends, MacArthur converted a national defeat in the Philippines into a personal triumph.[62] But even one of the most naturally reserved of the British generals, Sir Bernard Montgomery, deliberately cultivated relations with the press for much the same reasons as did Eisenhower and MacArthur. Montgomery's chief of staff included an entire chapter in his memoirs on military relations with the press and their importance.[63]

The Second World War saw strong governmental and military institutional controls over the Western democratic press, combined with the broad assumption that, in a war of national survival, excessive press criticism would amount to a form of treachery. The role of the war correspondent was correspondingly downgraded, despite the new immediacy of radio and the apparent friendliness of some generals and admirals. Generally, the level of cooperation between Western correspondents and their military authorities was high; the correspondents seldom complained or broke bad news stories, and they had little recourse or support if they considered themselves unfairly treated. For these reasons, quite soon after its end, the military-media relationship in the Second World War assumed the halo of a "golden age" from the government and military perspectives. From the 1950s onward, quite elaborate theories of "limited war" evolved to describe the use of military force and warfighting under very different circumstances, including the presumed role of the press. But in practice, governments and their armed forces involved in almost any military operation continued to evoke the myth of the cooperative Second World War correspondent through to the end of the twentieth century.[64]

Morale, Maneuver, Airpower, and the Media

A problem that had affected both warfare and the news media before 1914 was that of developing speech radio sets robust enough to be easily portable and to be taken onto a battlefield. The absence of such communications when directing mass armies with heavy firepower had a fundamental effect on the art of war and contributed much to the shaping of First World War battles, particularly on the deadlocked Western Front.[65] One of the leading German pioneers of armored warfare between the wars, Heinz Guderian, who had himself served in radio communications during the First World War, made a point in his 1937 book *Achtung—Panzer!* of stressing the critical importance of radio communications to mobile armored warfare.[66] The possession of radios by every German tank and of backpack radios at platoon level among the infantry was a significant advantage to the Germans in their dramatic defeat of the Allies in France in 1940.

The widespread use of radios also meant a fundamental change to command style between the wars. Whereas, in the past, commanders had watched a battle, now they commanded by *listening* to it taking place, in the form of radio traffic. Also, in a command style pioneered by Guderian, among others, it was possible for senior officers to go forward while remaining in communication with their headquarters via radio. One of the most publicity conscious of German generals of the Second World War, Erwin Rommel, also made a point of taking his own photographs as an aid to his propaganda image.[67] Increasingly, as armed forces became more reliant on electronic communications, the metaphor became accepted of military organizations resembling human bodies, with the headquarters and staff as a brain and the communications as the nervous system. This idea probably first occurred to J. F. C. Fuller as early as 1918, during the German offensive and British retreat. He later wrote, "I was watching the *déroute* of our troops in March 1918. Why were they retiring? Because our command was paralyzed."[68] This idea of psychological or strategic paralysis among both enemy troops and higher command became important to the aggressive style of land warfare known in the Second World War as "blitzkrieg," or in its Soviet variant as "deep battle." Developed further in the 1970s and 1980s as "maneuver war," this became the dominant war-fighting style of advanced industrialized states in the second half of the twentieth century.[69]

Maneuver war began very much, as in the case of Fuller, as a reaction to the Western Front in the First World War. As a systematic study of war became more common during the nineteenth century, and in particular, as reliable statistics on battles and campaigns began to be collected, military theorists confronted a major issue, perhaps first identified by the French colonel Ardent du Picq in his posthumous *Battle Studies* of 1870: armies and their commanders considered themselves defeated when only a minority of soldiers had become physical casualties.[70] From this observation came the military aphorism "Battles are decided by the number of men on both sides who run away." This observation was extended in the twentieth century to mass-industrialized war and its impact on civilian populations as well as military forces: in both world wars, the losing side capitulated having lost no more than a fraction either of its total population or of its military forces. From this came the search for methods of warfare that might induce or increase the likelihood of capitulation by the enemy without heavy losses on either side.

As has been shown, an awareness of the relationship between warfare, mass society, and mass media had been a developing feature of military art since the early nineteenth century. Concepts of individual and mass psychology, the mass media, and propaganda should have played an important role in maneuver war and similar new ideas, and in attempts to put them into practice. In fact, despite frequent reference to psychology, military morale, and other essentially abstract notions regarding human behavior, such theories seem to have developed almost without reference to mass society and the media, at least in the West. It has been rightly pointed out that the *British Journal of Psychology* was founded only in 1904, and that early thinkers such as Fuller were forced to improvise.[71] In contrast, Soviet views of maneuver war always placed an emphasis on propaganda and its importance.[72]

Among twentieth-century British military thinkers, it was Basil Liddell Hart who would emerge as the supreme exponent of war being waged for the mind of the enemy commander and his troops, notably with his concept of the "indirect approach." Like Clausewitz before him, Liddell Hart's popularity has waxed and waned, but he remains probably the most influential twentieth-century theorist on warfare, particularly on mechanized land war. Although Liddell Hart was himself a prolific writer and accomplished journalist, the single most clear exposition of his theories of war, *Strategy* of 1954, which dwells heavily on the importance of disrupting communications and influencing the minds of the commander and his troops, has effectively nothing to say about propaganda, the media, or other nonmilitary methods by which this might be achieved. Instead, his writings appeal to the self-evidence of the proposition that a physical attack on enemy communications can itself cause psychological effects:

> While a stroke close to the rear of the enemy force may have more effect on the minds of the enemy troops, a stroke far back tends to have more effect on the mind of the enemy commander. . . . These deductions were confirmed by the experience of the Second World War—above all by the catastrophically paralysing effect, physically and psychologically, that was produced when Guderian's panzer force, racing far ahead of the main German armies, severed the Allied armies' communications [in France in 1940].[73]

The explanation for Liddell Hart's development of a theory of maneuver war without a corresponding theory of the significance of mass media appears to lie in his rejection of the very concept of mass warfare and his search for a form of "bloodless victory," coupled with a strong belief in the fighting value of elites and in military professionalism.[74] Rather than control and direct mass populations as the basis of military force, as Moltke had sought to do, Liddell Hart and his followers hoped to substitute machines and much smaller armed forces.

The same argument applies with even more force to the development of theories of air warfare after the First World War, with which Liddell Hart associated himself as early as 1925, arguing that "a decisive air attack would inflict less total damage and constitute less of a drain on the defeated country's recuperative power than a prolonged war of the existing type."[75] Just as the tank was meant to substitute for the manpower of massed armies, so the aircraft was meant to provide a new technological means of war. The idea of organized air warfare on any scale was also scarcely conceivable without first wireless and then radio communications. Using airpower to assist the land battle was always an important part of mechanized or maneuver war. But a variety of theories of independent air warfare also developed in the 1920s, each based on the identical premise to that of maneuver war: defeating the enemy quickly, without the need for mass armies. The variety came in the choice of target: the enemy armed forces, military and political leadership and communications, or civilian population. Despite changes in technology during the twentieth century, these basic choices of target and targeting priorities have remained almost unchanged.[76]

Flowing naturally from such ideas, just as from the trends of militarism in the second half of the nineteenth century, was an increasing belief in the vulnerability of undisciplined civilians to attack from the air. Such beliefs, as the basis of mass bombing campaigns from the Second World War onward, were based on military conviction rather than historical analysis and took virtually nothing from studies of mass media and propaganda. By the end of the twentieth century, "in the more than thirty strategic air campaigns that have thus far been waged, air power has never driven the masses into the streets to demand anything."[77]

Like theories of mass warfare of the late nineteenth century, many of the more radical ideas of both maneuver war and airpower developed in a long period of comparative peace between major industrialized powers. The period 1945–1990 saw only a few small wars fought using mechanized armed forces, of which the most important were the Korean War of 1950–53 and the Arab-Israeli wars in 1967 and 1973. During this time, Western armed forces moved increasingly toward ideas of military professionalism, mechanization, and technological solutions to their problems (although most still maintained some form of conscription).[78] Particularly in the case of theories of nuclear warfare and deterrence, military thought virtually rejected public involvement and public opinion in war as irrelevant or even an irritant.[79] There was the potential in such developments for friction between military views of the nature of warfare and governments harking back to the perceived role of the press derived from Second World War experiences, especially given the changes both to society and to civilian communications and the media that had taken place since 1945. After the first real test of late-twentieth-century theories of airpower and maneuver war, the 1991 Gulf War, one senior British commander, whose brigade had both disrupted and destroyed their Iraqi enemies with great effectiveness, wondered "if commanders can now be ruthless enough, in a television age, to pursue the enemy to the limit, if the stakes are anything less than national survival."[80]

Cold War, Information War, Media War

If the period 1945–1990 saw no major wars between industrialized powers, it was dominated for the industrialized world by two political and military global strategic issues: the cold war and decolonization.[81] Both were envisaged on all sides as long-term and large-scale campaigns, to be fought by propaganda as well as military and diplomatic means. The media and propaganda issues of the cold war were so large and complex that they have only just begun to be evaluated by historians. But it is a matter of record that, for example, during wars of decolonization up to the 1970s, successive British governments conducted campaigns of propaganda and manipulation of the media in Great Britain itself, as well as in the countries in which war was taking place.[82]

The same period saw major changes in developed industrialized society, comparable to those that had taken place in the late nineteenth century. These included a trend toward globalization in which new communications technology played a significant part, prompting the emergence of what the American cultural theorist Marshall McLuhan called in the 1960s "the global village."[83] Television, emerging after the Second World War and expanding its coverage throughout the United States and Europe in the 1950s, had by the end of the 1960s replaced newspapers as the dominant news medium, as well as becoming the dominant medium of entertainment, with considerable social consequences.

Once more, developments in military communications in this period are hard to separate from developments in civilian communications. In 1962, the United States produced the world's first global communications system, based on orbiting satellites, for diplomatic and military purposes; this was followed about a decade later by equivalent systems available for civilian use. Western domestic ownership of televisions and telephones became virtually universal in the same period. Electronic computers, which had received a major boost during the Second World War as part of signals intelligence for their code-breaking properties, began to be used routinely for both military and commercial purposes. This was the media and communications context for the United States' involvement in the Vietnam War (or Second Indo-China War) of 1961–75. This was an encounter of highly professionalized, mechanized, and industrialized armed forces with "people's war" in the form of the twentieth-century revolutionary war theories of the East, a style of warfare that took propaganda and the deliberate manipulation of the media virtually for granted.[84] It was the first major defeat for the United States in the twentieth century, and also the first American war to be televised on a regular basis, leading almost inevitably to the claim that the defeat had in some way been caused by television, or at least by the United States news media failing to support the government as in the Second World War. This issue was made even more sensitive by President Ronald Reagan's assertion in 1981 that lack of political support at home rather than military failure had led to defeat, and that never again would the United States "send an active fighting force to a country to fight unless it is for a cause that we are prepared to win"[85]—a position inevitably compared by critics to the German stab-in-the-back claim after the First World War.

While it has been often claimed that the United States news media took the side of North Vietnam, arguments that television coverage or any other form of reporting was a direct cause of declining American popular support for the war have been repeatedly shown to be false by statistical analysis and other standard methods of historical investigation.[86] Against this must be set repeated assertions by contemporaries that a connection existed, even if it cannot be identified. One veteran American reporter, Robert Elegant, has impressed historians with the conclusion from his own experience covering the war:

> Despite their own numerous and grave faults, the South Vietnamese were, first and last, decisively defeated in Washington, New York, London, and Paris. Those media defeats made inevitable their subsequent defeat on the battlefield. Indo-China was not perhaps the first conflict to be won by psychological warfare. But it was the first to be lost by psychological warfare conducted at such great distance from the field of battle.[87]

The issue remains an extremely complex and controversial one, which will continue to attract much historical study. But the consequences to the art of war came less from what happened than from what was believed to have happened. As with the colonial wars of the late nineteenth century, part of the mythology of Vietnam was that it was "the uncensored war." This was, in practice, far from being the case, although war correspondents with American forces in South Vietnam were placed under comparatively few direct military restrictions. But the belief grew that public opinion in Western democracies had become overinformed about the nature of warfare by the news media, and that this posed a serious threat to those countries' ability to fight wars. This belief was shared not only by senior officers and politicians but also by many journalists, such as Robert Elegant or Britain's Robin Day, who in 1979 argued whether "in future a democracy which has uninhibited television coverage in every home will ever be able to fight a war."[88]

The result was an attempt in the 1980s, chiefly in the United States, to reach mutual agreement between government and news media organizations on their future relationship in war. Some input also came from Great Britain after its experience of the 1982 Falklands War.[89] Significantly, most of the effort was directed at regulating the behavior of war reporters in battle, as the most visible manifestation of the issue. These concerns provided the context for military-media relations in the 1991 Gulf War, in which a conspicuously high media profile was taken by the senior American officer in the Gulf, General H. Norman Schwarzkopf. Military behavior toward the media again produced considerable controversy, particularly from journalists and critics who argued that inappropriate measures had been taken against the traditional role of the media in civil society.[90]

While these issues of military-media relations in wartime were being debated, advances in technology once more changed the range and nature of communications available to all sides. The first twenty-four-hour television news service based on direct satellite broadcasting, Cable News Network (CNN), opened in Atlanta, Georgia, in 1979. Portable telephones, fax (facsimile) machines, and satellite-based systems, including global positioning systems and direct satellite broadcasting, were all features of the Gulf War. As such trends continued and accelerated at the end of the century, some theorists began to argue that what was happening was the military aspect of an emerging "postindustrial" society, in which knowledge and information were replacing physical creations as the basic objects of value.[91] Guided weapons had become so sophisticated and mechanized, industrialized armed forces had become so dependent on their electronic communications that a new form of warfare, called information war, could be fought based on paralyzing the enemy's ability to respond while causing little physical damage. Unsurprisingly, the theoretical basis for information war has been found to stretch at least as far back as suggestions made in 1932 by J. F. C. Fuller, as a variant on his "strategic paralysis" concept.[92]

Finally, after a generation in which military relations with the media in war were regarded as secondary or peripheral to the art of war, after the Gulf War another new term began to appear, *media war,*[93] reflecting the manner in which all sides interacted with the global mass news media as part of their military strategies. In the course of the 1990s, military interventions by the United States in such episodes as the Yugoslavia war (1991–1999) were accompanied by highly sophisticated information strategies, including once more an awareness of the importance of the mass media.[94] In the same period, emerging "new media" in the form of the Internet and other developments began to challenge conventional terrestrial television as the dominant source of news. It appeared that Western civil society had reached or passed a level of "media saturation" that made its response to military activities undertaken in its name potentially less well informed than in the past.

NOTES

1. Quoted in Brian Winston, *Media Technology and Society—A History: From the Telegraph to the Internet* (London and New York: Routledge, 1998), 331.

2. Susan L. Carruthers, *The Media at War* (London: Macmillan, 2000), 15–17.

3. Winston, *Media Technology and Society,* 337–42; Martin van Creveld, *Technology and War* (London: Brassey's, 1991), 311–20.

4. J. F. C. Fuller, *The Conduct of War 1789–1961* (London: Methuen, 1962), 78–79; T. N. Dupuy, *Numbers, Predictions and War* (Fairfax: Hero, 1985); Paddy Griffith, *Forward into Battle* (Chichester: Anthony Bird, 1981); B. P. Hughes, *Firepower: Weapons Effectiveness on the Battlefield 1630–1850* (London: Arms and Armour, 1974); Robert P. Scales Jr., *Firepower in Limited War* (Novato, Calif.: Presidio, 1995).

5. Fuller, *Conduct of War,* 93; M. S. Anderson, *The Ascendancy of Europe 1815–1914* (London: Longman, 1985), 150.

6. Daniel J. Hughes, ed., *Moltke on the Art of War: Selected Writings* (Novato, Calif.: Presidio, 1993), 14; Dennis Showalter, *Railroads and Rifles: Soldiers, Technology and the Unification of Germany* (Hamden, Conn.: Archon Press, 1986).

7. A. J. P. Taylor, *The First World War: An Illustrated History* (London: Hamish Hamilton, 1963), 15.

8. Richard Overy, *Russia's War* (London: Allen Lane, 1998), 73–98; Martin van Creveld, *Supplying War* (Cambridge: Cambridge University Press, 1977), 142–80.

9. Cyril Falls, *The Art of War* (Oxford: Oxford University Press, 1961), 67.

10. V. K. Triandafillov, *The Nature of the Operations of Modern Armies,* trans. Jacob W. Kipp (London: Frank Cass, 1994), 159–63; Shi-

mon Naveh, *In Pursuit of Military Excellence* (London: Frank Cass, 1997); Antoine Henri de Jomini, *The Art of War*, with an introduction by Charles Messenger (London: Greenhill, 1996), vii–viii; Michael I. Handel, *Masters of War: Classical Strategic Thought* (London: Frank Cass, 1996).

11. Arthur C. Clarke, *Profiles of the Future* (London: Pan, 1983), 72–93.

12. T. K. Derry and Trevor I. Williams, *A Short History of Technology* (Oxford: Clarendon Press, 1960), 622–23; Tom Standage, *The Victorian Internet* (London: Weidenfeld and Nicholson, 1998).

13. Paul Kennedy, *The Rise and Fall of the Great Powers* (London: Hyman, 1988), 225.

14. Hughes, ed., *Moltke on the Art of War*, 77.

15. Ibid., 13–14.

16. Martin van Creveld, *Command in War* (Cambridge: Harvard University Press, 1985), 264–68.

17. Bob Harris, *Politics and the Press: Britain and France 1620–1800* (London: Routledge, 1996), 107.

18. Carl von Clausewitz, *On War*, ed. and trans. Michael Howard and Peter Paret (London: Everyman, 1993), 3–68; Christopher Bassford, *Clausewitz in English* (Oxford: Oxford University Press, 1994).

19. Clausewitz, *On War*, 248.

20. Andrew Lambert and Stephen Badsey, *The War Correspondents: The Crimean War* (Stroud: Alan Sutton, 1994), 1–14.

21. Jay Luvaas, *The Military Legacy of the Civil War* (Chicago: University of Chicago Press, 1959); Paddy Griffith, *Rally Once Again* (London: Crowood, 1987).

22. Quoted in Brian Holden Reid, *The Origins of the American Civil War* (London and New York: Longman, 1996), 385.

23. Anderson, *Ascendancy of Europe*, 104.

24. Ibid., 24.

25. Peter J. Parish, *The American Civil War* (London: Eyre Methuen, 1975), 334.

26. Dee Brown, *The Fetterman Massacre* (Lincoln: University of Nebraska Press, 1984); Brian W. Dippie, *Custer's Last Stand: The Anatomy of an American Myth* (Lincoln: University of Nebraska Press, 1976).

27. Pat Hodgson, *Early War Photographers* (London: Osprey, 1974), 15; Pat Hodgson, *The War Illustrators* (London: Osprey, 1977).

28. Nicholas Hiley, "Audiences in the Newsreel Period," in Clyde Jeavons, Jane Mercer, and Daniela Kirchner, eds., *The Story of the Century* (London: British Film Institute, 1998), 59–62.

29. S. D. Badsey, "Battle of the Somme: British War Propaganda," *Historical Journal of Film, Radio and Television* 3, 2 (1983): 99–115; Nicholas Reeves, "Through the Eye of the Camera," in Hugh Cecil and Peter H. Liddle, eds., *Facing Armageddon: The First World War Experienced* (London: Leo Cooper, 1996), 780–800.

30. Shelford Bidwell and Dominick Graham, *Fire-Power: British Army Weapons and Theories of War 1904–1945* (London: George Allen and Unwin, 1982), 94–148; Peter Mead, *The Eye in the Air* (London: HMSO, 1983); Jonathan Bailey, "British Artillery in the Great War," in Paddy Griffith, ed., *British Fighting Methods in the Great War* (London: Frank Cass, 1996), 23–49.

31. Simon Singh, *The Code Book* (London: Fourth Estate, 1999), 107–16; Barbara Tuchman, *The Zimmermann Telegram* (London: Papermac, 1981); John Ferris, ed., *The British Army and Signals Intelligence during the First World War* (London: Army Records Society, 1992).

32. See, for example, Ralph Bennett, *Behind the Battle* (London: Sinclair-Stephenson, 1994); David Kahn, *Seizing the Enigma* (London: Arrow, 1996).

33. Norman Stone, *Europe Transformed 1878–1919* (London: Fontana, 1983), 31.

34. James Belich, *The New Zealand Wars and the Victorian Interpretation of Racial Conflict* (Auckland: Penguin, 1986), 60.

35. Donald R. Morris, *The Washing of the Spears* (London: Cardinal, 1973), 279–80.

36. Bill Nasson, *The South African War 1899–1902* (London: Arnold, 1999), 33–34.

37. George Orwell, "England, Your England," in *Inside the Whale and Other Essays* (London: Penguin, 1962), 84.

38. Correlli Barnett, *The Collapse of British Power* (Gloucester: Allan Sutton, 1987), 151–54.

39. John F. Marszalek, *Sherman's Other War: The General and the Civil War Press* (Memphis: Memphis State University Press, 1981).

40. Garnet J. Wolseley, *The Soldier's Pocket Book,* 2nd ed. (London and New York: Macmillan, 1871), 225.

41. For the "One hundredth year" phenomenon, see Dippie, *Custer's Last Stand,* 8–11.

42. Stephen Badsey, "War Correspondents in the Boer War," in John Gooch, ed., *The Boer War: Image, Experience and Direction* (London: Frank Cass, 2000).

43. Geoffrey Best, *War and Society in Revolutionary Europe 1770–1870* (Stroud: Sutton, 1998), 207–214.

44. Oliver Thomson, *Easily Led: A History of Propaganda* (Stroud: Sutton, 1999), 239–40.

45. Russell F. Weighley, *Towards an American Army* (New York: Columbia University Press, 1962), 200–206.

46. Hughes, ed., *Moltke on the Art of War,* 32.

47. John Terrain, *The Smoke and the Fire* (London: Leo Cooper, 1992), 22–29.

48. John Bourne, *Britain and the Great War 1914–1918* (London and New York: Edward Arnold, 1989), 210.

49. Brian Bond, *War and Society in Europe 1870–1970* (Stroud: Sutton, 1984), 188.

50. Leslie Midkiff DeBanche, *Reel Patriotism* (Madison: University of Wisconsin Press, 1997), 195–98; Clayton R. Koppes and Gregory D. Black, *Hollywood Goes to War* (Los Angeles: University of California Press, 1987).

51. John F. Williams, *ANZACs, the Media and the Great War* (Sydney: University of New South Wales Press, 1999); Jeffrey Grey, *A Military History of Australia* (Cambridge: Cambridge University Press, 1999), 80–118.

52. Stephen Badsey, "Haig and the Press," in Brian Bond and Nigel Cave, eds., *Haig: A Reappraisal Seventy Years On* (London: Leo Cooper, 1999), 176–95.

53. Reproduced in *A History of* The Times, vol. 4: *The 150th Anniversary and Beyond 1912–1948* (London: Times Printing House, 1952) part 1, chaps. 1–12, *1912–1920*; see also Holger H. Herwig, *The First World War: Germany and Austria-Hungary 1914–1918* (London: Arnold, 1997), 447–48; Lawrence V. Moyer, *Victory Must Be Ours: Germany in the Great War 1914–1918* (London: Leo Cooper, 1995), 335.

54. Bidwell and Graham, *Fire-Power,* 141–43.

55. Asa Briggs, *The History of Broadcasting in the United Kingdom, vol. 1: The Birth of Broadcasting* (Oxford: Oxford University Press, 1995), 54–57.

56. Asa Briggs, *The History of Broadcasting in the United Kingdom, vol. 2: The Golden Age of Wireless* (Oxford: Oxford University Press, 1995), 343.

57. Simon Callow, *Orson Welles: The Road to Xanadu* (London: Vintage, 1996), 399–408.

58. See, for example, Anthony Beevor, *Stalingrad* (London: Viking, 1998).

59. Alison B. Gilmore, *You Can't Fight Tanks with Bayonets* (Lincoln: University of Nebraska Press, 1998).

60. Quoted in Cornelius Ryan, *A Bridge Too Far* (London: Coronet, 1974), 72.

61. Philip M. Taylor, *British Propaganda in the Twentieth Century* (Edinburgh: Edinburgh University Press, 1999), 192.

62. Duncan Anderson, "Douglas MacArthur and the Philippines," in Brian Bond, ed., *Fallen Stars* (London: Brassey's, 1991), 181–82.

63. Michael Howard, "Leadership in the British Army in the Second World War," in G. D. Sheffield, ed., *Leadership and Command: The Anglo-American Military Experience since 1861* (London: Brassey's, 1997), 125–26; Francis de Guingand, *Operation Victory* (London: Hodder and Stoughton, 1947), 375–86.

64. Carruthers, *Media at War,* 12–14; Peter R. Young, ed., *Defense and the Media in Time of Limited War* (London: Frank Cass, 1992); Robin Brown, "Limited War," in Colin McInnes and G. D. Sheffield, eds., *Warfare in the Twentieth Century* (London: Unwin Hyman, 1988), 164–93.

65. John Terraine, *White Heat: The New Warfare 1914–1918* (London: Guild Publishing, 1982), 147–50.

66. Heinz Guderian, *Achtung—Panzer!* trans. Christopher Duffy (London: Arms and Armour, 1992), 197–98.

67. Len Deighton, *Blitzkrieg: From the Rise of Hitler to the Fall of Dunkirk* (London: Jonathan Cape, 1979), 203 (photograph).

68. Quoted in Brian Holden Reid, *J.F.C. Fuller: Military Thinker* (London: Macmillan, 1987), 48.

69. Colin McInnes, *Hot War, Cold War* (London: Brassey's, 1996); Robert Leonard, *The Art of Maneuver* (Novato, Calif.: Presidio, 1991).

70. Ardent du Picq, *Battle Studies,* translation in [author unknown] *Roots of Strategy 2* (Mechanicsburg: Stackpole, 1987).

71. Holden Reid, *J.F.C. Fuller: Military Thinker,* 24–25.

72. P. H. Vigor, *Soviet Blitzkrieg Theory* (London: Macmillan, 1983), 76–79.

73. Basil Liddell Hart, *Strategy* (New York: Praeger, 1954), 345.

74. Alex Danchev, *Alchemist of War: The Life of Sir Basil Liddell Hart* (London: Weidenfeld and Nicholson, 1998), 161.

75. Liddell Hart, *Strategy,* 364 (and quoting his own *Paris—Or the Future of War*).

76. Robert A. Pape, *Bombing to Win* (Ithaca: Cornell University Press, 1996), 55–86; John A. Warden, *The Air Campaign: Planning for Combat* (Washington, D.C.: Pergamon-Brassey's, 1989).

77. Pape, *Bombing to Win,* 68.

78. Creveld, *Technology and War,* 217–50.

79. Lawrence Freedman, *The Evolution of Nuclear Strategy* (London: Macmillan, 1981).

80. Patrick Cordingley, *In the Eye of the Storm* (London: Hodder and Stougton, 1996), 254.

81. Anthony Parsons, *From Cold War to Hot Peace* (London: Michael Joseph, 1995), vii–x.

82. Susan L. Carruthers, *Winning Hearts and Minds* (London: Leicester University Press, 1995); Paul Lashmer and Oliver James, *Britain's Secret Propaganda War 1948–1977* (Stroud: Sutton, 1998).

83. Philip M. Taylor, *Global Communications, International Affairs and the Media since 1945* (London: Routledge, 1997), 27–57.

84. Harry G. Summers, *On Strategy: A Critical Analysis of the Vietnam War* (Novato, Calif.: Presidio, 1982), 191–92; Chen-Ya Tien, *Chinese Military Theory, Ancient and Modern* (Stevenage: Spa, 1992), 223–64; Bernard B. Fall, *Ho Chi Minh on Revolution* (New York: Signet, 1968).

85. Quoted in Jon Roper, "Overcoming the Vietnam Syndrome: The Gulf War and Revisionism," in Jeffrey Walsh, ed., *The Gulf War Did Not Happen: Politics, Culture and Warfare Post-Vietnam* (Aldershot: Arena, 1995), 30.

86. The best short summary of the historical argument is Carlyle A. Thayer, "Vietnam: A Critical Analysis," in Young, ed., *Defense and the Media in Time of Limited War,* 69–88.

87. Quoted by Brian Bond, *The Pursuit of Victory: From Napoleon to Saddam Hussein* (Oxford: Oxford University Press, 1996), 190. See also Miles Hudson and John Stanier, *War and the Media* (Stroud: Sutton, 1997), 104–18.

88. Quoted in Philip M. Taylor, *Munitions of the Mind* (London: Patrick Stephens, 1990), 228.

89. Pascale Combelles Siegel, *The Troubled Path to the Pentagon's Rules on Media Access to the Battlefield: Grenada to Today* (Washington, D.C.: US Army War College, 1996); Robert Harris, *Gotcha! The Media, the Government and the Falklands Crisis* (London: Faber and Faber, 1983).

90. See W. Lance Bennett and David L. Paletz, eds., *Taken By Storm* (Chicago: Chicago University Press, 1994); Hamid Mowlana, George Gerbner, and Herbert I. Schiller, eds., *Triumph of the Image* (Boulder: Westview, 1992).

91. Alvin Toffler and Heidi Toffler, *War and Anti-War* (New York: Little, Brown and Company, 1993); Chris Hables Gray, *Postmodern War* (London: Routledge, 1997).

92. Stephen Badsey, "The Conceptual Origins of Information Warfare," Global Transformation Research Group, 1999 Series Paper 4, London, 1999; Paul Dickson, *The Electronic Battlefield* (London: Marion Boyars, 1976); John Arquilla and David Ronfelt, "Cyberwar Is Coming," *Comparative Strategy* 12 (1993): 141–65.

93. Stephen Badsey, "The Media War," in John Pimlott and Stephen Badsey, eds., *The Gulf War Assessed* (London: Arms and Armour, 1992), 222–28; Philip M. Taylor, *War and the Media: Power and Persuasion in the Gulf* (Manchester: Manchester University Press, 1992).

94. See Pascalles Combelles Siegel, *Target Bosnia: Integrating Information Activities in Peace Operations* (Washington, D.C.: National Defense University, 1998).

PART II

✦ ✦ ✦ ✦ ✦ ✦ ✦ ✦ ✦ ✦ ✦

THE MYRIAD FACES OF TOTAL WAR

CHAPTER 4

The Morale of the British Army on the Western Front, 1914–18

A Case Study in the Importance of the "Human Factor" in Twentieth-Century Total War

G. D. Sheffield

Armed with powerful weapons produced by modern industrialized economies, the armies of 1914–18 waged war of unprecedented destructiveness. The Western Front in particular has come to exemplify a form of warfare in which the individual combatant was helpless in the face of high explosives, machine-gun bullets, and poison gas. The fearsome weapons created by modern industry and technology seemingly had replaced the sinews and strength of the fighting man. Yet the "human factor" remained important, even if it has not always received the attention it deserved. The aim of this chapter is to examine one aspect of the human factor, the morale of British Expeditionary Force (BEF) on the Western Front, relating its "peaks and troughs" to success and failure on the battlefield. Using a broad range of sources, it seeks to build on and amplify the pioneering work of other scholars in the field, particularly two Canadian historians, J. B. Wilson and S. P. MacKenzie.[1]

Between August 1914 and November 1918, the British army evolved from a small regular force into a mass army. The original BEF of August 1914 consisted of professional soldiers supplemented by regular reservists recalled to the colors. By the end of 1914, the BEF's order of battle also included elements of the Territorial Force (TF), a part-time body of amateur soldiers originally intended for home defense. The Territorials were followed by the first New Army (or "Kitchener's army") units, which arrived in France in the spring of 1915. The New Armies were a mass volunteer force that by 1916 had transformed the nature of the BEF. No longer was it a small, elite, professional body; it was now a citizen army, the military manifestation of the British "nation in arms." The final act in the evolution of the British army was the introduction of conscription in 1916, conscripts being posted to existing units—regular, Territorial, and Kitchener. The extent of the evolution of the BEF over four years can be gauged by comparing its composition in August 1914 and November 1918. The original BEF consisted of one cavalry and four infantry divisions, all regular. At the Armistice, the BEF had sixty-one infantry and three cavalry divisions, including four Canadian, five Australian, and one New Zealand division; and the vast majority of British Empire officers and soldiers on the Western Front in 1918 were essentially civilians in uniform enlisted "for the duration."[2]

Morale, one of the most common terms in the modern military lexicon, is an imprecise term. A number of definitions have been attempted, ranging from the simplistic to the complex. One of the most useful that links the morale of the individual with the morale of the group is that of Irvin L. Child: "morale pertains to [the individual's] efforts to enhance the effectiveness of the group in accomplishing the task in hand."[3] The relationship between individual and collective morale can be described as follows: unless the individual is reasonably content, he will not willingly contribute to the unit. He might desert or mutiny but is more likely simply to refuse to work wholeheartedly toward the goals of the group. High group morale, or cohesion, is the product in large part of a good morale experienced by the members of that unit; and the state of morale of a larger formation, such as an army, is the product of the cohesion of the units that compose that army. The possession of individual morale sufficiently high that a soldier is willing to engage in combat might be described as "fighting spirit."

The work of Carl von Clausewitz gives valuable insights into the nature of collective military morale, and especially that of the BEF of 1914–18. He differentiated between professional armies, who possess such attributes as discipline, experience, and skill, and nonprofessional armies that have "bravery, adaptability, stamina and enthusiasm." Clausewitz divided morale into two components, "mood" and "spirit," and warned that one should never confuse the two. The mood of an army is a transient thing, which can change quickly; but an army with "true military spirit" keeps "its cohesion under the most murderous fire" and in defeat resists fears, both real and imaginary. Military spirit, Clausewitz argued, is created in two ways: by the waging of victorious wars and by the testing of an army to the very limits of its strength; "the seed will only grow in the soil of constant activity and exertion, warmed by the sun of victory."[4] The BEF is a prime example of a largely nonprofessional force that endured tremendous hardships but continued to fight effectively in a sustained conflict, in part because its morale was boosted at critical moments by a series of minor but significant successes.

Clausewitz's analysis is applicable to individual as well as collective morale. The mood of the soldier might fluctuate from minute to minute, affected by fairly mundane factors such as the state of the weather or the availability of food. An enlisted man of the Seventh Battalion "The Buffs," East Kent Regiment (7/Buffs), noted in 1916 that the spirits of the troops were depressed by wet weather but recovered when the rain stopped, while another soldier of First/Fifth London Regiment commented that "nothing changed one's spirits from buoyancy to utter despondency or vice-versa quicker than a shortage or surfeit of rations."[5] Thus it was perfectly possible for a soldier's mood to be poor but his military spirit to remain sound.

Interestingly, it seems that local conditions had a greater impact than wider events on the morale of the BEF. The working-class men that provided the BEF with the bulk of its soldiers did not shed their civilian identity on joining the army; their attitudes toward their officers and NCOs were profoundly influenced by their civilian experience, for instance. Leave and letters from home ensured that soldiers in the trenches were aware of developments on the home front, and factors such as food shortages among civilians undoubtedly influenced the mood of individual soldiers. Yet, having examined about 150 collections of unpublished soldiers' writings (letters, diaries, and memoirs), the author has discovered little evidence that incidents on the home front or wider political affairs such as Lord Landsdowne's "peace letter" of December 1917 or the Russian Revolution had a significant impact on the "spirit," that is, the willingness to fight, of the BEF. This is a topic that would undoubtedly repay further research.[6]

Therefore, although this chapter argues that the morale of the BEF remained fundamentally sound throughout the war, it is not suggested that individual soldiers or, indeed, entire units were ecstatically happy all the time. Rather, the combat performance of British soldiers reflected their commitment to winning. It is, of course, possible to find examples of groups or individuals who, at a particular time, lacked the willingness to fight. Some members of the First Gordon Highlanders were reportedly drunk and undisciplined during the retreat of March 1918.[7] The flight of an Irish battalion on the Somme in September 1916 and the rout of the Ninth Cheshires on 24 March 1918 provide even more dramatic evidence of the failure of military spirit of specific units at certain times. Cases such as these were, however, exceptional; 9/Cheshires, for example, fought effectively only one month after their rout.[8] The battles of the BEF were far from universally successful, and symptoms of poor morale were discernible on occasion, but the performance of British troops was rarely less than dogged. The British citizen army's mood fluctuated, but its determination to defeat the Germans remained essentially intact. In this respect, the morale of the BEF reflected the morale of the British nation in arms as a whole.[9]

The problems involved in attempting to assess the morale of a formation as large as the BEF are many and obvious. Ideally, an overall picture should be constructed from a series of case studies.[10] Enough evidence exists, however, to draw some tentative general conclusions about the morale of the BEF at certain points during the war. British high command had a strong belief in the importance of moral factors in war, drawn in part from a bastardized version of Clausewitzian theory,[11] and used an attritional strategy that had an explicit aim of destroying enemy morale. Unlike their American and French allies, the British army did not create a central organization devoted to the planning, direction, monitoring, and sustaining of the morale of their troops.[12] This relatively casual approach is partially explained by the assumption, held by many senior officers, that the morale of their men was fundamentally sound, and likely to remain so;[13] in the jaundiced words of the prime minister, David Lloyd George, generals believed that "Allied soldiers were infrangible steel, and enemy soldiers ordinary flesh."[14] This creed was not entirely founded on wishful thinking, for such assumptions were rooted in a belief in the "character" of the British soldier, and, what is more important, an intimate knowledge of the nature of the British regimental system and the paternalism of the British officer corps. These factors were, indeed, to prove vital in the maintenance of morale.[15]

Moreover, the British regimental system inhibited the establishment of a centralized body concerned with morale; as late as 1940, some senior officers resisted the creation of an Army Welfare Scheme and the use of extraregimental local welfare officers, on the grounds that the well-being of the soldier was the responsibility of the regimental officer.[16] It is typical of the British approach to morale that a campaign to mold the opinions of the soldiers of the BEF by a program of education sprang largely from the initiative of individual staff officers and the deputy chaplain-general (although with the encouragement of Field Marshal Sir Douglas Haig, the commander in chief of the BEF). These schemes were implemented in 1918, the last year of the war, in response to a perceived deterioration in morale. It is arguable that it would have been more sensible to introduce such a scheme earlier, to prevent morale declining in the first place.[17] This is not to argue that British High Command took no interest in the morale of their troops, rather that the information they received was not always reliable, and some was gathered in a haphazard and unsystematic fashion. Wilson has argued that high command attempted to ascertain the state of morale and discipline in units by the collection of quantitative data, especially figures for courts-martial offenses such as drunkenness, looting, and desertion and statistics for trench foot and shell shock. He concluded, however, that these indexes were not, on the whole, very reliable.[18] There was a tendency to confuse morale with discipline. This was misleading, for the two, although related, are not identical. Men serving sentences in military prisons are well disciplined but unlikely to have high morale.[19] Furthermore, not all regimental officers attempted to apply prewar standards of discipline to their troops, who were mostly civilians in uniform. In some Territorial, Kitchener, and Dominion battalions, failure to salute might have betokened neither slack discipline nor low morale. Applying criteria such as propensity to salute to the whole army was an inaccurate way of judging the fighting efficiency of some units.[20]

It is unwise to judge morale by the number of executions within a unit. Men sentenced to death seem to have faced something of a lottery as to whether their sentence would be commuted, for only 10.82 percent of death sentences were carried out.[21] The composition of the court-martial, whether or not the accused was represented, the attitude of the accuser's hierarchical superiors (from his battalion commander to the commander in chief), whether the accused was Irish or black: all these factors were important in deciding the fate of the individual.[22] The execution of malefactors, who were mostly deserters, was intended to deter others. The commanders of brigades in which condemned men served furnished a report "as to the state of discipline of the unit and his recommendations as to whether or not an example was necessary."[23] The decision of the commander in chief, who had the ultimate authority to confirm or commute a death sentence, was influenced by "the immediate needs of discipline," not necessarily that of the individual's battalion.[24]

The case of a private of the Twenty-Second Royal Fusiliers who was executed in December 1916 for deserting during the fighting on the Ancre in November illustrates the difficulties of generalizing about the state of morale of a unit from the execution of an individual. This action was a severe trial for the British infantry, yet his battalion performed well during the battle. The 22/Royal Fusiliers was characterized by an enlightened disciplinary regime, a high level of esprit de corps and excellent officer-man relations. The evidence suggests that the private was a poor soldier who had been lucky to escape a court-martial after his behavior in an earlier battle. It was possible that he was shell-shocked. In sum, he was an atypical soldier, and it would be unwise to generalize about the state of morale and discipline in 22/Royal Fusiliers from his fate.[25]

Censorship of soldier's letters represented the most systematic, if far from perfect, method used by the BEF for gauging morale. Reports based on such censorship were submitted at regular intervals to general headquarters (GHQ) and on occasion to the War Cabinet. Other ad hoc methods were also used. The war correspondent Philip Gibbs claimed to have been interviewed in late 1917 by Lord Milner and later by a British liaison officer serving with the French. Both questioned Gibbs as to his opinions of the morale of the BEF in general, and specifically, whether he believed the army would accept a compromise peace.[26] Another individual who advised the War Cabinet on British military morale was General Smuts, who submitted a report on his return from a tour of the front in January 1918.[27]

More junior officers also had quantitative methods of gauging morale. The brigade major of the Fourteenth Corps Heavy Artillery "learned to assess the morale of the infantry" by the number of complaints received from them of British shells dropping short. He believed that many of the reports of "short-shooting" were "entirely unreliable," and the better the morale of the infantry, the less likely they were to issue false claims about the inaccuracy of the gunners.[28] In December 1917, the officers of the First/Ninth King's Liverpool Regiment saw the willingness of troops to subscribe to War Savings Certificates as an indication of esprit de corps. The men were initially reluctant to part with their money but became more enthusiastic when the scheme was presented as a way of increasing the prestige of the battalion, and 1/9 King's eventually raised more money than any other unit in their division.[29]

Quite apart from these "statistical" means of establishing morale, officers used their experience and intuition to keep their finger on the pulse of units. Obvious signs of high morale included spontaneous humor and singing.[30] The junior officer soon learned that when marching men stopped singing and whistling they were growing weary or approaching danger.[31] R. W. F. Johnston, an officer of the Sixteenth Royal Scots, drew a sharp contrast between the demeanor of the battalion before an action of 25–26 August 1917 and one fought two months later. Before the first action, despite a "somber and dark" silence on the approach march, the men showed "keenness," and a degree of esprit de corps. By the time of the battle of 22 October, the men were "tired, dispirited and exhausted . . . without thought of victory." During the latter action, logistic arrangements had broken down, leaving the men short of food. It is interesting that Johnston used the absence of "jokes and singing in the ranks" as a criterion of low morale.[32] Contemporary advice for young officers laid heavy stress on the necessity for the officer to get to know his men.[33] While "grousing" was not necessarily a sign of low morale, the good officer was able to sense when the morale of his men was low.[34]

Despite the "scientific" nature of warfare in 1914–18, the gauging of morale remained an art rather than a science, as it had been throughout history. The use of statistics was fraught with danger, while, conversely, the opinions of regimental and even staff officers (who mostly had recent regimental experience), however subjective, cannot be lightly set aside. Many British officers developed close relationships with their men that made them sensitive to changes in mood and spirit among the rank and file. Likewise, the views of other well-informed individuals are worthy of attention such as General Gibbs, who built up a close relationship with an infantry battalion, the Eighth/Tenth Gordon Highlanders. Gibbs claimed that he had "complete liberty" to visit all parts of the front, spoke to men in the front line and gained much knowledge of "the spirit and personal experiences of the troops"; and one unit did record their satisfaction with his account of a visit to them in July 1915.[35] Finally, combat performance offers valuable evidence of morale, for troops that lack military spirit will not fight effectively. Such evidence, when used alongside the writings of other ranks and the findings of censorship reports, allows us to assess the morale of the BEF.

The first winter of the war was a terrible trial for the soldiers in the trenches, and this experience affected morale; the British official historians referred to the "depression" experienced by the men in the front lines in the winter of 1914–15.[36] Wilson argues that this depression was caused by a number of factors, including the harsh climatic conditions, the terrain, the primitive nature of the logistic infrastructure, poor quality of reinforcements for regular units, and frustration with trench warfare. All these things were important, but most important of all, Wilson suggests, was that the British army was inadequately prepared to conduct a campaign of static, trench warfare. The paucity of reserves and trench-fighting equipment is symptomatic of the lack of material preparation, but the effort needed to adjust mentally to trench warfare was also of importance. Wilson concludes that this depression, although serious in the short term, was short-lived. The first major British offensive of 1915, launched at Neuve Chapelle in March, demonstrated that the BEF's will to combat had not been undermined.[37]

There is considerable evidence to suggest that the British military leadership regarded the morale and discipline of the BEF with some concern in the winter of 1914–15. As early as 30 October 1914, a staff officer who had served with the Seventh Division delivered a lecture to the as-yet-unblooded Eighth Division, in which he warned that at Ypres the enemy had attempted to break the morale of the infantry by artillery fire. He also stressed the necessity of maintaining strict discipline, as did General Henry Rawlinson, who spoke after him.[38] In the same month, a staff officer treated soldiers arriving at Le Havre to "a homily upon morale."[39]

The average regimental officer was well aware of the need to make strenuous efforts to maintain the morale of the ordinary soldier. In 1938, an officer who had served with the Second Rifle Brigade gave a lecture to the staff college on the upkeep of morale in the winter of 1914–15. The practical steps that were taken by the officers of this battalion ranged from the enforcement of strict discipline to "the provision of amusements in the form of organized games, sports, concerts, boxing and horse shows."[40] Clearly, the depression of the winter of 1914–15 did not permanently erode the military spirit of the British soldier. Indeed, the fortitude of the enlisted men impressed contemporary observers.[41] Moreover, this period was not characterized by large-scale desertions or mutinies or by ineffective performances in combat. Morale began to improve in the spring of 1915, when the Fourth Middlesex, among other units, introduced vigorous training to shake off the sluggish ways of the winter.[42] According to Lieutenant General Sir William Robertson, then chief of staff of the BEF, the Battle of Neuve Chapelle, fought that spring, demonstrated that the BEF still possessed a considerable degree of offensive spirit.[43] In Clausewitzian terms, the mood of the BEF may have appeared depressed, but its military spirit remained intact.

A letter written by Private J. Allison in early 1915 perhaps offers a clue to the thinking of the regular enlisted man in this period. Allison, a soldier with many years' service, had escaped frontline duty and was working in a base hospital; he may have been wounded. After noting his "old mob . . . getting cut up." Allison opined that "this War is going to be a very long one so I have settled my mind down for it . . . the southafrican [*sic*] war was not a Patch to this one." There are a number of interesting points about this letter. Allison had deliberately (and illegally) posted the letter through the French civilian system, thus preventing an officer from censoring it, because, he said, one "carnt Put anything in it [a letter] to [*sic*] much" if an officer was going to read it. Even though no one in the military hierarchy would read his letter, Allison concluded with the words "God save the King."[44]

Allison's letter is evidence of a regular private, albeit one who had temporarily escaped frontline duty, coming to terms with the previously unanticipated reality of a long, static war. He accepted the fact with a certain amount of fatalism and even patriotism, tempered by a grumble. His views are consistent with those of other prewar regulars in this period. In his memoirs, Corporal John Lucy (Second Royal Irish Rifles) wrote of the ebbing of his personal morale and the stultifying effects of trench warfare on soldiers trained for mobile warfare, but his attitude to the war is epitomized by the title of chapter thirty-six: "Life Goes On."[45] Likewise, in his memoir Private Frank Richards (Second Royal Welsh Fusiliers) referred in passing to the lowering of morale caused by the harsh conditions of the winter of 1914–15, but the general tenor of his book is of stoic acceptance of his lot.[46]

Historians have paid little attention to the morale of the BEF from the spring of 1915 to the eve of the 1916 Somme offensive. Wilson's conclusion that there was an improvement in morale during 1915 and that spirits were generally high by June 1916[47] is broadly correct. There is also evidence, however, of a recurrence of depression among some troops during the winter of 1915–16. In November 1915, Robertson argued that "depression at home is beginning to be reflected in the Army in France."[48] This view was probably influenced by the reports on postal censorship. In November 1916, Captain Martin Hardie (Third Army censor) stated that a year earlier "Letters containing prolonged grousing" had been "fairly common."[49] At the other end of the scale, a Territorial private who had served in France for almost a year recorded in October 1915, "For the first time since the war began, I have heard soldiers say that we are losing." Although the writer retained his optimism, he believed the modest gains and heavy losses incurred at the Battle of Loos in September had undermined morale: "It gives one cold shivers to look at a map and see how far the Germans must be driven back."[50] Philip Gibbs believed that the winter of 1915–16 was worse even than the winter of 1914–15 and the one that was to follow the Somme offensive, for the sacrifices of 1915 appeared to have achieved nothing.[51]

This feeling of depression does not appear to have been common to the entire BEF. With the exception of a handful of formations that fought at Loos, few of the New Army units that had arrived in France by the early spring of 1916 had taken part in the battles of 1915. Their morale was generally high. It is also possible to find indications of high morale among troops who had been out in France for some time; a middle-class soldier of the First/Twenty-First Londons wrote in October 1915 that "it is only a matter of time before they [the Germans] give in,"[52] while Private Andrews, a professional man serving in a working-class unit, the First/Fourth Black Watch, believed that the men's "spirit . . . was still excellent," although men were more reticent about volunteering for "dangerous duty," than formerly.[53] The preparation for the Somme appears to have revived spirits in many units. The mood of some of the troops was depressed in the winter of 1915–16, but the BEF's military spirit remained essentially intact.

The Battle of the Somme was the largest single battle fought by the British army up to that point. From July to November 1916, the BEF advanced about seven miles, at the cost of 420,000 British casualties. Fifty-seven thousand casualties were incurred on the first day of the offensive alone, and the German army remained undefeated. Yet, in November 1916, a report on the morale of the Third Army based on the censorship of soldiers' letters (the only such records from the period to survive) could report that "the spirit of the men, their conception of duty, their Moral [*sic*], has never been higher than at the present moment." Not surprisingly, General Douglas Haig, the commander in chief of the BEF, commented on reading the report, "It is quite excellent." Although the Third Army did not play a major role in the battle after the initial stage, divisions from those armies that bore the brunt of the fighting were sent to Third Army in the course of the offensive. The Third Army's censorship reports were complemented by those of other armies.[54]

The censor's reports offer strong evidence of the reliability of soldiers' letters as indicators of morale and refute a recent suggestion that censors deliberately selected positive statements from soldiers' letters to produce falsely optimistic reports for submission to high command.[55] Captain Hardie, the censor for Third Army, was not a sycophant. His reports in the autumn of 1917 made no attempt to disguise symptoms of poor morale. In view of recent suggestions that official documents and some private papers were censored or falsified in an attempt to protect the reputations of senior commanders, it is noteworthy that these reports were retained in Hardie's private papers and emerged into the public domain only in the 1980s.[56]

Few consider poor morale a contributory factor in the failure of the British attack on 1 July 1916. Although a staff officer of the Thirty-Second Division suggested that a factor in the partial failure of that formation's assault was physical and moral exhaustion caused by excessive digging and a poor system of reliefs,[57] there is general agreement that morale, in the sense of willingness to fight, was high on the eve of the Somme offensive.[58] One artillery officer believed that the change of scenery, from the "dreary, drab and depressing surroundings of Flanders to the open plains of the Somme," lifted the spirits of the men. Certainly, in the first half of 1916, the Somme was a less active sector than the ever-dangerous Ypres Salient.[59] The obvious power and apparent effectiveness of the British guns also boosted confidence, leading to a widely held belief, which filtered down from senior officers to the rank and file, that the bombardment of the German positions would ensure the infantry's task would be an easy one, and that the forthcoming offensive would decide the war.[60]

John Keegan has written that the BEF of July 1, l916 was "a trusting army."[61] While there is a great deal of truth in this assertion, it would be wrong to exaggerate the extent to which ordinary soldiers shared their officers' optimism. It is instructive to compare an officer's and a private's recollection of General Aylmer Hunter-Weston's visit to the First Lancashire Fusiliers on 30 June. The officer recalled that Hunter-Weston's optimism was "naturally conveyed to my men, it gave us all good heart. In fact we thought that this must be the end of the war!!!" The private recalled "the ugly mutterings in the ranks" during the general's talk, which told a rather different story.[62] (The 1/Lancashire Fusiliers were a veteran regular battalion that was perhaps less impressionable than a green Kitchener unit.)

Rifleman Percy Jones, a soldier of a veteran Territorial Forces battalion, the First/Sixteenth Londons, was also unimpressed by a senior officer's assurances that casualties in the assault on 1 July would be low because of the effectiveness of the British artillery. Facing the formidable German defenses at Gommecourt, most of Jones's fellow enlisted men shared his skepticism about "the carefully drawn up plans." It is important to note that this cynicism did not undermine the willingness of the men of 1/16 Londons to fight. In two successive diary entries in late June 1916, Jones referred to the men's determination to go on "until something stops us."[63] During the week before the assault, only 7 out of 966 men of all ranks of 1/16 Londons reported sick, "a record," the battalion historian commented, "rarely beaten in peace time, even under the most favorable conditions," which indicated high morale.[64] Using the most important test of morale of all, both 1/Lancashire Fusiliers and 1/16 Londons fought well on 1 July, even though both their divisions' assaults ended in failure.

Many of the men who attacked on 1 July had never before taken part in a full-scale battle. In a letter of 7 July 1916, a private of the First/Thirteenth Londons, newly arrived from England, wryly recorded how his draft's unbridled enthusiasm, which provoked an amazed response from veteran soldiers, was quickly tempered after a few days in the trenches.[65] By July 1916, however, some New Army units had gained as much as nine months' experience of trench warfare, and on active service, raw troops experienced a steep learning curve.[66] To choose two regiments at random, of the fourteen battalions of the Northumberland Fusiliers that served with Fourth Army on 1 July 1916, eight had arrived in France in January 1916, two in November 1915, one in July 1915, and three in September 1915, while of the five New Army and Territorial battalions of the Royal Fusiliers serving with Third and Fourth Armies on that date, three had arrived in January, one in March, and one in July 1915.[67] There was no repetition of Loos in 1915, where inexperienced New Army troops were pitched into battle without first learning the rudiments of warfare on active service. Thus, in attempting to explain how British morale survived the disappointment and casualties of 1 July, it is necessary to dispense with the idea that the soldiers were "lambs to the slaughter."

Writing of the later stages of the Somme campaign, Haig's chaplain suggested that the army's "old 'death or glory' spirit" had largely disappeared. In its place, all ranks displayed "a quiet fortitude and a resolute determination to carry on to the end."[68] Captain Hardie, Third Army's censor, came to similar conclusions, writing in his report of November 1916 of a "dogged determination to see the thing through at any cost."[69] The endurance of the BEF is a theme that echoes through many reports from this period. In this respect, some comments made in September 1916 by B. H. Liddell Hart, then an enthusiastic young company commander of the Ninth King's Own Yorkshire Light Infantry, were typical. Liddell Hart wrote of the "wonderful courage and discipline" of the infantry. A few of the men were fearless; a few were too stupid to experience fear; but the majority, he stressed, were just ordinary men.[70]

That is not to suggest that morale was always high during the Somme campaign. Individual actions fought under difficult conditions could place the morale of units under some strain. The Thirty-fifth (Bantam) Division, a formation recruited from men under the normal height for enlistment, attacked unsuccessfully in August 1916. Some of the division's troops seem to have been little more than children and were found crying during the fighting.[71] In September, the Forty-ninth Division attacked Thiepval. It had already unsuccessfully assaulted this objective, and the evil reputation of Thiepval had adversely affected the division's morale. The Twelfth Division's action at Le Transloy in October, according to one survivor, caused a lowering of morale. Another officer of the Twelfth Division mentioned two factors that indicated the decline of morale: excessive straggling and a tendency for infantry to go to ground under hostile artillery fire. In retrospect, this witness claimed that fighting for "limited objectives" undermined the morale of the infantry: they were "murderous affairs to all infantry, with nothing to stir the imagination as to victory."[72]

This is not the whole picture. One battalion of the Twelfth Division, the Seventh Suffolks, suffered heavy casualties during July and August and during the fighting for Le Transloy had the handicap of understrength and untrained companies and the presence of many reinforcements. In addition, the battalion's battle began with an exhausting approach march along "muddy tracks." Yet, an officer of the battalion wrote, the morale of the battalion was "very good" considering all the difficulties.[73] Again, we have evidence that the spirit of the army was essentially sound, although external appearances suggested otherwise.

The ultimate test of morale is willingness to engage in combat, and the BEF's divisions continued to fight, with some degree of success, throughout the campaign. Conditions on the Ancre during the final operations of the campaign were exceptionally bad. The historian of Nineteenth Division declared that it had "never known greater exhaustion or discomfort than that experienced in November 1916." J. F. C. Fuller believed that the conditions were responsible for causing "considerable numbers" of British troops to the desert to the enemy, the only time in the war for which Fuller had evidence of this occurring.[74] In spite of these problems, the operation was a partial success, in which the Fifty-first (Highland) Division seized Beaumont Hamel. In sum, the evidence suggests that the BEF began the Somme campaign with a strong will to fight, and that the subsequent months of fighting, while imposing severe strains on individuals and units, did not erode the military spirit of the army. An artillery officer summed up the change when he wrote that it was not that the "will to win disappeared," but that the "spirit of adventurous participation" that had motivated the New Army in July 1916 "died away" during the Somme offensive.[75]

The year 1917 saw a repetition of the strategy of the previous twelve months. The British army engaged in two major offensives at

Arras (April–May) and at Ypres (July–November). Both battles became attritional affairs, and both resulted in heavy British losses: 150,000 at Arras, 250,000 at Ypres. Yet Captain Hardie's censorship reports and other evidence suggest that in the latter part of 1917, the BEF's morale remained sound, in spite of considerable strains.

The weather in the winter of 1916–17 was exceptionally cold. In January 1917, one battalion recorded temperatures of "20 to 25 degrees of frost."[76] The morale of the troops seems to have remained fairly good, however, although at least one unit received a lecture on esprit de corps before going into action, which suggests some doubts about their commitment may have existed.[77] Hardie's conclusions on the morale of Third Army in January 1917 are epitomized by the phrase "Tommy is still in the pink." Soldiers' letters did contain complaints, but there were "no indications" of a "wish for premature peace"; rather, there was a general acceptance that more sacrifices would have to be made before victory could be achieved. Hardie believed that the willingness of the ordinary British soldier to "submit without a murmur to guidance and authority, and be prepared simply to 'carry on' without comment or discussion," indicated confidence in the Allied cause, in the conduct of the war, and of "the justice and efficiency of our military training and methods."[78]

Haig, writing at the beginning of May 1917, was, perhaps predictably, at pains to stress the confidence of the BEF and the general belief that German losses were higher than those of the British.[79] Hardie's report of May 1917, the period of the Arras offensive, in which the Third Army took a prominent part, confirms this view, registering little change from the generally optimistic tone of previous reports.[80] In the second half of 1917, however, Hardie's reports were indicating a distinct change in the tone of soldiers' letters. In a report based on nine hundred letters read over the period 8 July to 24 August, he stated that "it must be frankly admitted that the letters show an increasing amount of war-weariness." He noted "a tinge of despondency which has never been apparent before" and considerable "unsettled feeling about the continuation and conclusion and after-effects of the war," leading to a replacement of "active enthusiasm" by "passive acceptance." Talk of peace, which had been rare earlier in the year, was now "frequent." Complaints about matters such as lack of leave outnumbered references to the strain of combat by about five to one, and the average soldier did not seek "peace at any price," but there was "an immense and widespread longing for any reasonable and honorable settlement that will bring the war to a close."[81] A further report of 19 October offered even more alarming conclusions: that the willingness of the soldier to abandon his rights as an individual and obey his military masters without question was beginning to end.[82]

Other evidence supports Hardie's views. Gibbs believed that the Third Battle of Ypres adversely affected morale:

> For the first time the British Army lost its spirit of optimism, and there was a sense of deadly depression among many officers and men. . . . They saw no ending of the war . . . , and nothing except continuous slaughter.[83]

The French official history also commented on the "weariness" of the BEF at the beginning of winter,[84] and a recent study demonstrates the extent to which Australian morale had been eroded by the end of the campaign.[85] Yet, as Gibbs himself stated, the discipline of the BEF remained intact.[86] Even Lloyd George, who denounced the "stupid and squalid strategy" of the last stage of Passchendaele, which, he believed, exhausted the BEF and destroyed its confidence in its leaders, commented on the dogged fighting of the army in this phase.[87] Perhaps the most important piece of evidence for British morale at the end of 1917 is a report to the War Cabinet, dated 18 December 1917. Compiled on the basis of seventeen thousand frontline soldiers' letters and general impressions formed during the previous three months, this report is "the nearest thing available to a gallup [*sic*] poll" for the period. The report stated categorically, "The Morale of the Army is sound." Positive and negative letters were about evenly balanced in the Second Army, which had taken the lead in the offensive at Ypres. Positive letters written by men of other armies considerably outnumbered negative communications, suggesting that the conditions at Ypres were indeed placing the men of the Second Army under considerable stress. Despite much unfavorable news (the "Russian debacle and the Italian setback") and considerable "war weariness" and "an almost universal longing for peace," this report suggested that the BEF remained willing to fight on to achieve victory.[88]

The success of the German counterattack at Cambrai on 30 November 1917 placed a question mark against the morale of some British formations. General Smuts was sent to France to report on the situation. In two memoranda submitted to the War Cabinet, he concluded that the "moral [*sic*] of the army is good." Smuts proceeded to highlight some major problems that were likely to affect the morale of the BEF. The men, particularly the infantry, were tired, a problem exacerbated by the need to prepare defensive positions. This also reduced the time available for training and "rest . . . a psychological factor of the utmost importance." Smuts foresaw that, should the British agree to the French request for the British to take over more trenches, "we shall be running serious risks. We shall be straining the army too far." In sum, Smuts did not believe that Passchendaele had significantly weakened the resolve of the BEF, although complacency was unwise.[89] The very honesty of censorship reports—that they included evidence that morale was less than perfect, caused some controversy among the British high command. In October 1917, the Chief of the Imperial General Staff (CIGS) recorded his belief that "by no means" every British soldier on the Western Front was "possessed of a good morale," and that this was "only natural and to a greater or lesser extent common to all armies."[90] Writing two days later, the ever-optimistic Brigadier General John Charteris, Haig's intelligence chief, complained that Robertson had underestimated "the extraordinary high morale" revealed by the monthly censorship reports.[91] While a brief report submitted to the cabinet on 13 September certainly supports Charteris's view, the rather more somber tone of other extant reports tends to support the CIGS.[92]

There are several factors that help to explain why the BEF's morale remained relatively high in early 1917, declined as a result of the offensive at Ypres, but did not collapse. The British won a series of minor but important victories in the first half of 1917. In January, the BEF renewed offensive operations on the Somme and gained some success. The German retreat to the Hindenburg Line in March 1917 appeared to be conclusive evidence that the Germans were losing the war and allowed the Somme to be presented as a victory because the Germans had abandoned the field of battle to the Allies.[93] The brief phase of mobile warfare as the British followed up the retreating Germans also seems to have provided a boost to the morale of some units.[94] The initial success in the Arras offensive in April and the seizing of Messines Ridge in June also offered evidence of British victories.

The failure of the initial stages of the Third Battle of Ypres, the heavy casualties sustained, and poor weather help explain the weakening in morale in August. There are several factors that may have helped offset this decline. First was the genuine success of the limited offensives fought by Plumer's Second Army at the battles of the Menin Road, Polygon Wood, and Broodseinde (20 and 26 September, 4 October). Second, the fact that the BEF was advancing, albeit slowly and at a high price, was important. On the Somme in October 1916, an officer of the seventh East Surreys had written that he was presently back with the battalion transport (i.e., some way behind the forward trenches), but "even where we are we are some way over where the German front line was before the 1st July which is very satisfactory."[95] Similarly, thirteen months later the capture of Passchendaele Ridge, although costly in casualties, was proof that the British were advancing steadily: "Troops were now resting where once they could not have stood up." The BEF's sacrifices had brought tangible reward.[96] Finally, the offensive at Cambrai in November 1917, although ultimately a failure, was initially brilliantly successful.

The evidence offered in this section could be supplemented by a host of other examples drawn from lower echelons of the army[97]

The most important testimony to the state of British morale during the Passchendaele offensive is that, with the atypical exception of the Etaples mutiny of September 1917,[98] the men of the BEF continued to obey orders and to fight reasonably effectively. Taken together, the evidence suggests that the morale of the British soldier was severely tested in 1917, but not to destruction.

In 1918, the BEF had to relearn how to conduct mobile operations, first in a major retreat, and then advancing to victory. The BEF lost 239,793 men in 40 days in the spring of 1918, compared with the loss of 244,897 in 105 days of the Third Battle of Ypres.[99] As has recently been pointed out, however, during the period of the Allied offensive (8 August–11 November 1918), combat fatalities of the BEF were remarkably low: only about twenty thousand for Fourth Army, the spearhead formation.[100] This hitherto little noticed factor might well have had an effect in maintaining British morale.

The initial success of the German spring offensive that began on 21 March 1918 has been traditionally attributed, in part at least, to "the poor and cowardly spirit of the officers and men" of the Fifth Army.[101] To some extent these accusations were politically inspired, but they have been echoed by some historians.[102] Did British military morale really collapse in the spring of 1918? A censorship report of July 1918, based on the study of 83,621 letters, suggests it did not. This report, covering the period from April, states that the "high moral quality" shown in March was "amply confirmed" in the latter part of the period. The comments on the morale of the Fourth (formerly the Fifth) Army are particularly interesting. The report frankly admits that it would have been "misleading" to suggest that this formation was "happy." Besides a general sense of war weariness, there was decline in confidence in politicians, and if not in the higher command, at least in the "higher administration of the Army." In spite of everything, however, the Fourth Army's "combative spirit" "remains very high," the letters of the ordinary soldiers containing unambiguous evidence of the men's "determination to stick it out to the end," to achieve the victory they still longed for. The report also contained comments of a similarly positive nature about the morale of other formations. "The persistence and determination" of the men of the Second Army, who fought a defensive battle on the River Lys in April, was described as "remarkable": "They were *very* tired but unbeaten."[103]

A soldier of the Second Devons perhaps caught the essence of the morale of the BEF during the spring battles when he wrote of the general belief that, despite all setbacks, they remained confident that everything would be all right in the end.[104] An artillery officer considered Haig's "Backs to the Wall" message of 11 April 1918 to be damaging to morale because many soldiers "had not admitted even to themselves" how serious the position was.[105] Furthermore, mere rumors that the British were striking back at the Germans was enough, in good Clausewitzian fashion, to boost morale.[106]

On 21 March 1918, the Germans took 98.5 square miles of ground from the British Third and Fifth Armies (all but 19 square miles from the latter). The British defenders suffered about 38,500 casualties, including twenty-one thousand men taken prisoner. The Germans captured approximately five hundred guns, mostly from the Fifth Army. Such heavy losses of prisoners and guns are usually the tokens of defeat, and there is also much anecdotal evidence of British soldiers surrendering without putting up much resistance. It would be unwise, however, to deduce from these facts and figures that British morale was low in March 1918. As Martin Middlebrook points out, many factors serve to distort the picture, not least the overall strategic plan and the unfamiliarity of the British forces with the concept of "defense in depth." Ignorance of this new doctrine resulted in the advancing Germans cutting off as many as one-third of the defenders, who were crammed into the forward zone. Middlebrook suggests that although the Fifth Army's morale was far from "excellent," the morale of at least some units was "steady."[107]

The records of stragglers collected by the military police also cast doubt on the traditional belief that the Fifth Army routed. A variety of provost sources agree that the majority of stragglers in the first days of the battle were not drawn from frontline units. Many of the stragglers that headed for the rear at the beginning of the battle appear to have been noncombatants employed in constructing defensive positions. Some statistical evidence has also survived. The busiest periods for the straggler posts of the Eighth, Twenty-fourth, Fiftieth, Sixty-sixth, and Sixteenth Divisions occurred during the 27–30 March, that is, at least six days after the battle began. These figures do not tell the whole story, for of these formations, the Sixteenth, Twenty-fourth, and Sixty-sixth Divisions were in action from 21 March, but only the Twenty-fourth Division records any figures for the first two days of the battle. Either the figures for the other divisions were incomplete, or they did not collect any stragglers, which is possible. The evidence of the Twenty-fourth Division certainly supports the contention that the worst period for stragglers came toward the end of March. Sheer exhaustion seems to have been a major factor in causing them to straggle.[108]

Had the morale of the Fifth Army indeed collapsed, the Germans would probably have won the war, for the autumn of 1918 was to demonstrate the serious consequences of a genuine weakening in an army's morale. The German spring offensives were, in Winston Churchill's words, "judged by the hard test of gains and losses . . . decisively defeated."[109] In short, the British army's morale was sound enough to allow it to fight the German army to a standstill. Even in a conflict as technologically based as the First World War, sound morale was an indispensable factor in military success.

Recently, David Englander has published an interesting analysis of British army morale and discipline in the last two years of the war. In this article he speculates whether the 1918 German spring offensive forestalled a major mutiny.[110] Moreover, he argues that, although the Fifth Army's figures for sick wastage were relatively low, which might be taken as an indicator of sound morale, it would be a mistake to assume

> that the failure of the 5th Army was more of a military defeat due primarily to enemy superiority in manpower and firepower rather than

> a moral collapse due to fatigue and depression. The evidence, from postal surveillance, *should it ever be located,* may well present a different story.[111]

The evidence presented in this chapter suggests that Englander's arguments are flawed. While the morale of the BEF in early 1918 was not high, it was a very long way from collapse; in March–April 1918, the Fifth Army certainly took a battering but was not defeated; and the postal evidence, which is referred to above and has actually been in the public domain for some years, testifies to the essential soundness of the Fifth Army's morale.

The events of the spring of 1918 cannot be divorced from those of the summer and autumn, when the Allies went onto the offensive and won a series of crushing victories. In the absence of a censorship report for the second half of 1918, a book by Major General A. A. Montgomery is probably the nearest we have to an "official" view of British morale in the final campaigns on the Western Front. Montgomery, the chief of staff of the Fourth Army, laid great stress on the increase in British morale and a simultaneous decrease in German morale when the BEF took the offensive in August.[112] More dramatically than during the Somme or Passchendaele offensives, the autumn of 1918 presented the British soldier with tangible evidence of success. The BEF was gaining ground, capturing large numbers of guns and prisoners, and the fighting spirit of many German units was perceptibly deteriorating. In the words of an officer of the Forty-sixth Division, the BEF "was at last obtaining a just reward for all its dogged and patient fighting."[113] It is interesting to compare these comments with those of a German infantry officer, who believed that

> the reason for the slow decline of morale within the German Army over the final months of the war was the feeling of the soldiers that they were being ground to pieces in one useless, pointless, and hopeless offensive action after another.[114]

It is likely, nevertheless, that the British infantry's morale was somewhat brittle by the time of the Armistice. R. H. Mottram argued that by late 1918 "a new spirit of taking care of one's self" had emerged among the infantry, who would have begun "not refusing but simply omitting to do duty" if the war had continued beyond 11 November.[115] In the autumn of 1918, there were several mutinies in the Australian Corps, which played a major role in the Allied offensive. Some were protests against the disbandment of units, but two mutinies, that of the Fifty-ninth and First Battalions in September, amounted to combat refusals caused by excessive weariness and a sense that the soldiers were being asked to do more than their fair share.[116] One should be wary, however, of drawing firm conclusions about the state of the BEF's morale from such evidence. Mottram was not serving with an infantry battalion at the end of 1918. His views were undoubtedly exaggerated and can profitably be contrasted with a host of other evidence, drawn from the contemporary writings of regimental soldiers and enlisted men, that suggests morale was high.[117] While the Australian mutinies suggest that these particular troops had reached the end of their tether, it is far from clear that morale in the Australian Corps as a whole was on the point of collapse. Indeed, one can tentatively suggest that much evidence points in the opposite direction.[118] Above all, the ultimate evidence of high morale is willingness to engage in combat. The irrefutable facts are that British Empire troops continued to advance and win battles, taking heavy casualties in the process, until the very end of the war. After the Armistice came into effect, there was a distinct change in the attitude of many soldiers toward their employment—in military terms, there was a collapse of morale and discipline—but this occurred after, not before, the end of hostilities.

In conclusion, it would be difficult to improve on Charles Douie's assessment of British morale in 1918. Douie served as a temporary subaltern in the First Dorsets, and ten years after the war he wrote (in response to the "disillusioned" school) of the generally "magnificent" state of morale in the BEF in the last year of the war. He argued that the BEF of 1918 was inferior in quality to its predecessor of 1916, yet it was sufficiently skilled to fight a numerically superior enemy to a standstill in the spring in two great actions, to counterattack

> and remain continuously on the offensive from August to November. The infantry at least had no doubt that they were winning, and their faith was justified when the greatest military Power of modern times finally collapsed in disordered retreat.[119]

The BEF of the autumn of 1918 consisted of a blend of veterans of earlier campaigns and young, fresh conscripts who had been spared the rigors of trench warfare and the defensive battles of the spring. This was a winning combination of enthusiasm and experience.

Generally, a rapid and decisive victory can be achieved only if one side is greatly inferior to the other in terms of fighting power. On the Western Front, the two sides were roughly equally matched. In the March 1915 to March 1918 period, British offensive operations frequently began with an attempt to reopen mobile warfare, but when the hoped-for penetration failed to materialize such battles were continued in an attempt to "wear out" the enemy's strength and morale. Thus, although the opening phases of the Somme in 1916 and the Third Battle of Ypres in 1917 were designed to achieve a breakthrough, they evolved into attritional battles. In early 1915, Haig became convinced that German morale would shortly collapse,[120] and thus continued to fight attritional battles, designed in part to deplete enemy morale. With some justice, in his final dispatch Haig claimed that the battles of the previous four years were "a single continuous campaign" that had contributed to weakening the German army.[121] Certainly, for a variety of reasons, German military (and civilian) morale began to crumble in the summer and autumn of 1918.[122] In retrospect, the Allied victory of 1918 owed much to the morale of the British soldier remaining intact while that of his German counterpart crumbled. This is not to deny the importance of factors such as improvements in British tactical and operational methods and the arrival of a large United States army on the Western Front, but these would have availed little had the morale of the BEF collapsed. In this respect, the "human factor" was at least as important as technology in determining the outcome of the First World War.

NOTES

This is an updated version of Occasional Paper no. 2 issued by the Institute for the Study of War and Society, De Montfort University, Bedford, in 1995. It is based on G. D. Sheffield, "Officer-Man Relations, Morale and Discipline in the British Army, 1902–22" (unpublished Ph.D. thesis, University of London, 1994), chap. 3. This thesis, minus this chapter, forms the basis of *Leadership in the Trenches: Officer-Man Relationships, Morale and Discipline in the British Army in the Era of the Great War.*

The author is grateful to Peter Simkins, Keith Simpson, Chris McCarthy, and Paul MacKenzie for their help and advice, and to Ray Sibbald for permission to reproduce the paper.

Thanks are due to the following for allowing me to quote from material for which they hold the copyright: Gertrude Hardie (Martin Hardie papers); J. B. Gregory (J. Allison papers); R. G. S. Johnston (R. W. F. Johnston papers); Paul P. H. Jones (P. H. Jones papers); the Liddle Collection, University of Leeds (G. Banks-Smith papers and R. B. Marshall papers). It has proved impossible to contact the copyright holder of the C. E. L. Lyne papers, with whom the Imperial War Museum would like to restore contact. Crown Copyright material in the Public Record Office is reproduced by permission of the Controller of Her Majesty's Stationary Office. I would like to thank the trustees of the Imperial War Museum and the trustees of the Royal Military Police for permission to consult material in their possession, and to offer my sincere apologies to anyone whose copyright I have inadvertently infringed.

1. J. B. Wilson, "Morale and Discipline in the British Expeditionary Force, 1914–18" (unpublished M.A. thesis, University of New Brunswick, 1978); S. P. MacKenzie, "Morale and the Cause: The Campaign to Shape the Outlook of Soldiers of the British Expeditionary Force, 1914–18," *Canadian Journal of History* 25 (1990): 215–31; idem, *Politics and Military Morale* (Oxford: Clarendon Press, 1992); J. G. Fuller, *Troop Morale and Popular Culture in the British and Dominion Armies 1914–1918* (Oxford, 1991). A concise statement of the author's views on the role of paternalism in the upkeep of the BEF's morale can be found in his chapter in H.

Cecil and P. Liddle, eds., *Facing Armageddon: The First World War Experienced* (London: Leo Cooper, 1996), 413–24. Paternalism, logistics, popular culture, and other important factors in the maintenance of British morale that have already received attention from the present author and others are not discussed here.

2. For the expansion of the British army from 1914 to 1918, see P. Simkins, *Kitchener's Army* (Manchester: Manchester University Press, 1988); I. F. W. Beckett and K. Simpson, *A Nation in Arms* (Manchester: Manchester University Press, 1985); I. F. W. Beckett, "The Real Unknown Army: British Conscripts 1916–19," *The Great War* 2, 1 (1989): 4–13.

3. See J. H. Sparrow, *The Second World War 1939–45 Army: Morale* (London: His Majesty's Stationary Office, 1949), 1–2; T. T. Paterson, *Morale in War and Work* (London: Max Parrish, 1955), 99. Child's definition is quoted in I. McLaine, *Ministry of Morale* (London: Allen and Unwin, 1979), 8.

4. C. von Clausewitz, *On War,* ed. M. Howard and P. Paret (Princeton, N.J.: Princeton University Press, 1976), 187–89

5. Unpublished account, 52 (based on a diary entry of 8 Nov. 1916), R. Cude papers, Imperial War Museum (IWM); [A. Smith], *Four Years on the Western Front* (reprint, London: London Stamp Exchange, 1987), 100.

6. For some indications of possible avenues of research, see D. Englander's important article "Soldiering and Identity: Reflections on the Great War," *War in History* 1, 3 (1994): 300–318.

7. Lt. D. D. A. Lockhart's account in War Diary (WD) 1/Gordons, War Office (WO) 95/1435, Public Record Office (PRO).

8. Unpublished account, 41, H. D. Paviere papers, 81/19/1, IWM (Paviere erroneously states the division was disbanded a little later); 24 Mar. 1918, WD 17/Royal Fusiliers, WO 95/1363, PRO.

9. For British civilian morale, see note 6, above; N. Reeves, "The Power of Film Propaganda—Myth or Reality?" *Historical Journal of Film, Radio and Television* 13, 2 (1993): 181–201; J. M. Bourne, *Britain and the Great War* (London: Edward Arnold, 1989), 199–214.

10. For an example, see G. D. Sheffield, "The Effect of War Service on the Twenty-second Battalion Royal Fusiliers (Kensington) 1914–18, with Special Reference to Morale, Discipline, and the Officer-Man Relationship" (unpublished M.A. thesis, University of Leeds, 1984).

11. K. R. Simpson, "Capper and the Offensive Spirit," *Journal of the*

Royal United Services Institution (JRUSI) 118, 3 (1973): 51–56; G. F. R. Henderson *The Science of War,* ed. N. Malcolm (London: Longmans, 1906); Field Service Regulations, part 1, "Operations," 1909 (reprinted with amendments, 1914), 13, 138, 142; T. Travers, *The Killing Ground* (London: Unwin Hyman, 1987), 37–82.

12. T. M. Camfield, "'Will to Win'—The U.S. Army Troop Morale Program of World War I," *Military Affairs* 41, 3 (1977): 125; D. Englander, "The French Soldier, 1914–18," *French History* 1, 1 (1987): 50–51; E. L. Spears, *Prelude to Victory* (London: Jonathan Cape, 1939), 102–3.

13. See the comments of Haig's director of military operations, J. Davidson, in *Haig: Master of the Field* (London: P. Nevill, 1953), 64–65, 125.

14. D. Lloyd George, *War Memoirs* (London: Oldham, 1938), 1: 825,

15. See G. D. Sheffield, "Officer-Man Relations, Morale and Discipline in the British Army," 1902–22" (unpublished Ph.D. thesis, University of London, 1994), passim.

16. M. C. Morgan, *The Second World War 1939–1945 Army: Army Welfare* (London: Her Majesty's Stationary Office, 1953), 1.

17. MacKenzie, "Morale and the Cause," passim; idem, *Politics and Military Morale,* 7–31.

18. Wilson, "Morale and Discipline," 311.

19. Sparrow, *Second World War,* 2.

20. Sheffield, "Officer-Man Relations," 37–40; idem, "Effect of War Service," passim.

21. *Statistics of the Military Effort of the British Empire, 1914–22* (London: His Majesty's Stationary Office, 1922), 649.

22. J. Putkowski and J. Sykes, *Shot at Dawn* (Barnsley: Wharncliffe, 1989), 9; G. Oram, *Worthless Men: Race, Eugenics and the Death Penalty in the British Army during the First World War* (London: Francis Boutle, 1998), 18.

23. W. Childs, *Episodes and Reflections* (London: Cassell, 1930), 142.

24. A. Babington, *For the Sake of Example* (London: Leo Cooper, 1983), 16–17, 18–19.

25. 26 Dec. 1916, WD 22/Royal Fusiliers, WO 95/1372, PRO; WO 93/49, PRO; H. Berrycloath, interview; H. E. Harvey, *Battleline Narratives* (London: Bretano's, 1928), 135, 157–8.

26. P. Gibbs, *The Pageant of the Years* (London: W. Heinemann, 1946), 210.

27. "Memorandum of a Visit to the Western Front by General Smuts," 27 Jan. 1918, GT 3469, Cabinet Papers (CAB) 24/40, PRO.

28. J. H. Bateson to J. E. Edmonds, 26 May 1934, CAB 45/132, PRO.

29. E. H. G. Roberts, *The Story of the Ninth King's' in France* (Liverpool: Northern, 1922), 94.

30. W. Slim, *Unofficial History* (reprint, London: Transworld, 1970), 12.

31. "Mark VII" [M. Plowman, pseud.], *A Subaltern on the Somme* (London: J. M. Dent, 1928), 66.

32. Unpublished account, 75, 87–88, R. W. F. Johnston papers, 82/38/1 IWM.

33. E.g., *Notes for Young Officers* (London: HMSO, 1917), 3.

34. E. L. M. Burns, *General Mud* (Toronto: Clarke, Irwin, 1970), 63.

35. Gibbs, *Pageant,* 201; P. Gibbs to J. E. Edmonds, 26 July 1930, CAB 45/134, PRO; O. Rutter, ed., *The History of the Seventh Service Battalion of the Royal Sussex Regiment, 1914–19* (London: The Times, 1934), 21.

36. J. E. Edmonds and G. C. Wynne, *Military Operations, France and Belgium, 1915* (London, 1927), 1:2–3.

37. Wilson, "Morale and Discipline," 67–118.

38. Lecture delivered by Brigadier General R. A. K. Montgomery, C.B., D.S.O., at the camp of the Eighth Division near Winchester, on 30 October 1914, 5, 9, Department of Printed Books, IWM.

39. "Around Armentieres and the Ypres Salient," in *Twenty Years After,* supplementary volume (London, 1938), 503.

40. Col. M. G. N. Stopford, M.C., "Trench Warfare—General—Winter 1914–15," (lecture to Junior Division, 1938), 13, Conf. 3898, Staff College Library (SCL).

41. See B. O. Dewes diary, 17, 21 Nov. 1914, 84/22/1, IWM.

42. T. S. Wollocombe diary, 103, 18 Mar. 1915, Royal Military Academy Sandhurst Library.

43. W. R. Robertson, *From Private to Field Marshal* (London: Constable, 1921), 229–30.

44. Letter, 7 Jan. [?] 1915, J. Allison papers, 85/15/1, IWM.

45. J. Lucy, *There's a Devil in the Drum* (London: Faber and Faber, 1938), 255, 267, 311–12.

46. F. Richards, *Old Soldiers Never Die* (reprint, London: Faber and Faber, 1965), 60, 97–98.

47. Wilson, "Morale and Discipline," 139, 157.

48. W. R. Robertson, *Soldiers and Statesmen 1914–1918* (London: Cassell, 1926), 1:206.

49. "Report on Complaints, Moral etc.," 3–4 [Nov. 1916], M. Hardie papers, 84/46/1, IWM.

50. Letter, 28 Oct. 1915, P. H. Jones papers, P.246, IWM.

51. P. Gibbs, *The Battles of the Somme* (London: W. Heinemann, 1917), 14–15; idem, *The Realities of War* (London: W. Heinemann, 1920), 169–70.

52. Letter, 26 Oct. 1915, G. Banks-Smith papers, Liddle Collection, University of Leeds.

53. W. L. Andrews, *The Haunting Years* (London: Hutchinson, n.d.), 180.

54. "Report on Complaints, Moral etc.," 6, 12 [Nov. 1916], and "Summaries of censorship Reports on General Conditions in British Forces in France," M. Hardie papers, 84/46/1, IWM.

55. G. De Groot, *Douglas Haig 1861–1928* (London: Unwin Hyman, 1988), 235–36.

56. D. Winter, *Haig's Command* (London: Viking, 1991), 303–15.

57. A. C. Girdwood to J. E. Edmonds, 30 June 1930, CAB 45/143, PRO.

58. See, for example, letter to J. E. Edmonds [signature of writer illegible], 5 Sept. 1930, CAB 45/137, PRO; W. Turner, *Pals: The Eleventh (Service) Battalion (Accrington) East Lancashire Regiment* (Barnsley: Wharnecliffe, n.d.), 131–32.

59. Letter [signature of writer illegible], n.d., CAB 45/134, PRO.

60. Letter, 6 Nov. 1929, C. Howard, CAB 45/134, PRO; M. Middlebrook, *The First Day on the Somme* (London: Somme, 1971), 96–97, 116.

61. J. Keegan, *The Face of Battle* (reprint, Harmondsworth: Penguin, 1978), 218–19.

62. J. Collis Browne to J. E. Edmonds, 12 Nov. 1929, CAB 45/132, PRO; G. Ashurst, *My Bit: A Lancashire Fusilier at War 1914–18* (Ramsbury: Croward, 1988), 95–96.

63. Diary, 220–22, 26, 27 June 1916, P. H. Jones papers, P.246, IWM.

64. J. Q. Henriques, *The War History of the First Battalion Queen's Westminster Rifles 1914–1918* (London: The Medici Society, 1923), 84.

65. P. D. Munday to father, 7 July 1916, 80/43/1, IWM.

66. C. E. Carrington, "Kitchener's Army: The Somme and After," *JRUSI* 123, 11 (1978): 17. See also Col. W. Robertson to J. E. Edmonds [n.d. but ca. 1936], CAB 45/137, PRO; R. H. Mottram, *Journey to the Western Front* (London: G. Bell and Sons, 1936) 144. For an example of

the Thirty-sixth (Ulster) Division's learning experience, see P. Orr, *The Road to the Somme* (Belfast: Blackstaff, 1987), 118–19.

67. E. A. James, *British Regiments, 1914–18* (London: Smason Books, 1978), 46–47, 113; H. C. O'Neill, *The Royal Fusiliers in the Great War* (London: W. Heinemann, 1922), 6.

68. G. S. Duncan, *Douglas Haig as I Knew Him* (London: Allen and Unwin, 1966), 45.

69. Censorship Report, Third Army [Nov. 1916], 6, M. Hardie papers, 84/46/1, IWM.

70. B. H. Liddell Hart, "Impressions of the Great British Offensive on the Somme," 51, CAB 45/135, PRO.

71. Diary, 26 Aug. 1916, Hugh Dalton papers, Dalton I/I 130, British Library of Political and Economic Science.

72. E. Skinner to J. E. Edmonds, 30 Apr. 1936, CAB 45/137, PRO; letter to J. E. Edmonds [signature of writer illegible], 13 Apr. 1936, CAB 45/132, PRO.

73. L. A. G. Bowen to J. E. Edmonds, 31 Mar. 1930, CAB 45/132, PRO.

74. *A Short History of the Nineteenth (Western) Division 1914–18* (London, n.p. 1919) 29; evidence of J. F. C. Fuller, personal papers (PP), 1922, Cmd. 1734, "Report of the War Office Committee of Enquiry into 'Shell-Shock,'" 22.

75. Unpublished account, 176, C. E. L. Lyne papers, 80/14/1, IWM.

76. C. A. C. Keeson, *The History and Records of Queen Victoria's Rifles* (London: Constable, 1923), 212.

77. War Diary, 1/6 Londons, 17 Feb. 1917, WO 95/2729, PRO.

78. "Report on Morale etc III Army," l, 5, 7, January 1917, M. Hardie papers, 84/46/1, IWM. See also Charteris to Macdonogh, 24 Feb. 1917, WO 158/898, PRO, for favorable comments on the Second Army's morale.

79. D. Haig to W. R. Robertson, 1 May 1917, WO 158/23, PRO.

80. "Report on Morale etc" [May 1917], M. Hardie papers, 84/46/1, IWM.

81. "Report on Moral etc." [Aug. 1917], M. Hardie papers, 84/46/1, IWM.

82. "Report on Peace" 19 Oct. 1917, M. Hardie papers, 84/46/1, IWM.

83. Gibbs, *Realities of War,* 396.

84. Quoted in Lloyd George, *War Memoirs,* 2:1468.

85. A. Ekins, "The Australians at Passchendaele," in P. H. Liddle, ed.,

Passchendaele in Perspective: The Third Battle of Ypres (London: Leo Cooper, 1997), 244–46.

86. Gibbs, *Realities of War,* 396.

87. Lloyd George, *War Memoirs,* 1467.

88. "The British Armies in France as Gathered from Censorship," 18 Dec. 1917, GT 3044, CAB 24/36, PRO; G. Blaxland, *Amiens 1918* (reprint, London: W. H. Allen, 1981), 7. For a useful overview, see P. Scott, "Law and Orders: Discipline and Morale in the British Armies in France, 1917," in Liddle, ed., *Passchendaele in Perspective,* 349 70.

89. General Smuts, "Memorandum," 3 Jan. 1918, GT 3198, CAB 24/37, PRO (see also app. B.1, "Report by Third Army Commander" [Byng]"); "Memorandum of a Visit to the Western Front by General Smuts," 27 Jan. 1918, GT 3469, CAB 24/40, PRO.

90. W. R. Robertson to D. Haig, 18 Oct. 1917, WO 158/24, PRO.

91. "Remarks of B.G.G.S., 'I,'" 20 Oct. 1917, WO 158/24, PRO.

92. "Note on the Moral of British Troops in France as Disclosed by the Censorship," 13 Sept. 1917, GT 2052, CAB 24/26, PRO.

93. The final frame of a 1917 version of the film *The Battle of the Somme* (IWM video, 1987) makes this point by showing a map of ground abandoned by the Germans in early 1917.

94. *The History of 2/6 Battalion the Royal Warwickshire Regiment 1914–19* (Birmingham, n.p. 1929), 47.

95. Capt. R. B. Marshall to mother, 3 Oct. 1916, LC.

96. Blaxland, *Amiens 1918,* 8.

97. For an example, see A. R. Armfield, letter, Oct. 1963, in correspondence concerning *The Great War* BBC television series, held at IWM.

98. The Etaples mutiny was largely a product of local circumstances. See Sheffield, "Officer-Man Relations," 295–98; J. Putkowski, "Toplis, Etaples and 'The Monocled Mutineer,'" *Stand To!* 18 (1986): 6–11; G. Dallas and D. Gill, *The Unknown Army* (London: Verso, 1985), 63–81.

99. J. Terraine, *The First World War* (London: Papermac, 1984), 168.

100. R. Prior and T. Wilson, *Command on the Western Front* (Oxford: Basil Blackwell, 1992), 391.

101. H. Gough, *Soldiering On* (London: A. Barker, 1954), 176–78.

102. E.g., Keegan, *Faces of Battle,* 276. See also the correspondence on the subject (including from the author and John Keegan) in the *Times Literary Supplement,* 13 May; 3, 17 June; 8 July 1994.

103. "The British Armies in France as Gathered from Censorship," app. to July 1918, WO 256/33, PRO. For an example of a soldier's "grousing" masking a readiness to fight, taken from the 23 Mar. 1918 diary entry of Lt. F. Warren (17/King's Royal Rifle Corps), see A. Bird, ed., *Honour Satisfied* (Swindon: Crowood, 1990), 87.

104. R. A. Colwill, *Through Hell to Victory* (Torquay: Devonshire, 1927), 103.

105. "A Heavy Gunner Looks Back," in *Twenty Years After,* 1:12.

106. Unpublished account, 33, C. J. Lodge Patch papers, 86/9/1, IWM; Bird, ed., *Honour Satisfied,* 77.

107. M. Middlebrook, *The Kaiser's Battle* (London: Allen Lane, 1978), 105, 308, 322–39, 341.

108. "Straggler Posts," 25, 30, Royal Military Police Archives. This document is a short history of the subject, apparently compiled shortly after the Armistice, which quotes at length from contemporary documents, many of which have now disappeared. For a further discussion of the subject, see G. D. Sheffield, "The Operational Role of British Military Police on the Western Front, 1914–18," in P. Griffith, ed., *British Fighting Methods in the Great War* (London: F. Cass, 1996).

109. W. S. Churchill, *The World Crisis 1911–1918* (reprint, London: Odhams, 1938) 2:1289.

110. D. Englander, "Discipline and Morale in the British Army, 1917–1918," in J. Horne, ed., *State, Society and Mobilization in Europe during the First World War* (Cambridge: Cambridge University Press, 1997), 141.

111. Ibid., 136; emphasis mine.

112. A. A. Montgomery, *The Story of the Fourth Army in the Battles of the Hundred Days, August 8 to November 11, 1918* (London: Hodder and Stoughton, 1920), 1, 5, 9, 145–6, 237.

113. R. E. Priestly, *Breaking the Hindenburg Line* (London, n.p. 1919), 87.

114. G. Ritter, *The Sword and the Sceptre* (London: Allen Lane, 1973), 4:232.

115. R. H. Mottram, J. Easton, and E. Partridge, *Three Personal Accounts of the War* (London: Scholartis, 1929), 127.

116. C. E. W. Bean, *The Official History of Australia in the War of 1914–1918,* vol. 6: *The A.I.F. in France: May 1918–The Armistice* (reprint,

St. Lucia, Queensland: University of Queensland Press, 1983), 875–76, 933–34, 939–40; J. G. Fuller, *Troop Morale and Popular Culture in the British and Dominion Armies 1914–1918* (Oxford: Oxford University Press, 1991), 24–28.

117. See, for example, the optimistic view of a private of 1/23 Londons: A.E. Abrey, letter, 7 Oct. 1918, 84/4/1, IWM.

118. This comment is based on the examination of primary source material in Australian archives. I hope to return to this topic at a future date.

119. C. Douie, *The Weary Road* (London: John Murray, 1931), 15–16.

120. D. Haig to Lady Haig, 1 Apr. 1915, Haig papers, National Library of Scotland.

121. J. H. Boraston, ed., *Sir Douglas Haig's Despatches* (reprint, London: Dent, 1979), 319–21.

122. "Why Germany Capitulated on November 11 1918—A Brief Study Based on Documents in the Possession of the French General Staff," app. 2, SCL; R. H. Lutz, *The Causes of the German Collapse in 1918* (Stanford: Stanford University Press, 1934), 176–77.

CHAPTER 5

Not All Beer and Skittles?

Everyday Life and Leisure on the Western Front

Chris McCarthy

Industry and technology had a powerful impact on the armed forces of the British Empire during the Great War. At the outset of the conflict, the British Expeditionary Force (BEF) was an effective but tiny army that mainly served as a colonial police force. By November 1918, the BEF had become one of the most mechanized of all combatants on the Western Front. Railways emanating from the Channel ports threaded their way across the fields of France and Flanders like a giant spider's web, bringing supplies to the railheads in towns just a few miles from the front line: heavy ammunition for the artillery; stores of cement, barbed wire, and corrugated iron for the engineers; fodder for the horses. From there, light railways bore them to the rear of the trenches.

Motorized vehicles, too, were a common sight. The BEF had arrived in France in 1914 with only 950 motor vehicles, mostly requisitioned from shops and removal firms around London. By the end of the war, it had many thousands. The tasks undertaken by these vehicles varied widely. Ambulances ferried the wounded to base hospitals; staff cars moved officers between their various headquarters; and dispatch riders on motorcycles carried orders between units.

The crucible of war forced the BEF to adapt in other ways as well. To undertake battle against a continental foe, Britain had to expand its armed forces. As a result, 5 million men were under arms over the course of the war. To feed, clothe, and equip them was a triumph of logistics for the British army, occasioned by the very scale of a conflict that had troops stretched out along a 126–mile[1] front. It was made possible only by the industrialization that had begun over a century before.

As battles progressed, the relationship between industry and the military became more pronounced. The tactical need to innovate and develop existing weaponry—the *science* of war, if you will—moved crucial industrial processes on apace. They, in turn, had a direct impact on the specialties of the Army, in particular. Take, for instance, the evolution of signals into wireless communication and the development of new weapons such as trench mortars, gas, tanks, and fighter aircraft. Each of these technological developments necessitated dramatic changes in tactics and the very structure of the BEF.

Yet, while industrialization improved the killing capabilities of the army, in terms of both hardware and the wherewithal to keep its troops fighting fit, it did little to influence the way in which the average soldier spent his time, whether in or out of the line, largely because of the stationary nature of the war. When movement did take place, the ordinary soldier considered it a red-letter day to be transported by anything other than his own two feet.

Of greater significance to the life of the infantryman in the First World War was the recognition by high command of the need to keep gainfully occupied or entertained the vast number of troops under its control, especially when out of the line. It was a matter of both welfare and discipline: the better you look after a soldier's well-being, the better he fights; the more time you occupy with organized activity, the less he has to spend on undesirable pursuits.

This chapter looks at just how the average British soldier on the Western Front was occupied when not engaged in battle: from the routines laid down by the army, through organized activities he might choose to get involved in, to those he engaged in when left to his own devices. It looks at where he lived, what he did on duty, and how he spent the endless hours with time on his hands.

It may have already come as a surprise to some that soldiers did not spend the entire four years of the war in a frontline trench, rifles trained over the parapet, at the ready to repel an attack at any moment from the enemy beyond. Compelling images of men struggling through mud with blood and gore all about them have insinuated themselves into our collective memory and are regularly reinforced by feature films and TV documentaries dwelling on the same. Intellectually, we know that is not the whole picture, but less dramatic conditions are easily pushed aside by the more powerful imagery. We forget that the war was fought in summer as well as in winter. We forget that wet, windy, cold days were followed by dry, still, hot ones. We imagine the gut-wrenching fear that undoubtedly seized men at the sharp end, but we forget the humor that followed close on its heels and which is the natural reflex of humans in adversity. In short, we lose sight of the middle ground.

When, in the course of the last seventy years, veterans have been asked of their overriding memories of the war, many have cited boredom as the predominant impression. They refer to the repetitive routine and endless, confined days when nothing much happened. Ennui and lethargy were as much their enemy as the men in the trenches opposite, making it all the harder for them to react when required.

If troops were not in the front line continuously, how long were they there, and how often? The answer is, it varied. The ideal was that, of the three infantry brigades in a division, one brigade would be at rest, often at the railhead some four or five miles behind the line. The remaining two brigades in the divisional area would place one battalion each in the line, one in the support trenches, one in reserve, and one at rest. Each battalion in the front line put three companies forward and kept the fourth as battalion reserve. In turn, each company placed two platoons in the fire bays of the frontline trench, plus one in support and one in reserve.

The theory was that battalions would relieve each other every four days, and brigades every few weeks. The entire division would change from the line to corps rest area, some ten or so miles away, every three months. In practice, brigade and divisional reliefs worked in quiet spells, but an offensive or bad weather would force changes to this system.

For the brigade at rest, the system meant pulling back to a town to sleep during the day and, often, moving up to the line at night to work. Battalions at rest, by contrast, seldom got back farther than to one of the camps two or three miles away, and these were often still under shell fire. Working parties were more frequently sent out from here, as the distance was short, and most nights would be spent working.[2] Thus, taking all the variations into account, troops spent approximately one week in three actually in the front line.

When in the trenches, the boundaries of time were marked by "stand-to," which took place daily one hour before both dawn and dusk,[3] considered to be the most likely times for enemy attack. Every available man was required to be present: it was a major opportunity for rifle, ammunition, and clothing inspection; checking gas masks; issuing orders; and informing men of their sentry duty. Breakfast was cooked after morning stand-to. The usual fare was bread, bacon and tea,[4] but in winter an issue of rum might be offered, and was rarely refused.

After stand-to frontline vigilance became the duty of sentries, who were always posted in pairs. During the day, one would watch for movement in no-man's-land while his partner rested close by. At night, though, both kept watch together, mostly to keep one another alert. Each sentry or pair was on duty for two hours, except in bad weather, when this could be shortened to an hour or less. The remainder of the men spent the time resting and awaiting their stint of guard duty. Sleep was always highly prized: during the day, most men would take every opportunity to catch as much as they could. In the frontline and support trenches, equipment had to be worn at all times, so even in sleep, the soldier would have his arm through the sling of his rifle.

Work started after evening stand-to. Trenches, and in particular breastworks, which were used in places where digging was made impractical by marshy ground, required a great deal of maintenance. A single shell could ruin trenches and parapets, but on a more mundane level, sandbags rotted and fell apart. There were new trenches to be dug, wire to be erected and repaired, patrols to be sent out and orders to be brought up along with rations, water, and ammunition. In fact, the whole front line came to life after darkness fell.

During the four days in the line, each platoon always had a non-commissioned officer (NCO) and officer on duty whose job it was to patrol the trench line, making sure the sentries were alert and paying attention. Officers also had administrative work to do, in particular the completion of the trench log book, in which all work was recorded, ready for the commander to hand over to his counterpart on relief. Similarly, the trench stores book, containing an inventory of barbed wire, pickets, duckboards, nails, picks, shovels, and so on, was kept up to date so that the stores could be handed over to relieving troops in exchange for a receipt.

Reliefs were nearly always carried out at night, and it could be midnight before the incoming troops, starting out at dusk, would reach the front line, where the outgoing troops waited anxiously. The handover was performed as quickly as possible: it was a dangerous time when the enemy could easily get wind of what was happening and put an artillery barrage on the front line, causing casualties to both incoming and outgoing troops.

On coming out of the line, the chances were the soldiers would be billeted in a farm. Farms were almost all of universal design: buildings on four sides with a dovecote for winter food and a void in the middle that held a midden—graphically described by the humorist Bruce Bairnsfather as a "rectangular smell."[5] An account in the history of the New Zealand Rifle Brigade tells how, one dark night, a company commander "made an involuntary reconnaissance of the ancestral midden and met his C.O. the next morning, wearing an ill-fitting suit of velveteen provided by his obliging host."[6]

An officer would usually be housed in a room in the farmhouse itself, sleeping on the bed provided or on his cot in a sleeping bag. He mostly took his meals in the kitchen. Assuming the owners had not been evacuated, meals were often provided by the woman of the house. Alternatively, the officer's servant would share the kitchen with her.

The men usually occupied the hayloft on the second floor of the barn, and it was not unknown for the farm animals to be still in residence on the first floor. The rustic accommodations forced most soldiers to rely on straw as a mattress. Farmers were paid to provide fresh straw, although this was rarely available. As a rule, the older the straw, the flatter and lousier it was, leaving many enlisted men to suffer through an uncomfortable and somewhat itchy night. Barns were sometimes fitted out with tiered wire-netting bunks, sometimes four high and often as deep. The trick was to get a bunk on the top tier, as the mud and rubbish accumulated from days in the front lines would fall through the wire netting, giving the man on the bottom bunk a constant dust bath.

The wealthier Belgian farmers had sometimes built good brick outbuildings with concrete floors, while further south in France the barn was usually a wattle-and-daub structure. These structures, although having stood in some cases for hundreds of years, did not take too kindly to a constant stream of soldiers passing through them and quickly fell into disrepair: walls collapsed and floors were soon deep in mud. The farmers were unwilling or unable to maintain them, and soldiers hung empty corn sacks over the holes in the walls to defeat the constant drafts. It seems the notion of patching them with timber did not occur to anybody,[7] and in any case, all spare and not-so-spare wood was used to make fires. It is likely, too, that troops worked on the premise they would be there for only a few days and left problems for the next occupants to sort out.

If rest was taken in a town, then factories, schools, warehouses, in fact, any large buildings were utilized as billets. These were often less comfortable than farms but had the advantage of keeping the whole unit under one roof. In the Loos sector in the summer of 1915, billets in the houses around the pitheads were eagerly sought: the parlor or spare bedroom of a modern house was considered infinitely preferable to a drafty barn or factory.[8]

Relations between local inhabitants and the army were more or less cordial throughout the war, partly because the army chose not to live off the land—a policy instituted by the duke of Wellington during the Napoleonic wars. The reception given to the troops was not uniform, however. In some areas the troops were welcomed warmly at first, but the novelty wore off as the war dragged on. Feelings hardened toward what were, after all, occupying troops, even if they were allies, and they began to be regarded cynically as "money on the hoof." R. J. Russell of Eleventh (Lewisham Battalion) Royal West Kents had this to say of a billeting village in August 1915: "Brucamps was a small straggling village, which offered but little comfort. The inhabitants seemed to resent our intrusion . . . we found them unwilling to sell anything, whilst even the pump handles had been removed to prevent our drawing upon their water-supply."[9] In other areas, the troops were treated well throughout the war.

Farmers were not expected to put up with the intrusion of the troops for nothing, and a rent of 5 centimes was paid for each soldier accommodated, and 1 franc for each officer.[10] Stabling and grazing land was also paid for, and compensation of up to 125 francs[11] was handed out for damage, real or imaginary. An example of this is reported by First Battalion, New Zealand Rifle Brigade.[12] As they left for the front line, the enterprising owner of the ground occupied by Regina Camp charged them for a tree, which he claimed had been cut down by men of the battalion. The account was paid. The Fourth Battalion had a similar experience over the same tree a few days later, again paying the bill. The owner tried it, too, with the Working Battalion as they left the area, but this time the New Zealanders finally realized what was happening and the game was up.

When there was insufficient accommodation for resting troops in barns, as in the congested area around Ypres, or when the ground behind the line had been razed, camps were used as an alternative. Camps were naturally preferred by the army, as they kept the unit together and out of mischief. They rejoiced in names such as Red Rose, Toronto, and Piccadilly, usually named by the units who built them. The Ypres area had a famous one known as Dirty Bucket camp, which by all accounts lived up to its name, though how it got it is a matter of conjecture. Camps were hutted or tented or sometimes both. The hut used in 1915 was known as the Armstrong—named, like all huts, after its engineer inventor. It was a prefabricated structure with wooden framing of four-by-one-inch timber, covered with canvas. The official line said that it stood up to the weather conditions well but was found to be cold in winter. The reality was a bit different, according to an officer in the Fifth Sussex:

> The huts are covered in a poor sort of material, so when there is rain outside there is a sort of Scotch mist inside, and everybody is always damp.[13]

After the autumn of 1915, Armstrongs were mostly replaced with Tarrant Huts, which measured fifteen-by-twenty-five feet and were made of double-thickness boarding. Later, in 1916, the Nissan Bow Hut—a sixteen-by-twenty-seven-foot corrugated-iron affair—became the most common hut in France.

Camps of tents were also common. These were invariably bell tents, so called because of their shape, and they slept twelve to sixteen men per tent with their feet to the central pole. Alarmingly, they were issued in easy-to-spot white canvas. The First/Fourth Gordons were given a free hand in camouflaging theirs, with interesting results: tents were painted in disruptive patterns of yellow, red, and green; the chaplain had a fetching black number; one was turned into a wigwam, with figures in brown and blue walking on a background of verdant green; and there was at least one red-and-blue checkered pattern. The prize, however, must go to the minimalist who simply pinned a sprig of hawthorn over the door to comply with an army order stating that all tents must be disguised either by painting or by covering with green stuff. It appears, however, that the army was not entirely at fault in sending out white tents in the first place. They had actually ordered the tents painted brown, but some bright spark had painted the inside by mistake.[14]

When first out of the line, the men were often allowed to sleep late. On rising, the soldier's first job was to clean his equipment and uniform ready for inspection. The first parade was normally at 8.45 A.M., when the morning work was assigned. According to the orders read out by the orderly sergeant the previous night, this could mean a route march, a company drill, a field-craft exercise, fatigues, or general training. Like the carrying parties that blighted most rest periods, these parades were regarded as needless work by the troops and were bitterly resented by officer and enlisted man alike. Specialist troops, such as signalers, bombers, and machine gunners, were usually spared laboring in order to practice their specialty instead.

Rest was an opportunity to tighten up on discipline and keep the men fit after the sedentary life in the trenches. It was important for an army that marched everywhere to keep its feet in good condition, so one or two mornings would be spent in route marches.

If working parties were loathed, bath parades were welcomed; a chance for a wash and some clean clothing was a high spot. The process of bathing seven or eight hundred men was fairly complicated: battalions were given specific time slots, and with the number of troops away on working parties, it was hard to assemble the group in the time available. Breweries were often turned into bathhouses, as the vats were large enough for several men to bathe in. Pithead showers were a luxury for troops lucky enough to be in the right area, having unlimited hot and cold water. Bath water was usually tepid and could be up to fifth-hand. The whole process took ten minutes of actual bathing time. While the men bathed, their underclothes and uniforms went through a steam disinfector and, according to one reference at least, were dipped in creosote.[15] On average, the troops might get a bath every three to four weeks.

After the morning's work, the troops were dismissed sometime between 11.30 A.M. and noon to have their "dinner" and were usually left to their own devices afterward. How they spent their free time was largely determined by how much money they had in their pockets.

Just how much the infantryman received as pay is a cloudy issue. Frank Richards, author of *Old Soldiers Never Die,* said that at the time of the Battle of Loos in 1915, the pay of a "first-class" soldier—meaning a proficient soldier and a first-class shot—"was eighteen pence[16] a day, with some were receiving fifteen pence and some a shilling. A few of us were on nineteen pence. The latter end of 1917, a first-class soldier's pay was increased to two shillings a day, and the others in proportion."[17]

A pay book dated 28 August 1917, however, shows one particular private in the London Regiment getting 1 shilling per day basic plus 3 pence proficiency pay, less 6 pence stoppages, leaving 9 pence take-home pay per day. His pay increased from 29 September 1917 to 1 shilling and 9 pence, made up of 1 shilling regimental pay plus 3 pence proficiency pay, 3 pence war pay, and a further 3 pence representing a penny per year of active war service.

In accordance with the King's Regulations of 1912, proficiency pay was awarded when the soldier fulfilled annually tested criteria:

> Cavalry: To be a good horseman and man at arms.
> Artillery: To reach the standard required by the Corps regulations.
> Infantry: Physical endurance in marching with skill at arms.[18]

Soldiers in the Artillery, Engineering, and Army Service Corps all received more money than the infantryman, much to his chagrin. When in France, the men were paid directly in francs, usually at two-week intervals. The rate of exchange for the issue of cash to troops in the Expeditionary Force was fixed at 3 shillings and 7 pence for 5 francs.[19] To give an indication of the purchasing power in 1914, beer was 2.5 pence[20] a pint; twenty cigarettes cost 4 pence a packet; a loaf of bread 5 pence 3 farthings.[21] Most staggering of all, £1 in 1914 was equivalent to £50 today.[22]

At the start of the war, an average civilian engineering worker earned a weekly wage of £1, 2 shillings, and 10 pence, and a construction worker £2 and 7 pence.[23] At the end of 1917, the soldier received 12 shillings and thruppence[24] all told.

Compared with the French soldiers, however, the British were millionaires. In 1914, the French soldier earned one sou (equivalent to a halfpenny) a day.[25] The Australians, though, were the real winners, receiving a massive 5 shillings a day—and as a consequence, earning the sobriquet "the fucking five-bobbers" from the British soldier. One suspects that this huge pay differential was a strong contributory factor in the Australians' notorious lack of discipline out of the line. For single men especially, with no family commitments, there was little on which to spend their fortune and "high living" had obvious attractions. In this respect they could be likened to the Western "yuppies" of the 1980s—hard at work but even harder at play.

Money for most, however, was in short supply, and men tried various means to obtain it. Collecting and selling souvenirs was popular throughout the war. Shell fuses, the German *pickelhaube* helmets, and Luger pistols were all highly sought-after and found a ready market with the men back at base. Other keepsakes also had their vogue. At one stage, anyone who was anyone had a shell ring filed from an aluminum nose cap, though this trend ended when an enterprising soul began casting them out of lead. Shell cases, too, had their turn, being fashioned into flower vases and ashtrays, while rifle cartridges made good paper knives and model airplanes. Salvage, however, made shell cases increasingly difficult to obtain, and the more earnest collector refined his search in favor of items with a specific history: the nose cap of the shell that buried him, say, or one that nearly killed the commanding officer. Nose caps, indeed, had an immense vogue. According to a Private Bruce:

> There were times when No Man's Land on a misty morning resembled nothing so much as Margate sands in August, only instead of happy children building sand castles there were men digging for nose caps.[26]

Those really desperate for money resorted to selling off their equipment, leather jackets, harness, and even rations of bully beef, which was quite sought-after by the local peasants.

But as every soldier has known since the beginning of time, the real road to immense riches is gambling. Gambling was strictly against King's Regulations. In fact, the only game officially allowed was "housey-housey," or bingo, as it is now more familiarly called; probably allowed because it is almost impossible to "fix." Here's how the game was conducted:

> It costs a penny for a green card with numbers on. When fifty or sixty cards are sold, the game starts. A man picks numbered discs out of a cloth bag, one at a time, and shouts the number. If it is on your card, you mark it off. The first one to have all his numbers marked off shouts "House!" and wins the prize.
>
> It is good fun, but has a special language of its own which takes a lot of understanding. Nearly every number has a name—"Kelly's eye", "legs eleven", clickety-click", top-o'-the-house.[27]

Inevitably, the owner or backer of the game deducted a percentage of the kitty "for his trouble."[28]

The most popular gambling game by far, however, was Crown and Anchor:

> The paraphernalia used in "Crown and Anchor" consists of a piece of canvas or oil cloth two feet by three feet. This is divided into six equal squares. In these squares are painted a club, diamond, heart, spade, crown, and an anchor, one device to a square. There are three dice used, each dice marked the same as the canvas. . . .
>
> . . . the Tommies place bets on the squares, the crown or anchor being played the most. The banker then rolls his three dice and collects or pays out as the case may be. If you play the crown and one shows up on the dice, you get even money, if two show up you receive two to one, and if three, three to one. If the crown does not appear and you have bet on it you lose, and so on. The percentage for the banker is large if every square is played but if the crowd is partial to, say, two squares, he has to trust to luck. The banker generally wins.[29]

Gambling was an organized business among the troops. One man might run half a dozen "Crown-Anchor" gambling boards in different parts of the camp. Two men worked each board, and dozens were employed as lookouts to report the approach of strangers. It is said that some NCOs were bought off by the gambling-school boss.

These games were a great temptation for some men, despite the distinct possibility that they might lose all their pay within an hour. Frank Richards describes what the lucky ones could avail themselves of:

> There was a large concentration of troops around this village and the cafes were always full of troops. It was not an uncommon sight to see a dozen Crown and Anchor boards outside each café, and one evening we won over a thousand francs—about £40 which we invested in champagne.[30]

Cards were also much in evidence. Pontoon, or blackjack as the Americans knew it, was ever popular with enlisted men, and so was Brag. Another game called Nap held an attraction for some, though patently not for an American serving in the British army, by the name of Empey who experienced the game and quipped: "It is well named. Every time I played it I went to sleep."[31]

By far the most popular game among British officers was bridge, although poker is occasionally mentioned.

The Australians were keen on the gambling game two-up, known also to miners in the north of England. The rules couldn't be simpler. It consisted of two coins, usually laid on the index and second finger, which were then thrown up into the air. Bets were placed on whether the coins come down heads, tails, or "odds." Money was paid out at "evens."

One pursuit requiring no money at all was writing home, and this the men did with a vengeance—or so some of the officers felt, for every letter had to be censored by them for anything that might be of use to the enemy. This task was as onerous to the officers as working parties were to the men. In the first place, most disliked it because it was intrusive—at least one officer had to censor letters from a man in his company addressed to a young lady he himself was holding a torch for. Second, the letters were, on the whole, dead boring. After all, if the average junior officer knew little of what was going on, the men knew even less.

Among the items prohibited from mention was the name of the

place in which the troops were staying. Still, the British soldier, being a cunning and resourceful devil, did his best to circumvent this restriction by trying to encode his words in some form or other. Here is an example of such a letter sent home by a soldier in the Eleventh Royal West Kents:

> Dear Mother,
> We have moved. I must not tell you where, but we are not far from a town the same name as myself.
>
> Your loving son
> Albert[32]

Not surprisingly, this particular piece of deft prose did not escape the censor's blue pencil.

Letters were sometimes sent in regulation green envelopes. These could be sealed uncensored, but the soldier was required to sign a declaration on the back stating that the information inside would be of no use to the enemy. A random selection would then be opened at base.

Officers, by contrast, seem to have far fewer restrictions in this respect, if the number of explicit letters surviving today is anything to go by. Some even sent home entire diaries.

Walks and, if there was a nearby stream, swimming were popular activities for the impecunious. Fishing, too, had its following, but it was not always the pastoral idyll that one might expect:

> So they chucked a Mills bomb into the Scarp and when it went off, salmon floated to the top. Now this went on for some time until one day some fool brought up a trench mortar bomb and blew such a hole in the river bank that it actually changed the direction of the bloody river![33]

Soldiers never wasted an opportunity to delouse their clothes while they talked and smoked. Some of the most popular cigarettes of the day were Ruby Queen, Red Hussar, 'Arf a Mo, Gold Flake, and, topping the list, Woodbines. The delousing technique was to run either a thumbnail or a candle flame along the seams of the clothing. It did little to arrest the lice but much to avenge the soldier's greatest source of permanent discomfort.

There were those, too, who undertook quite individualistic pursuits: a Captain Gosse amused himself by stuffing small French mammals and sending them to the Natural History Museum in London.[34]

Naturally enough, sports played a large part in keeping the troops occupied, with by far the most popular being football[35]—or, as the French knew it, "le fool-ball," as it seems the average French farmer was very reluctant to give up his good grazing land for such a frivolous reason.[36] There were football matches in the thousands. Officers were amazed by the readiness of their men to start an impromptu football match when, moments before, they had appeared exhausted. The general staff considered sport a good way of letting off steam, particularly in the officers-versus-men matches, where the rank and file could point out the deficiencies and incompetence of the officers with no recriminations.

Inter-company/battalion/brigade/division matches were taken extremely seriously. Battalion teams usually featured a number of peacetime professionals. The Sixteenth Royal Scots had the entire Heart of Midlothian, a professional football club, in their ranks—a point nimbly glossed over by the divisional history when later celebrating the team's inevitable victories.

A typically important match is described in the Eighteenth Division's history:

> "A" battery of the 82nd Brigade RFA were playing the DAC in the Divisional Cup, and their players were brought to the field in a newly-painted and varnished G.S. wagon, drawn by six grand-moving blacks—the battery's showiest gun team. Also there were enough chargers, ridden by brigadiers and brigade-majors and staff-captains and ordinary officers come to see the tie, to make one think it was a point-to-point meeting.[37]

Boxing was far more popular than it is now and, like football, was well supported; no doubt, the opportunity to gamble was not overlooked. Both officers and men boxed. In fact, Walter Long, the commander of the Fifty-sixth Brigade, was twice middle-weight champion of the British army.[38]

Horse shows, too, were common, giving officers in particular a rare chance to show off their riding proficiency; though a temporary Kitchener army officer hailing from deepest suburbia might have regarded the opportunity with rather less enthusiasm. Naturally, the mounted troops took these events very seriously, and there was fierce competition for the best turned out team and the like. Troops not directly involved in the shows viewed them like a day at the races and thoroughly enjoyed them as such.

Another chance to poke fun at the officers and gamble was provided by good old-fashioned sports days, just like at school, with "three-legged," "egg 'n' spoon" and "sack" races, but lent a martial air by events such as long-distance bomb throwing, stretcher-bearer races, and accurate bomb throwing.[39]

Next to football, cricket was the most popular peacetime sport of the day. Indeed, the war interrupted what has come to be known as "the golden age" when county cricket clubs could regularly fill the stands for ordinary midweek matches. Such enthusiasm for the game was naturally transferred, along with many players, to the BEF. The Rifle Brigade's history describes how the Tenth Battalion "took violently" to cricket and remarks that it was "a welcome change from the inevitable football." The Eighteenth Division records a cricket match between the divisional headquarters and the Sussex Pioneers, where a certain Sergeant Caesar—a member of the Surrey Ground Staff[40]—"took the Divisional Staff apart."

Rugby football had a strong following particularly in Welsh regiments, although most battalions could turn out a team. With Canadian and American troops, baseball was the natural favorite. Even this rather foreign sport found a following in the BEF, as the First Essex Regiment took on the Third Canadian Infantry Regiment during the Battle of the Somme in 1916 and eventually beat them.[41]

Although there was a great deal of sport going on, as many have experienced at school, it was left to a chosen few to play: the rest were encouraged to watch.

No account of soldiers' free time would be complete without mention of drinking. Most drinking took place in *estaminets,* a sort of cross between a pub and a café—something like a wine bar, in fact. At best an *estaminet* would be like a café; at worst, it would be someone's living room. Generally it amounted to a plain, somewhat spartan room with a stove at one end and furnished with either long, refectory-type tables and benches or smaller tables and chairs. A small counter or bar would be in one corner, often a piano and, in the truly well-appointed ones, a billiard table.

Drink was the main but not sole attraction of *estaminets.* Perhaps as important, they offered a tentative link with civilian normality. With nearly all French men at the front, *estaminets* were usually run by a woman, with a couple of daughters or waitresses to help out and a regular posse of children. According to the author Saki,[42] it was "a fixed rule that *estaminet* children should be big enough to run about and small enough to get between one's legs."[43]

In addition to beer at a penny a glass and "van blonk"[44] at 1 or 2 francs a bottle, there was an opportunity to eat egg and chips or an omelet, play "house," have a singsong, and exchange banter with the waitresses. The beer, though, was often a quite flat concoction that did not live up to the standards of most members of the BEF. According to one officer:

> Belgium water, by the way, is light brown to look at, and hardly distinguishable either in color or taste from their beer.[45]

This was an opinion shared by most soldiers and extended to all beers available along the Western Front.

Estaminet hours were strictly controlled by the army. They were open from 11 A.M. until 1 P.M., and in the evening they opened at 6 and closed at 8. The limitation of drinking hours, however, was not entirely successful. General Birdwood, commander of the Anzac Corps, believed drink was responsible for "85% of the trouble over here." The problem from the army's point of view was that alcohol encouraged insubordination and absences without leave. The army distinguished, though, between two forms of drunkenness: "simple drunkenness" and "drunkenness on duty."

A private soldier would not be tried by court-martial for an act of "simple drunkenness," which was deemed to be "an act of drunkenness committed when not on active service, when the soldier had not been warned for duty, nor had by reason of the drunkenness rendered himself unfit for duty—unless four instances of drunkenness [had] been recorded against him within the twelve months preceding the offense." This offense would be dealt with by his commanding officer unless the defendant elected for a court-martial. The punishment for a first offense was a reprimand, for the second a fine of 2 shillings and 6 pence, for the third and fourth a fine of 5 shillings, rising to 10 shillings if more frequent still.[46]

By far the more serious crime, though, was "drunkenness on duty" when in the line, for which an automatic court-martial was given. Only some 35,313 field general courts-martial for drunkenness took place throughout the war.

It was not unknown for officers to indulge in a drop, too. An officer of the Tank Corps admitted to entertaining two Australian officers rather well and waving them off at 2:30 A.M. He found them some hours later, going around and around in a circle, following the tapes used to fence off the officers' tents, convinced that eventually they would be led back home.[47]

"Van blonk" was a little more alcoholic than and of a comparable price to beer, though Frank Richards maintained that the wine was much better at the start of the war than at the end. Canteens often had bottles of Guinness, and sometimes, when the "van blonk" turned out to be immature champagne, Black Velvets would be the order of the day.[48] Enlisted men were not allowed to buy spirits over the counter, but the more determined found ways around this rule.

In hutted camps "wet canteens" provided an alternative to *estaminets*. This description of a wet canteen in England gives an impression of the conditions:

> As he was carried through the doorway, the stifling, vitiated air caught his throat, and for some time he coughed and choked, with his eyes running and his cheeks burning. But beyond the mingling smells of wet clothing, stale beer, tobacco smoke and sweat, there was a pungent chemical odor that he could not place. . . .
>
> . . . And in a flash it came: it was the smell of a urinal. . . . Long before closing time the puzzle was resolved. Once inside the canteen, the only warm place in the camp, one stayed until closing time; to leave it was to fight one's way out through a packed crowd and to find a return hopeless. All the men inside drank steadily, and they remained there several hours, urinating where they stood or sat, with, if sober, as little inconvenience to their neighbors as could be managed.[49]

There are some references to soldiers drinking themselves to death, though this was not so much normal *estaminet* drinking than abuse of the rum ration, where, for instance, men might find a gallon jar of rum and drink it at one sitting.

Before the outbreak of war, music halls flourished in towns all over Britain, and this type of entertainment easily translated into concert parties on the Western Front. The first concert party was started by the Fourth Division in late 1915 and was called The Follies. By the end of the war, most divisions had one. The parties were encouraged by high command, who considered them good for morale. As the ranks of any battalion could produce professional music-hall turns, it was hardly surprising that many of the acts were quite good, and a first-class show could be seen for the cost of a penny or so. There were singers (serious), singers (comic), magicians, acrobats, and comic turns. Their efforts were appreciated by officers and men alike.

All concert parties had one or two female impersonators who, if surviving photographs are to be believed, could make a pretty convincing job of it. For the length of the show at least, troops were prepared to suspend their belief entirely and accepted the performers as women. The Fifty-sixth (London) Division's concert party, The Bow Bells, was generally regarded as the best.

Cinemas, too, were run by divisions and were immensely popular throughout the war. They were usually set up in a suitable building but occasionally out in the open. Comedy films were the most popular, especially those of Charlie Chaplin. Roland Fielding, in his *War Letters to a Wife,*[50] tells of watching the acclaimed British documentary on the *Battle of the Somme* on the battlefield itself, with real guns providing unintented sound effects in the background. He thought the film "a really wonderful and most realistic production" but questioned the wisdom of showing it to troops about to attack. He did not question the entertainment factor.

> Who was it had told us French women were beautiful, or knew how to wear clothes? Bah! these women of Amiens with their flat peasant faces and their shuffling walk were drab bundles of humanity possessing none of those feminine attractions of which we had our visions.

So said Sydney Rogerson in his book *Twelve Days.*[51]

The subject of women is a tricky one. Most soldiers were quite young and relatively inexperienced in the ways of the world—then again, there were those who were not. Very little is written about the subject, but we can assume that the same multiplicity of standards applied then as does today. Certainly, the occasions on which a frontline soldier could form any relationship at all with a woman were quite limited—farm girls, waitresses, and the women of their billeting village providing the only opportunities. The majority of such relationships were probably quite innocent—the men welcoming the chance to talk and flirt with a member of the opposite sex and one not dressed in khaki.

The army was fully aware of the temptations to young men and tried to prevent subsequent problems. Frank Richards recalled:

> Each man had been issued with a pamphlet signed by Lord Kitchener warning him about the dangers of French wine and women; they might have well not been issued for all the notice we took of them.[52]

We should not be too naive about these things, however. Brothels were taken over from the French and were officially unofficial.

Mostly in back-area towns, they enjoyed the seductive name of Approved Houses. They were similar in design to *estaminets,* but with an upstairs. The bar downstairs sold drinks at over-the-odds prices and acted as a waiting room.

A vivid description is provided by Alfred McLelland Burrage, known as the author Ex Private X:

> At the back of the barracks [at Hesdin] was a soldier's brothel. It had not been put out of bounds and even if it had been I should have felt compelled to have a look at it. . . .
>
> The interior was rather like that of a Nonconformist meeting house, and never before or since have I seen commercialized vice look so unattractive. The "characters" might all have taken part in one of Shakespeare's tavern scenes or stepped straight out of *The Beggar's Opera.* The *patron,* a hulking bully, and his obscene old consort would have made a perfect Mr and Mrs Peacham. There was an appalling serving wench who could not bring a drink without making some Rabelaisian remark or an indecent gesture. But to me the most amusing of the whole circus was the spectacled young woman in a severely cut black dress who sat at the receipt of custom and took the wages of sin. You felt that you couldn't say anything out of place to *her.* . . .
>
> As for the resident ladies, the least said of them, poor things, the better. They were old and worn and hideous, with death's heads instead of faces.
>
> We sat for a half hour drinking and resisting invitations to "come upstairs." I noticed that the very bad *vin blanc* was five times the price charged at the more virtuous establishments, and since *they* profiteered unconscionably, Monsieur and Madame must have done very well out of the drinks alone.
>
> I saw only one man plonk down his five francs and march brazenly upstairs. . . .
>
> Why were not the "red lamp" houses put out of bounds? A man was given every encouragement to contract a contagious disease, and once he had caught it he was treated like a criminal.[53]

The rooms used by the girls had little more than a bed and washbasin in them. This lack of comfort was not a drawback, as the entire process took only ten minutes—which, given the following description, was probably quite long enough:

> Half the places are so stinking, and the wenches such old cows, that a man's got to be canned well up for his stomach not to retch.[54]

Prices and services at such establishments varied. For instance, the price at Bethune in 1915 was quite reasonable at 2 francs.[55] Some of the girls employed there, it appears, were quite imaginative:

> In one of [the brothels in the Rue des Gallenne, le Havre], it was said, one of the sirens wore the uniform of a British captain and was much in demand among the disrespectful Tommies. If this were true it was a brain wave on the part of the lady.[56]

Just before the Battle of Loos, a line of three hundred, mainly young men, was seen outside a brothel in Bethune.[57] It was then, and still is, a common sentiment among soldiers that they did not want to die a virgin.

Some girls, of course, worked as independent operatives and not under the auspices of a brothel:

> Of course, whenever the soldiers were away from the line at rest amongst civilized surroundings there was a rush for a "bit of skirt." One "Lady" at Berchem set up in business in a field to attend to the wants of the men. She took about 485 francs from 162 men in three days.[58]

The presence of brothels and prostitutes presented the army with moral and physical problems. The most bothersome problem, it would seem, was the possibility that men would contract venereal disease (VD) and have to be put on sick leave. Partly in an effort to relieve the VD problem in 1915, the medical officer of health and the police authorities at Havre[59] decided to open, under medical supervision, a street of brothels that had previously been out of bounds. Over a period of fifty-seven weeks, a total of 171,000 men visited houses in this street (460 troops a day). Only 243 stated that they had been infected there. Known as Maisons de Tolerance, these controlled brothels were set up in other towns, but in April 1918, they were put out of bounds in deference to strong public feeling in the United Kingdom.[60]

The possibility of contracting VD was enough to keep a lot of men out of the brothels. There was a great social stigma attached to the disease at the time. Even so, the army considered it counter-productive to punish patients, as this would encourage concealment. Officers were, however, charged 2 shillings and 6 pence per day for hospital stoppages, and the men 7 pence. Proficiency pay could also be lost at the discretion of the commanding officer. In fact, this was the case for all diseases not contracted through service, though, during the war, only alcoholism and venereal diseases fell into this category.

In the First Essex Regiment, having VD was classed as "sick through negligence," and pay and allowances home were stopped, thus ensuring one's nearest and dearest knew all about it. Contraceptives, known as "Red, White and Blues," were available, but then, as now, availability and use are vastly different things:

> They'd be all right if a man used 'em, and kept sober, but most of the chaps are soused, and don't give a dam' then.[61]

VD, however, was a serious problem, particularly among Empire Troops, who suffered at a ratio of thirty-two cases out of every thousand men. In 1918, sixty-thousand men in the British army had a form of VD. During the course of the war, some four hundred thousand troops were treated. For the army on all fronts at the time of the Armistice, 23,900 beds were set aside for venereal cases—gonorrhea accounting for 66 percent of cases treated and syphilis 24 percent. For gonorrhea, the soldier was hospitalized and treated by irrigation. Syphilis was treated with injections of "606," a mixture of arsenobenzol and mercury. This "casualty" rate effectively took six divisions out of the line. In fact, it was said that there was a division out of the line on every day of the war through VD.

The colonial troops suffered more because they had more money. When they went on leave, they visited London or Paris and were seduced by big-city temptations. In fact, Britain was the main source of VD, having no controls on prostitution. The War Office, not wishing to bite off more than it could chew, ignored the problem and passed it on to the Home Office, and it ping-ponged between the two for the rest of the war, much to the increasing annoyance of the Dominions. The New Zealanders, for instance, in the first six months of 1917, had 1,138 men admitted to hospital with VD—915 in England and 223 in France—making up over 10 percent of their one division on the Western Front. Even so, they balked at giving free contraceptives to men going on leave, as the Canadians did; but contraceptives were on sale cheaply at canteens.

Throughout the modern era, ongoing advances in technology and industry have changed war and the military in myriad ways. War itself has become more widespread and lethal. In the Great War, increasing demands placed on armies to supply their troops with food, water, and ammunition led to innovation and a miracle of modern organization. There can be no doubt that industrialization greatly assisted British High Command in this respect during the First World War; but as we have seen in the course of this chapter, the daily life of the ordinary soldier was little affected by it. His purely military needs may have been fulfilled more quickly and reliably, but his social needs were fundamentally simple and remained largely unchanged, save for the measures introduced by the high command to maintain discipline among the large number of men under arms. Indeed, the lives of soldiers while not in battle changed but little due to the modern nature of the Great War.

NOTES

1. James E. Edmonds, *Short History of World War I.* (Oxford: Oxford University Press, 1951).
2. R. H. Mottram, John Easton, and Eric Partridge, *Three Personal Records of the War* (London: Scholartis Press, 1929), 19–20.
3. This section has been based on the official publication *Divisional Trench Orders 1917* of the Forty-second Division.

4. Basil Liddle Hart once remarked to the historian, Peter Simkins, then his protégé, that the smell of bacon cooking always reminded him of the trenches.

5. Bruce Bairnsfather, *Bullets and Billets* (London: Grant Richard, 1914), 66.

6. W. S. Austin, *The Official History of the New Zealand Rifle Brigade* (Wellington: Watkins, 1924), 71.

7. Various references: see Mottram, Easton, and Partridge. *Three Personal Records of the War,* 96; and R. A. Lloyd, *A Trooper in the Tins* (London: Hurst and Blackett, 1938), 160.

8. *A War Record of the Twenty-first London Regiment (Surrey Rifles) 1914–1919* (published by the battalion, 1928), 37.

9. Capt. R. O. Russell, *The History of the Eleventh (Lewisham) Battalion, the Queen's Own Royal West Kent Regiment* (Lewisham: Lewisham Newspaper, Co., 1934), 57.

10. General Routine Orders 1547—29/4/16.

11. General Routine Orders 1239—31/10/15.

12. Austin, *Official History,* 182.

13. A. J. Sansom, *Letters from France* (London: Melrose, 1921), 70.

14. Cecil Sommers, *Temporary Heroes* (London: Bodley Head, 1918), 95.

15. C. H. Dudley Ward, *History of the Welsh Guards* (London: The London Stock Exchange, 1988), 72.

16. Before the decimalization of UK currency in 1970, there were twelve pennies, or "pence," to a shilling and twenty shillings to £1 sterling.

17. Frank Richards, *Old Soldiers Never Die* (London: Faber and Faber, 1933), 120. Frank Richards was a private in the Second Battalion, Royal Welch Fusiliers.

18. King's Regulations of 1912, paragraph 683.

19. General Routine Orders 1332.

20. Two and one-half pence.

21. Five and three-quarter pence.

22. "Cost of Living Index," Eightieth Anniversary of the Great War, *The Telegraph,* 28 April 1994.

23. Ibid.

24. Three pence.

25. Martin Windrow and Frederick Wilkinson, eds., *Universal Soldier* (Enfield: Guinness Superlatives, 1971), 220.

26. The Thirty-seventh Division, *The Golden Horseshoe* (London: Cassell, 1919), 47.

27. Bombardier "X," *So This Was War!* (London: Hutchinson & Co, 1930), 65.

28. Arthur Guy Empey, *Over the Top* (London: Putnams, 1917), 147.

29. Ibid., 145.

30. Richards, *Old Soldiers Never Die,* 194.

31. Empey, *Over the Top,* 149. Empey was a private in the Machine Gun Corps.

32. Russell, *History of the Eleventh (Lewisham) Battalion,* 73. Albert was the main town within the Somme battlefields.

33. J. Nicholls, *Cheerful Sacrifice* (London: Leo Cooper, 1990), 63.

34. Philip Gosse, *Memoirs of a Camp Follower* (London: Cassell, 1934), 30.

35. Soccer.

36. Mottram, Easton, and Partridge, *Three Personal Records of the War,* 102.

37. Capt. G. H. F. Nichols, *The Eighteenth Division in the Great War* (London: William Blackwood & Son, 1922), 191.

38. G. D. Sheffield, ed., *Leadership and Command* (London: Brasseys, 1997), 95.

39. J. D. Hill, *The History of the Fifth Leicestershires* p.124. (Echo Press).

40. This was probably W. C. Caesar, who played for Surrey as a bowler after the war, grandson of Julius Caesar, the Surrey batsman of 1850s–60s. William Seymour, *The Rifle Brigade,* vol. 2 (London: The Rifle Brigade Club, 1936), 146.

41. J. W. Burrows, *Essex Units in the Great War 1914–1919* (J. H. Burrows, 1931), 212.

42. "Saki" was the pen name of Lance Sergeant (L/Sgt.) Hector Hugh Munro, Twenty-second Battalion, Royal Fusiliers, killed in action 14 November 16 on the Somme.

43. "The Square Egg," in *The Complete Saki* (Harmondsworth: Penguin, 1976), 541.

44. The soldiers' usual rendition of *vin blanc.*

45. Sommers, *Temporary Heroes,* 25.

46. King's Regulations of 1912, paragraphs 509–511.

47. Sound archive, Imperial War Museum.

48. Richards, *Old Soldiers Never Die,* 94; and Eric Hiscock, *The Bells of Hell Go Ting-a-ling-a-ling* (London: Arlington Books, 1976), 39.

49. Eric Partridge, ed., *A Martial Medly* (London: Scholartis Press, 1931), 39.

50. Roland Fielding, *War Letters to a Wife* (London: Medici Society, 1930), 109.

51. Sydney Rogerson, *Twelve Days* (London: Arthur Barker, 1930), 146.

52. Richards, *Old Soldiers Never Die,* 11.

53. Ex Private X [Alfred McClellan Burrage], *War Is War* (London: Gollancz, 1930), 26.

54. Partridge et al., *A Martial Medly,* 29.

55. Richards, *Old Soldiers Never Die,* 118.

56. Ex Private X, *War Is War,* 17.

57. Richards, *Old Soldiers Never Die,* 119.

58. Extracts from A. H. Davis. *The Diaries of a Tommy(1916–19)* (London: Cecil Palmer, 1932), 107.

59. Today known as Le Havre.

60. *Medical Services: Diseases of War* (London: HMSO, 1923).

61. Partridge, *A Martial Medly,* 29.

CHAPTER 6

✦ ✦ ✦ ✦ ✦ ✦ ✦ ✦ ✦ ✦ ✦

The Indian Corps on the Western Front

A Reconsideration

Robert McLain

When the Great War began in 1914, few in British military circles envisioned that it would be of such a long or sanguinary nature.

Similarly, no one realized that the empire would be forced to look eastward, to India, to offset its tremendous losses. British High Command balked at employing Asian troops in Europe, yet the long casualty list, rather than delicate colonial sensibilities, dictated the decision to do so. By the end of October 1914, the first elements of a twenty-four-thousand-man force, the Indian Corps, or Indian Expeditionary Force (IEF) as it was also termed, began to filter into the trenches that snaked along the Western Front. Within one year, war on the Western Front had reduced its two divisions, the Third Lahore and Seventh Meerut, to skeletons of their former selves. While the IEF represented only a small fraction of the more than 1 million Indians who were to be recruited during the war, it nonetheless possessed a significance greater than its numbers.[1] Indian nationalists hoped that the Indian Corps would be the lever they needed to gain political autonomy. Militarily, the unit provided the razor-thin margin between a stalemate in the autumn of 1914, which led rather tortuously to victory, and an outright Allied defeat. In societal terms, the IEF stood as a microcosm of the colonial encounter, fraught with those elements of race and caste that informed the overall Anglo-Indian relationship.

Despite its importance, the IEF has garnered scant historical notoriety, particularly when compared to the array of material dedicated to the empire's other forces.[2] The purpose of this chapter is thus to offer a timely and appropriate reconsideration of the Indian role in France, not only awarding overdue credit but also affirming, modifying, or challenging the most commonly held notions regarding the Indian Corps. The essay raises a number of questions: How does one assess the IEF's first two months in the trenches, a period that laid waste to the Kiplingesque likeness of the stalwart Indian sepoy, or enlisted man, and replaced it with an unflattering depiction of the Indian troops as panicky and inclined to self-inflicted wounds? How did the Indian army's reliance on the peculiar "martial races" recruiting system, an ideology that maintained the military superiority of particular castes and tribes of northern India, affect its ability to recoup its losses? Were the heavy casualties among white officers the essential cause of the IEF's supposedly poor performance, as the British surmised, or did deaths and wounds among Indian officers also play a significant role? Last, how did the attitudes of the IEF's contemporaries, who saw the Indian Corps as a perilous, misguided failure, affect the historical record?

Prior to the Great War, Indian troops had rarely served outside South Asia, and only once had the army sent units west of the Suez; they occupied Malta and Cyprus during the Balkan Crisis of 1878.[3] Not until August 1913 did the Committee of Imperial Defense reconsider plans for the Indian army should England become embroiled in a continental war. The government of India agreed in principle to deploy Indian troops overseas, but with the understanding that they would serve merely as garrisons to release British units for the main fighting.[4] These best-laid plans hardly survived the start of hostilities. By 23 August, the British Expeditionary Force (BEF) and Alexander von Kluck's First Army had locked in a bloody death grip after Kluck had blundered into General Horace

Smith-Dorrien's Second Corps at Mons. Just two days later, the Second Corps blunted another German assault at Le Cateau, but at a distressing cost. In the span of one week, the BEF had suffered over fifteen thousand killed, wounded, and missing, approximately 10 percent of its total force. September offered no respite as Douglas Haig's First Corps lost thirty-five hundred men in one day along the Aisne, while in October, the beleaguered Second Corps incurred another fourteen thousand casualties.[5] It took no great feat of statistical genius to realize that the BEF had to have immediate relief or it would soon exist only on paper. British High Command quickly reached a rather unsettling conclusion: of all the empire's forces, only India could provide immediate reinforcements.[6]

Sir Frederick Smith, a staff officer attached to the IEF, noted that many fellow officers believed they were "taking part in a very hazardous experiment" by sending Indian troops to the Western Front.[7] Smith feared, as did anyone with even a sliver of political awareness, that denuding India of troops would incite sedition among the colony's radicals. If nationalist agitation became too pronounced, the Raj would have little chance of maintaining control over the vast subcontinent.[8] Nor was apprehension over India's potentially troublesome domestic situation the only concern. The IEF presence in France violated a tacit agreement among the European powers forbidding the use of colonial troops outside their respective realms. Engaging in geographically remote proxy wars with "colonials" was one thing, bringing them to the continent quite another. Moreover, the vast majority of British officers agreed with the aged Lord Roberts, former commander in chief of the Indian army (1885–1893): "No one," Roberts told IEF commander Sir James Willcocks, "has a higher regard for them [the sepoys] than I have; but they have their limits. Up to that they will do anything . . . beyond that they will not go." Roberts continued, giving Willcocks a final admonition: "With British officers they fight splendidly; without them they cannot do much."[9] Simply put, the supposedly inferior sepoy, even with the benefit of British leadership, stood a less than even chance against the superb German troops.

The prospect of an Allied defeat, however, quickly overcame the litany of reservations. By late September, a large portion of the IEF had arrived at the Marseilles docks, with individual battalions and companies first entering action on 23 October as reinforcements for General Edmund Allenby's Cavalry Corps. Within one week the IEF began taking over its own eight-mile stretch of the British front, an area Willcocks aptly described as "a dismal dead plain."[10]

The events that followed the Indian arrival in France seemed to confirm the worst fears of those who believed that the Indian Corps could not succeed. On three separate occasions from late October to mid-December, IEF battalions either retreated without orders in the face of powerful German attacks or hesitated to attack when ordered to do so. On the night of 29 October, a battalion of Gurkhas fell under concentrated fire from German howitzers and heavy guns. Their trenches, already a veritable swamp, became increasingly untenable as the men began to run short of ammunition. Between 8 A.M. and 1 P.M. the next day, the Gurkhas repulsed repeated German attacks, only to retreat without orders after most of their British officers had been killed or wounded.[11]

On the 24 November, another unit seemed to falter when the Fifty-eighth Battalion (Vaughn's Rifles) hesitated to carry out orders to retake trenches that had just fallen into German hands. Three and a half weeks later, on 20 December, the most serious incident took place when the Germans exploded a mine under the Indian trenches, entombing half of a company of the Highland Light Infantry and a double-sized company of Gurkhas. The entire Indian front came under extremely heavy artillery and machine-gun fire. The Germans pressed home their attack, routing the already depleted 129th Baluchi Battalion. In a subsequent inquiry, one officer reported encountering "about 300 men of different regiments, mostly 129th Baluchis. . . . Many had thrown away their rifles and said that all their officers had been killed."[12] Willcocks, realizing his entire front might collapse, immediately requested that Sir Douglas Haig's First Corps relieve his outmanned, outgunned, and exhausted men.

While IEF officers found these incidents deeply disconcerting, they also knew that their men had been severely bloodied and had received few reinforcements. As early as 3 November, the corps had sustained roughly two thousand casualties out of its twenty-six-thousand–man contingent. The units that had "broken" had suffered especially severe losses. The 129th and 57th (Wilde's Rifles) Battalions, for example, had lost 579 men up to 3 November, a number made more significant when one considers that each Indian battalion consisted of only 764 men, in contrast to their one-thousand-strong British counterparts. Excluding support troops, the two units had lost half of their effective combatants in two days, a pattern that repeated itself throughout the corps. Casualties up to 31 December amounted to 9,579. The Indian Corps had endured its first tour in the trenches, but just barely. Over 40 percent of the IEF's original complement lay dead or wounded.[13]

Assessing the First Two Months: Martial Races, Officers, and Self-Inflicted Wounds

The Indian Corps' first eight weeks of combat present the historian with two rather striking contrasts, one confirming the prejudices of the IEF's critics, the other reinforcing the "traditional" image of the faithful sepoy. Indeed, many IEF contemporaries seized on the events of November and December with a dogmatism clearly indicative of their ingrained belief in Indian inferiority. A more careful consideration, however, reveals that on many occasions the IEF actually performed well, exhibiting incredible bravery on a personal and unit level. Moreover, despite what the British officers believed, their Indian counterparts could, and often did, perform effectively on the battlefield.

In the fighting of 31 October, during the First Battle of Ypres, the men of the Fifty-seventh Battalion (Wilde's Rifles) fought as ferociously as any Tommy or German soldier of the war. Havildar Gagna Singh held his position against overwhelming odds, killing five Germans with his bayonet before collapsing from wounds. Shell fire wounded Sepoy Usman Khan three times before he agreed to be evacuated. Faced with capture or suicide, Jemadar Kapur Singh used his last cartridge to shoot himself.[14] In the same action, Khudad Khan of the 129th Baluchis won the first V.C. (Victoria Cross) awarded to an Indian in France; he manned his machine gun until his crew had been killed and he himself seriously wounded. Three weeks later, Naik (Corporal) Darwan Sing Negi won the V.C. for his role in clearing German trenches, refusing evacuation for his wounds until his men had finished their task.[15]

The fortitude of the maligned 129th Baluchis, who retreated on December 20, provides yet another such example. Between October 23 and November 3, a mere ten days of fighting, the regiment had lost 50 percent of its British officers, 30 percent of its Indian officers, and 33 percent of its men. The 129th had endured four miserable days before 20 December, supporting a French attack so "hopeless in its conception" that it resulted in "a useless slaughter at a time when economy in both men and materials was of paramount importance."[16] The night before the crisis, torrential rains had washed away the fire-steps and begun collapsing the walls of the trenches, creating a mire of icy water and knee- and waist-deep mud and jamming well over half the rifles. The next morning, a German mine wiped out the units on the Baluchi left flank, leaving it exposed. The constant German artillery fire had also destroyed communication lines, affording the IEF no way to summon artillery support as the enemy shelled their front with impunity. It is small wonder that the 129th, and indeed the entire corps, had reached a crisis point. Indeed, it can hardly be said that the 129th and other units "cracked" on 20 December. Rather, the shattered remnants of the IEF, with most battalions numbering between two and three hundred exhausted men, could stand no more.

Here we come to the crux of the matter, one discernable only if one is prepared to contemplate the influence of colonial ideology on Great War historiography.[17] There existed, and continues to exist, virtually no restriction on criticizing the Indian Corps. Contrarily, any reproof directed at the BEF bordered on defeatism or, worse, treason; eighty years later, the suggestion that the BEF experienced its own lapses would garner a cool response at best.[18] It is indeed the pinnacle of historical conceit to suppose that BEF, though arguably one of the best armies of the century, never came apart under the intense violence of the Western Front. Lord Rawlinson, the Fourth Army Corps commander, wrote that the Twenty-eighth Division "had to be broken up and distributed among the 3rd and 5th Divisions" as a result of their poor performance.[19] Douglas Haig privately told a shocked king of the "crowds of fugitives who came down the Menin Road from time to time during the Ypres battle having thrown away everything they could, including their rifles and their packs."[20] In his autobiography, Robert Graves recalled the frank discussions that he and other instructors had at the British training ground for new draftees, the "Bull Ring" at Harfleur:

> It seemed to be agreed that about a third of the troops in the British Expeditionary Force were dependable on all occasions: those always called on for important tasks. About a third were variable. The remainder were more or less untrustworthy: being put in places of comparative safety, they lost about a quarter of the men that the best troops did.[21]

Haig and Graves both understood that one could make such comments only quietly and under certain circumstances—in a private audience with His Majesty or in the company of a few infantry instructors.

As Graves unwittingly revealed, the IEF and BEF shared a variation in the quality of battalions. Yet there existed other important similarities. Both were both small and highly professionalized forces, augmenting their strong esprit de corps through appeals to regimental history and, often, a tradition of family military service. While this trait made the units very skilled initially, it also had an opposite effect in that mounting losses hampered their cohesion and potency—neither expeditionary forces could quite match the level of efficency they had enjoyed in the opening days of the war. A significant portion of the BEF of 1916 had so little training that planners for the Somme offensive relied solely on the weight of artillery fire to carry the men of the New Armies into and across the German lines.[22] Indeed, the British army really earned its reputation as a well-oiled fighting machine later in the war. When at Cambrai in November 1917, for example, the BEF deployed over three hundred tanks, foreshadowing the mechanized, combined-arms warfare of the future.

Obviously, the most critical problem for both the IEF and the BEF centered on how to replace the heavy losses among their long-serving soldiery. The BEF could at least muddle through by drawing on the Indian army until aid arrived from the Dominion forces and newly raised home formations. The Indian Corps, however, faced a far more difficult situation. As part of the Indian Army, the IEF depended entirely on the theory of the "martial races" to guide its recruiting policy. The martial races idea had developed gradually in the course of the later nineteenth century, before Lord Roberts formalized the process in the 1890s. In the opinion of Roberts and his successors, "the martial value of a regiment recruited from the warlike races of northern India" brooked no comparison to "one recruited among the effeminate races of the south."[23] The inhabitants of the south had lost their fighting edge, and their masculinity, through too many years of peace and the enervating effects of a blistering hot climate. Furthermore, they lacked the presumably shared Aryan ancestry of India's northwestern frontier tribes and the Anglo-Saxons who had swept into medieval England, a trait that added an important ethnological element to the martial races equation. For Indians in the military, Roberts's scheme made perfect sense. As the heirs of India's warrior caste, the *Kshatriyas,* they could protect their privileged position in the Raj's power hierarchy. The martial races idea, by dint of imperial ideology, caste right, and religious duty, had the sanction of both ruler and subject.

On the surface, Roberts presented the concept as a matter of military necessity, yet the theory demonstrated an amazing ability to conform to British security needs. The Sikhs and Punjabi Muslims, both of whom the British heavily recruited, could act as bear-baiting reserves, holding up the anticipated Russian thrust through Afghanistan until help arrived from England. The Sikhs, like the Gurkhas of the northeast, fortuitously dwelt on the outskirts of British political power, isolated from seditious schemes yet excellently placed to respond to both internal and external threats.

The peculiar symbiotic martial races notion thus became the cornerstone of the Indian army, functioning effectively as long as the military limited its roles to domestic peacekeeper and reserve force. Yet, as early as 1892, military authorities had warned the Indian army that a major conflict would collapse the highly decentralized recruiting system.[24] Indeed, each regiment and battalion selected its own men, usually sending out an officer and a few members of the other ranks to obtain a predetermined number of men from particular villages, clans, and families. Each officer kept with him an *Umdewar* book, a list of "hopefuls" who received special consideration as kinsmen of troopers already in the regiment.[25]

Indian units, bound as they were by unique ties of kinship, religion, and caste, could under no circumstances keep pace with the Western Front's furnacelike consumption of men. The martial races had compartmentalized the Indian army to a high degree. At least to a company level, religious, caste, and class restrictions applied. British officers simply could not send replacements to any unit but had to observe caste or tribal affiliations, creating in effect an exaggerated version of the British "Pals Battalions," in which one joined the same unit as one's neighbors, coworkers, or friends. While the martial races idea made for extremely close-knit units, it also contained the germ of the IEF's destruction.

The obvious solution to the IEF's manpower shortage, a drawing down of existing army units, proved unpalatable as long as colonial authorities perceived a potential for internal unrest. Rather than relying on what they already had, the Indian army enrolled new recruits and recalled its most decrepit reservists, men who represented the worst that India had to offer in the way of manpower. One survey deemed 876 of 5,250 replacements unfit for duty; 50 of them lacked a simple musketry course, one of the most basic training procedures. Staff officers for the corps opined that the replacements were "not only a source of actual danger to themselves, but tended to lower the efficiency of those with whom they were placed in combat." George MacMunn recalled that "the first parties of reservists . . . were the laughing stock of the depot, feeble old men who were of no use and of whom a large number were rejected."[26] To provide an alternative to the wretched reserve situation, the Indian army exempted reservists of over fifteen years service from European duty. The fear of internal unrest had also subsided, allowing the British to begin tentatively detaching men from existing units for service in France. By the end of January, Willcocks was able to report that his new drafts were a vast improvement over those of the previous winter, but even these adjustments could not sustain the IEF indefinitely.[27]

The problem of replacing British officers, unfortunately, proved even more acute; the Western Front demonstrated an insatiable appetite for this precious commodity. At start of the war, the Indian army had over 2,500 officers, 257 of whom were on leave in England. Kitchener ordered these officers to remain in place so that they could help in the training of the New Armies. The IEF battalions serving in France had only twelve officers each and only ninety-nine Indian army officers on reserve to provide replacements. Indeed, as early as mid-November, General Willcocks was warning Kitchener's staff that "if the Corps is to be maintained as such, we shall need all the officers we can [get]." Willcocks's fears came to fruition all too quickly. By 31 December, the Indian Corps had lost 292 British officers, killed, wounded, and missing.[28]

British commanders drew the inescapable conclusion that the losses among their own ranks had multiplied the baneful effects of the Western Front. After all, the sepoy and his commander shared a unique bond, whereby the former became the *jawan,* or "youth," under the protection of the officer. Another officer simply could not step into this simulacrum of the father-son relationship. Even if it were possible, there existed only a limited number of Indian army officers with the necessary language skills for each particular unit.[29] Nor could Indians take their place. They were believed inferior, and the system of command refused Indians the king's commission, granting in its stead a commission from the viceroy. The viceroy's commissioned officer (VCO) commanded only to the platoon or company level, and his authority technically extended only to Indian soldiers.

Certainly the close bonds that existed between the sepoy and British subaltern provided the Indian Corps with a strong cohesive element. Yet contemporaries and historians of the IEF have overemphasized the essentiality of the British officer at the expense of his Indian counterpart. The losses among the VCOs offer an equally viable, yet neglected, explanation for the IEF collapse in December. The Indian officer not only brought valuable experience to the regiment, but he also acted as a role model to the young sepoy. The highest ranking VCO, the formidable subadar-major, served as an in-house village elder and acted as an adviser to the British commander on unit matters.[30] His lower-ranking counterpart, the subadar, acted as the company commander in the field. Once either of these trusted subordinates joined the casualty list, his unit lost a vital element of morale and control. The IEF in fact differed little from other armies who shared similar officer losses.[31]

Obviously, the fact that Indian officers obtained a modicum of command in itself suggests that they possessed leadership skills, a contradiction not lost on British commanders. Willcocks surmounted the discrepency by explaining that the VCO, though "well fitted to fill *temporarily* the place of the British officer in the field," could by no means replace him.[32] Similarly, the Indian, though "gallant and staunch," was "feeble without the unlimited initiative and fearlessness" of the white commander. The ideal of the "officer and gentleman," articulated as a form of class superiority within the British army, had found a more idiomatic expression in India, one redolent with late-nineteenth century colonial ideology. Only with the aid of the British, it was believed, could decadent India recapture its past grandeur. The English had both a right and a duty to remain on the subcontinent until they had completed this task. Likewise, the indigenous officer, even though a product of the martial races, had also degenerated, though much less so than the masses of Indians who "once were white" but had suffered too long from the "deteriorating effects of aeons of tropical sun."[33]

In this ideological framework, instances where the VCO seized the initiative took on an anomolous quality, a fleeting moment in which he touched ancient glory and reached par with his English superior. Command fell on the Indian officer by chance rather than design, a de facto result of the Western Front's voracious appetite for men and material. Even so, few authors have made a serious effort to find out what occurred in cases where the VCO obtained this temporary status. In fact, the leadership and bravery of the VCO stands out on these occasions. The following example of an officer who won the Military Cross suffices to make the point effectively:

> *2nd Lt. Rána Jodha Jang Bahádur (a), 1st Battalion, France (4.11.15)*
> During a feint attack made by the Indian Corps to the north of La Bassée Canal on October 13, 1915, this officer commanded a double company with great ability and conspicuous gallantry in the face of fierce fire from rifles, machine guns, grenades, and bombs, and was severely wounded in the neck. On the previous evening this very gallant officer was wounded in the arm by a rifle bullet but . . . returned to the firing line to see his company through the engagement which was due to commence the next day. . . . The bravery of Rána Jodha Jang Bahádur was previously observed on September 25th last, when he led his company right up to the German wire under heavy rifle and machine gun fire.[34]

Official regimental histories are so replete with similar accounts that one reaches the natural conclusion that the Indian VCOs fulfilled a vital role in the functioning of the IEF. Should one assume, for example, that the loss of a subadar-major, who often had served with the same unit for years, had little or no effect on morale, wheras British officer casualties virtually paralyzed the Indian troops? If, as Indian Corps diarists maintain, British losses decapitated the corps, then losses among Indian officers broke its spine. In the first ten days of action, the maligned 129th Baluchi Battalion lost six of its twelve British officers and four of its Indian subadars, the latter having often served as the company commanders in the field.[35] By 20 December, the battalion had none of its original British officers, only three Indian officers, and 214 enlisted men. The loss of the VCO broke a vital link in the Indian unit. Once a significant number of Indian officers had joined the ranks of the dead and disabled, unit cohesion suffered, just as in the case of the British officer.

There remains one issue, a very difficult one, to examine before we move forward. After the first weeks of combat, the corps' own medical officer issued a report confirming that some sepoys had engaged in self-maiming to avoid the trenches. According to the report, of the 1,848 Indian soldiers who had been admitted for treatment up to 3 November, 1,049, or 57 percent, had hand wounds. By contrast, the British units that had come from India as part of the corps, about one-third of the IEF, had only 140 wounds to the hand. Willcocks, hoping to "put a stop to this idiotic and dangerous thing," had two men shot. "I ask what it means," he continued, "to be owing hundreds of men who cannot be replaced for weeks and even months while their comrades . . . do double duty in those horrible trenches."[36]

While the corps unquestionably had a problem with self-maiming, there are nonetheless lingering questions regarding just how prevalent the practice really was. According to the statistics compiled by the IEF's medical officers, over one-half of the hand injuries suffered by Indian troops came within the first two weeks of battle. If one accepts that most of the hand wounds up to 3 November were self-inflicted, it would mean that, on average, nearly *one hundred men a day* either shot themselves or simply raised their hand over the parapet of their trench and let the Germans do the job for them. The conclusion that

Indian troops engaged in a sudden and massive outbreak of self-maiming simply boggles the mind, for it presumes that the war dealt them such profound psychological shock that they immediately cast aside deeply held convictions of personal and unit integrity. The spate of self-inflicted wounds defies the strong esprit de corps of the Indian army's regiments and battalions, as well as the sepoy's intense concern with personal honor, or *Izzat*.

The medical officer's report comparing British and Indian units of the corps presents a problem as well, since Indian battalions outnumbered their British counterparts ten to six in the first few weeks of the war. Moreover, the British battalions had not yet been as heavily engaged as the Indian units and had had fewer casualties for the time period covered in the medical officer's report. Part of the explanation may lie in the possibility that inadequate trenches had exposed some units to greater enemy fire, especially the trenches that characterized the early fighting.

One common thread, however, offers a feasible explanation for the problem of self-inflicted hand wounds, namely, that the units seemed most affected possessed a large number of the soldiers originating from areas on the fringes of British power, regions with far more tenuous ties to the Raj and its associated traditions of military service. The 129th Baluchis, Fifty-seventh and Fifty-ninth Rifles, for example, recruited heavily from the Pathan tribes that dwelt along the border region between British India and Afghanistan. By way of further contrast, we can note that the only case of mass desertion on the Western Front came in March 1915, when twenty-four Pathans of the Fifty-eighth Rifles crossed over to German lines, while the statistics of hand wounds among the Gurkha and British battalions hardly differ.

The obvious question, though—what percentage of the hand wounds were self-inflicted as opposed to genuine—will remain a puzzle until more evidence comes to light. It may be safest to conclude that the Indian Corps had a moderate problem with the practice, confined to certain units. Suffice to say, one should use great caution in drawing conclusions from the medical officer's statistics.

Neuve Chapelle and the Official History of the War

After enduring a horrendous winter, the IEF finally gained time to lick its wounds and train in the nascent art of trench warfare. In their next major action, the March 1915 Neuve Chapelle offensive, the Indian Corps performed superbly, meeting all of its first-day objectives. Despite its early success, the assault stalled in the face of German machine-gun and artillery fire, an all-too-common occurrence on the Western Front. IEF and BEF commanders quickly blamed one another for the failure, igniting a debate that has a continuing resonance for interpreting the Indian part in the overall BEF narrative. For the Indian Corps officers, the lack of recognition merely confirmed what they already believed—that their men were not getting the credit they deserved. Indeed, the Neuve Chapelle offensive presents critical issues regarding the historiography of the IEF and the writing of the official history of the war, *Military Operations in France and Belgium* (1926).

The offensive, though small when compared to later endeavors, arguably stands as the most important engagement of the war for Indian forces, demonstrating rather clearly what the IEF could and could not do. Given the proper training and weaponry, the Indian soldiers performed as well as any in the field. Nonetheless, the IEF could not consistently sustain heavy losses, even with an amended recruiting policy in place.

The Indian participation in the operation also underscored the BEF's continuing need to flesh out its depleted ranks with colonial troops. Indeed, the circumstances that had compelled the War Council to dispatch the IEF to France still applied. Britain had already combed out its remaining regular troops, forming the Twenty-seventh, Twenty-eighth, and Twenty-ninth Divisions from its overseas garrisons. The Germans had immediately and severely bloodied first two of these hastily formed units at St. Eloi in mid-February, while the Twenty-ninth had been detached for the misconceived Dardanelles scheme. As for the Territorial divisions that had begun to appear in France, even the 46th (North Midland), "the best Territorial division then available," would require "some weeks training in trench warfare before it could be asked to take over a section of the battle front."[37] As late as eight months after the war began, the crimsoned BEF still exhibited a critical need for Indian Corps.

As the day for the assault approached, IEF commanders carried with them an awareness that not only their own reputations but also the reputations of their men hung in the balance. General Willcocks thought Neuve Chapelle represented "an experiment which might have momentous consequences," comparable to that faced by Japanese commanders in the Russo-Japanese War. "It was," said Willcocks, "a matter of East versus West." Major General H. D'U. Keary, head of the Lahore Division, exhibited a more cautious attitude, one indicative of the heavy losses of his unit had incurred in the winter. "How the Indian troops will do, I don't know," he wrote; "most of my div[ision] got very severely handled in the early part of the war and consist of a good deal of new drafts."[38] Keary had good reason for concern, for planners had slated his old unit, the Garhwal Brigade, along with Rawlinson's Fourth Army Corps, to lead the attack.

Planners had originally conceived the Neuve Chapelle scheme as part of a joint Anglo-French offensive. The British part in the assault called for the capture of Aubers Ridge as its primary objective. Although the ridge was only forty feet high, it dominated the surrounding countryside and allowed direct observation of the British lines. On 7 March 1915, only three days before the attack was to begin, General Joffre, commander of the French army, notified Sir John French that he was calling off the operation in his sector. French nevertheless decided to assail the German lines independently, mainly to answer criticism from his ally that the British had not contributed enough in the December fighting around Ypres.

In the weeks before the battle, the Royal Flying Corps had gained local supremacy over German airpower, using their advantage to map the German trenches in detail and preventing the enemy from discovering the massing of matériel and men. Planners also buried some of the communication lines and ran them as far forward as the battalion command posts. The offensive also drew on the guns of four divisions, in addition to some heavier pieces from England and other sections of the line. Once completed, the artillery plan contained a complement of 340 weapons, forming a gradual arc to the west and southwest of Neuve Chapelle. The total frontage for the assault was two thousand yards, allowing one gun for every six yards of the enemy line, an exceptional feat at this time of the war. Artillery crews registered their targets surreptitiously, hoping to avoid alerting the Germans, and planned a short, violent bombardment, lasting for only thirty-five minutes. The first wave of the infantry would leave the trenches while the guns were still firing, rather than waiting for the shelling to stop. In this manner, it was hoped that the lead troops would encounter only stunned Germans and destroyed positions.[39]

These painstaking preparations boded well for the Indian Corps, as Keary's old Garhwal Brigade and Rawlinson's Twenty-third and Twenty-fifth Brigades moved up quietly to the front trenches. At 7:30 A.M., the BEF began artillery began pounding the German trenches. Twenty minutes later, the four battalions of the Garhwal Brigade left the positions on their six hundred–yard front. Three of the Indian Corps battalions, the Second of the Thirty-ninth Garhwal, Second of the Third Gurkha, and the Second Leicester, easily met their first objectives. All along the German line the wire had been cut and the breastworks collapsed. The enemy soldiers who had survived the bombardment emerged stunned and incapable of offering significant resistance.

Yet German opposition stiffened as the assault carried into successive defensive lines. The performance of the Second Garhwal Battalion in these early stages was especially notable, and the Indian troops rose to the occasion, once again taking the initiative. A Garhwal rifleman, Gobar Sing Negi, took over his platoon after German fire killed its commander. In the subsequent trench-clearing operation, Gobar Sing was the first man around each traverse, bayoneting and shooting enemy soldiers along the way and in the process earning a posthumous Victoria Cross. In addition, two other Garhwal officers won the Military Cross, another the Order of British India, and two the Indian Order of Merit.

The 2/39th Garhwal's companion unit, the First Battalion of the Thirty-ninth Garwhal Brigade, met with misfortune immediately when they drifted right and struck an undamaged portion of the German line at the Port Arthur salient. Tragically, the enemy machine-gunned the two lead companies in front of uncut wire. The survivors desperately tore at the entanglements with their hands and forced their way into the German position. Although they had fought gallantly and achieved a measure of success, a two hundred–yard section of enemy-occupied trenches separated two companies of the 1/39th from the remainder of their unit. On the left of the Garhwals, the British Twenty-fifth Brigade of Rawlinson's Fourth Corps had also captured its objectives. Rawlinson's Twenty-third Brigade, however, met the same fate as the 1/39th Garhwals, with far higher casualties. One of the Twenty-third's regiments came out of action under the command of a second lieutenant and down to 150 men. Despite the heavy casualties in the errant British and Indian units, the assault had met all its initial objectives by 10 A.M.

According to the official history, *Military Operations in France and Belgium,* the Dehra Dun Brigade of the Meerut Division and the Twenty-fourth Brigade of Rawlinson's Eighth Division were to move up and implement the next phase of the British attack. The Twenty-fourth Brigade planned to pass through the newly captured German lines and seize Aubers Ridge, while the Dehra Dun Brigade drew the task of capturing the Bois de Biez, an overgrown wooded area to the southeast of Neuve Chapelle. Sir James Edmonds, author of the official history, states that the offensive then faced two critical delays: heavy German fire, which slowed the Dehra Dun Brigade's advance across no-man's-land, and the excessive caution of the Meerut Division commander in dealing with the uncaptured portion of the German trench. The attack stalled for five precious hours, allowing the Germans to establish a strong secondary line of defense with interlocking fields of machine-gun fire.

Edmonds further claimed that Rawlinson could not renew his assault until the Indian Corps had cleared the Port Arthur position, a task the IEF troops accomplished at 5 P.M. Not until nightfall did lead elements of the Dehra Dun Brigade move into the edge of the Bois de Biez, the flames of a burning farmhouse lighting the way. The Dehra Dun commander, Brigadier General C. W. Jacob, faced a difficult decision. The brigade was now in a position to capture the wood, but both flanks were "in the air," since the rest of the Indian and the Fourth Corps were on the opposite side of the Layes River, and it was doubtful that they could hold the position against a concerted German attack. Jacob expected help from two of Rawlinson's brigades, but both were stopped at the river's edge by strong German resistance. Jacob's after-action report for 10 March confirms that "if the Eighth division had been able to cooperate with me I would have been able to maintain myself on the edge of the wood."[40] In a move that enraged General Douglas Haig, the First Army commander, Jacob decided to withdraw his men back across the Layes River.

The delay of the 1/39th Garhwals and Rawlinson's Twenty-third Brigade of is important for three reasons. First, it reveals the tendency of the *Military Operations* author to exonerate some commanders at the expense of others. Second, Edmonds placed the blame for the delay primarily on the shoulders of the Indian troops, specifically the 1/39th Garhwal battalion and their unfortunate drift into the undamaged German trenches. Third, Edmonds overlooked the fact that IEF commanders built a solid case against Rawlinson for his failure to move up the supporting brigades in a timely manner. The problem, in fact, lay in Rawlinson's own corps, so the reverse was true: the Indian Corps was delayed because of events in the Fourth Corps on their left.

Rawlinson could not have advanced even if the Indians had cleared the Port Arthur position earlier. The Fourth Corps commander had already cast three of his five reserve battalions into the fray, while the remainder were employed in carrying stores or were simply placed too far in the rear to be of assistance. In short, this meant the troops intended for the second phase of the attack were already engaged in other endeavors. Moreover, the attack simply could not proceed until the Fourth Corps cleared the first line of German defenses, for they formed the left flank of the Indian Corps.[41] Indeed, Rawlinson's mishandling of reserves and the slaughter of his Twenty-third Brigade along the German wire offer a far more compelling explanation for the attack's delay than the destruction of the two Garhwal companies at the Port Arthur redoubt. Sir John French, in a dispatch written three weeks after the battle, concurred that the "difficulties . . . might have been overcome if the General Officer commanding the 4th Corps had been able to bring his reserve brigade more speedily into action." Indian Corps staff officers likewise determined that "the check sustained by the 23rd brigade had thrown the machinery out of gear," thereby necessitating a reorganization of the attack.[42]

Further evidence in favor of the Indian Corps comes from an examination of Rawlinson's actions after the fighting had ended. Rawlinson tried to sack his Eighth Division commander, Joey Davies, for his supposed dereliction in bringing up his reserve brigades. Davies countered by gathering evidence on his own behalf, saving his job in the process. More significant, he identified Rawlinson's handling of the reserves as the real cause of the lag, as well as the subsequent stall in the Indian Corps' advance. The bid to remove Davies demonstrates that Rawlinson had attempted to shift responsibility for the crucial delay from himself to his subordinate Davies, as well as to the Indian Corps. *Military Operations in France and Belgium* similarly tends to place the preponderance of blame on the IEF.

The real significance of Edmonds's account of the Neuve Chapelle fighting lies not only in the short shrift given to the Indian soldiers but also in revealing the conviction of British army officers that the Indian Corps had no business on the Western Front. Indeed, British army "regulars" positively relished their long tradition of regarding "nonregulars," that is, everyone else, much like an unattractive cousin that one has been forced to escort to the ball. In the estimation of Sir John French and his successor Sir Douglas Haig, the Indian soldier was simply part of a misguided, grand failure. Indian Corps officers had no illusions regarding their brethren in the British army. They clearly recognized the personal animus that existed toward the IEF and greatly resented the tendency of BEF commanders, and later Edmonds, to blame the Indians for particular setbacks. In a private letter to his brother, Lahore Division commander H. D'U. Keary thundered against the highest echelons of the BEF: "the truth is that [Field Marshal] French and [General] Haig hate the Indian Corps and want to get rid of the whole thing." Keary continued his diatribe, angrily noting that "no one in the Indian Corps feels either safe or induced to do his best, there has been so much injustice done and said." Willcocks joined the fray as well, noting that his only fault had been to fall "under command of a man [French], with whom, notwithstanding every possible endeavor I could not hit it off."[43]

The stridency of both Keary and Willcocks is telling, for it demonstrates that both men believed Neuve Chapelle had expiated the supposed sins of the Indian Corps the previous winter. Keary expressed pride in his men: "[The] Indian Corps operation was completely successful. My old brigade . . . forming the front line of our attack and carrying everything before them." Keary's reference to his former unit, the Garhwal Brigade, is instructive as well, for it confirms that the Indian soldiers could do as well as anyone in the field, given proper training, equipment, and artillery support. At the same time, it reinforces our earlier point that the IEF, like the BEF, had units of varying efficacy. The Garhwal Brigade, for example, unquestionably stands as the best that the IEF had to offer, winning two Victoria Crosses out of the five awarded to the corps. A colonel who served with the Garhwals recalled that they "never had a bad show and so kept their original morale and prestige. . . . They never lost their form though they had their full share of casualties."[44]

The exuberance displayed by Keary and Willcocks proved fleeting, for Neuve Chapelle had cost the Indian Corps heavily. Three of the Meerut Division's regiments had lost 84 percent of the troopers. The Forty-Seventh Sikh Battalion alone lost seven British and eleven Indian officers, along with 80 percent of its men.[45] The Meerut's companion unit, the Lahore Division, remained relatively intact, though not for long.

On 22 April 1915, the Germans launched the Second Battle of Ypres, marking the first use of poison gas in the war. The French front collapsed as chlorine fumes asphyxiated men in their trenches. The British quickly began a series of helter-skelter counterattacks, hoping to salvage the situation. Keary's Lahore Division, after undertaking a three-day forced march to Ouderdoum, reached the front early on 26 April, just in time to take part in the worst fighting it would experience in France. The result can be described only as disastrous. German gunners poured fire on the assaulting brigades, who had to cross one thousand yards of open ground just to reach their assembly points.[46] British artillerymen, in the midst of a severe shortage of ammunition, could do little to help. Indian and British troops fell in heaps as they crested the last slope in front of the German lines, advancing into a hail of machine-gun and artillery fire. Within three days, Keary's unit had ceased to exist as an effective force, losing 3,889 men, or nearly 30 percent of its ranks.[47] Keary saw the episode as a "slaughter," even as he acknowledged that "something had to be done, and done quickly" to maintain the front; it was "the very devil of a fight much worse than Neuve Chapelle."[48]

On 9 May, the remainder of the Meerut Division met an identical fate, losing 2,629 men in one day during the Battle of Festubert, a subsidiary action in the Second Ypres engagement. Indian and British troops again met machine-gun and heavy artillery fire as soon as they left their positions. Men tumbled back into the trenches, hit before they had stepped off the scaling ladders, or dropped dead on the parapet. Whole lines of soldiers, Highlanders and Gurkhas alike, fell as machine guns traversed their ranks.[49] The Meerut Division, like the Lahore, suffered devastating losses. Individual battalions emerged with 50 to 80 percent casualties. The

Forty-first Dogras lost 401 out of 645 men, and the British Second Black Watch 270 of 450, while the Fifty-eighth lost another 252 men from its already depleted companies.[50]

When the fighting at Second Ypres finally slowed to a halt on 25 May, British commanders recognized that the IEF could no longer remain on the Western Front. The balance of its time in France passed quietly, with the exception of some diversionary attacks as part of the Loos offensive in September. By nightfall of 10 November, the IEF completed its withdrawal from the British line.[51]

The devaluation of the IEF, though lamentable, is hardly unforgivable. By the end of hostilities, the BEF had grown into an imposing force of more than 6 million men, casting the earlier efforts of the relatively tiny Indian Corps into pallid insignificance. Later campaigns, such as Passchendaele and the Hundred Days, could draw on dozens of divisions and span weeks rather than days.[52] Nonetheless, it is important to resist the urge to become overawed by the eventual scale of the conflict if one is to grasp the full significance of the Indian Corps' role in Flanders. If we truly hope to gain a fair picture of the Indian role on the Western Front, we must take a shorter historical view, that of the first ten months of the war. During that period, the Indian Corps forestalled disaster for British arms on more than one occasion, buying time for the first of the Territorial Armies to fill the void left by the demise of the original BEF. It is difficult indeed to disagree with the assertion of one staff officer that the empire was saved "first by the Expeditionary Force, secondly by the Indian Corps, thirdly by the Territorial Divisions, and fourthly by the Overseas and Kitchener Armies."[53]

One final cloud has hung over the IEF for the last eighty-six years, namely, the belief that they failed. If the term *failure* denotes that the Indian Corps floundered in its attempts to advance beyond its original trenches, then one could apply the designation to all the Western armies for the majority of the war. The Indian Army in fact accomplished one of its primary missions: to act as a sacrificial body that would defend the empire until help could arrive from

Britain. The irony lay only in the enemy and the location—the Germans in Flanders rather than encroaching Russians in Afghanistan. The Indian Corps, in its own peculiar and tragic fashion, stand as India's "Old Contemptibles."[54]

NOTES

1. See F. W. Perry, *The Commonwealth Armies: Manpower and Organisation in Two World Wars* (New York: St Martin's Press, 1988), 96–97. Perry notes that approximately 826,000 of the troops acted as combatants, with just over 552,000 serving outside of India. Over 138,000 Indians served on the Western Front.

2. While there are some studies related to the Colony's role, such as Philip Mason's *A Matter of Honour,* T. A. Heathcote's *The Indian Army,* Byron Farwell's *Armies of the Raj,* and more recently David Omissi's excellent *The Sepoy and the Raj,* they generally cover a broad chronological span. The most specific work relating to the Indian Corps itself is an unpublished dissertation by Jeffrey Greenhut, "The Imperial Reserve: the Indian Corps on the Western Front, 1914–1915" University of Kansas (1980), and Greenhut's excellent article with the same title, published in the *Journal of Imperial and Commonwealth History* 12, 1 (October 1983): 54–73.

3. Byron Farwell, *Armies of the Raj: From the Great Indian Mutiny to Independence, 1858–1947* (New York: W. W. Norton and Company, 1989), 75.

4. Sir James Edmonds, *Military Operations in France and Belgium, 1914* 2 vols. (London: MacMillan and Company, 1926) 1, 12–13. Hereafter referred to as *Military Operations.*

5. Ibid., 1:224, 361–362; II:222.

6. Greenhut, "The Imperial Reserve," 54–55.

7. Lt. Col. J.W.B. Merewether and Sir Frederick Smith, *The Indian Corps in France* (London: John Murray, 1919), 479.

8. Nationalists became increasingly frustrated with the pace of reform as the war progressed, particularly as India's financial burden increased. See Algernon Rumbold's *Watershed in India* (London: Athlone Press, 1979) and Percival Spear's *Oxford History of Modern India, 1740–1947* (Oxford: Oxford University Press, 1965).

9. Sir James Willcocks, *The Romance of Soldering and Sport* (London: Cassell and Company Ltd., 1925), 282.

10. Sir Sames Wilcocks, *With the Indians in France* (London: Constable and Company, 1920), 81.

11. Merewether and Smith, *The India In Corps*, 70–75; Greenhut, "The Imperial Reserve," 56.

12. Cited in Greenhut, "The Imperial Reserve," 59–60

13. Merewether and Smith, *The Indian Corps*, 40, 206; "The Imperial Reserve," 54–55.

14. Philip Mason, *A Matter of Hounor: An Account of the Indian Army, its Officers and Men* (New York: Holt, Rinehart and Winston, 1974), 416. Merewether and Smith, *The Indian Corps*, 36–37.

15. Merewether and Smith, *Indian Corps*, 41, 131.

16. W. S. Thatcher, *The 4th Battalion D.C.O. Tenth Baluch Regiment in the Great War (129th Baluchis)* (Cambridge: Printed by W. Lewis at Cambridge University Press, 1932), 19, 27.

17. For the development of imperial racial concepts in the nineteenth century India see Thomas R. Metcalf's *Ideologies of the Raj* (Cambridge: Cambridge University Press, 1994).

18. Perhaps the best analogy lies in the history of the American Civil War. Americans from the South are loathe to admit that Robert E. Lee's Army of Northern Virginia ever turned its back to a Union assault. Indeed, some of the best Confederate units faltered at times.

19. General Sir Henry Rawlinson to Lord Kitchener, 23 December 1914, WB–11, Kitchener Papers, Public Records Office [PRO], London.

20. Cited in Charles Cheverix Trench, *The Indian Army and the King's Enemy's, 1900–1914* (London: Thames and Hudson, 1988), 43.

21. Robert Graves, *Goodbye to All That* (reprint, New York: Anchor Books, 1985), 181–182.

22. John Keegan, *The Face of Battle* (reprint, New York: Penguin Books, 1978), 226. Keegan notes that the 30th, 32nd, and 34th Divisions lacked in equipment and training, as well as experienced officers and N.C.O.s.

23. Cited in Mason, *A Matter of Honour*, 347.

24. Stephen P. Cohen, *The Indian Army, Its Contribution to the Development of a Nation* (Berkeley: University of California Press, 1971), 70.

25. Farwell, *Armies of the Raj*, 188.

26. George MacMunn, *The Martial Races of India* (London: Sampson Low, Martin, 1933), 328.

27. Merewether and Smith, *Indian Corps*, 463.

28. Ibid., 206; Sir James Willcocks to Capt. Oswald FitzGerald, 11 November, 1914, Kitchener Papers; Merewether and Smith, *The Indian Corps*, 206. Captain Oswald FitzGerald, a captain in the 18th Bengal Lancers, became Kitchener's Aide-de-Camp during Kitchener's tenure as Commander-in-Chief of the Indian Army (1902–1909). FitzGerald drowned with Kitchener in June 1916, when their ship the *Hampshire* struck a mine off the Orkney coast. See Trevor Royle's *The Kitchener Enigma* (London: Michael Joseph, 1988) for a detailed account of the sinking.

29. See Mason, *A Matter of Honour*, 341. The language problem could be especially acute in "mixed" battalions, in which one company might come from a different region and linguistic background than its cohorts.

30. Cohen, *The Indian Army*, 50.

31. John Terrain has touched upon this phenomenon in his book *To Win a War*.

32. Willcocks, *With the Indians*, 6.

33. Quoted in Mason, *A Matter of Honour*, 348.

34. Exerpted from the London Gazette, November 4, 1915, in *Historical Record of the 39th Royal Garhwal Rifles, 1887–1922* (Aldershot: Cale and Polden, ltd., 1922), 184. The volume was compiled from various sources by Brigadier J. Evatt, D.S.O.

35. Thatcher, *Tenth Baluch Regiment*, 246.

36. Willcocks to Kitchener, 12 November 1914, Kitchener Papers.

37. Edmonds, *Military Operations*, 1915, 1, 53, 73–73.

38. Wilcocks, *With the Indians*, 207; H. D'U. Keary to Capt. F. W. Keary, 7 March 1915, Keary Papers, Imperial War Museum, London.

39. Robin Prior and Trevor Wilson, *Command on the Western Front: The Military Career of Sir Henry Rawlinson, 1914–1918* (Oxford: Blackwell, 1992), 24, 32, 33.

40. Excerpt from Jacob's report to Willcocks, cited in Willcocks, *With the Indians*, 217, 218.

41. Prior and Wilson, *Command in the Western Front*, 52, 55.

42. Merewether and Smith, *Indian Corps*, 249.

43. Wilcocks to FitzGerald, 5 October 1915, Kitchener Papers. See also Tim Travers' *The Killing Ground* for an excellent account on the treatment

the Somme and Paschendaele campaigns received in the offical history. Edmonds ordered the first draft of the Paschendaele volume to be rewritten so that Sir Douglas Haig would be cast in a more positive light. Edmonds also altered the accounts of the Somme to satisfy commanders who took offense to the handling they received in the first draft.

44. Cited in Mason, *Matter of Honour,* 418.

45. Greenhut, "The Imperial Reserve," 65.

46. Merewether and Smith, *The Indian Corps,* 297–312.

47. Edmonds, *Military Operations, 1915,* 1:258–61; Greenhut, "The Imperial Reserve," 65–66; Merewether and Smith, *The Indian Corps,* 340.

48. H. D'U Keary to Capt. F.W. Keary, 30 April 1915, Keary Papers.

49. Edmonds, *Military Operations, 1915,* 2:21–22, 38; Greenhut, "Imperial Reserve," 65–66.

50. Merewether and Smith, *The Indian Corps,* 360–362.

51. Ibid., 460–465.

52. Paddy Griffith, *Battle Tactics on the Western Front: The British Army's Art of Attack, 1916–1918* (New Haven: Yale University Press, 1994), 14.

53. Merewether and Smith, *The Indian Corps,* xxii.

54. The Original BEF had adopted the moniker "Old Contemptibles" in response to Kaiser Wilhelm's reference the expeditionary force as a "contemptible little army."

CHAPTER 7

The Somme in British History

Brian Bond

What's in a name? For most French, the Somme is simply the name of a river, like the Thames or the Severn; but for the British, "the Somme" has become a potent myth—grim, dark, somber—epitomizing the "horror" and "futility" of the First World War and, more specifically, the tragically ludicrous belief that numbers and valor alone could somehow overcome the awesome defensive power of modern military technology. Ironically, although the "battles of the Somme" were mentioned from 1916 onward, none of the British offensives took place on or across that river. Indeed, the official designations were more accurate, stretching from the initial Battle of Albert (1–13 July) through Flers-Cource Lette (15–22 September) to the final muddy phases of the Ancre and the Ancre heights in November. The Somme could just as well be associated with the Anglo-French retreat in March 1918 or the victorious Allied advance a few months later. But who, except a handful of misguided military historians, would dream of linking the words *Somme* and *Allied victory*? The Somme must evoke notions of incompetent generalship, horrendous losses, and failure. There is a further enormous distortion between myth and historical reality: the campaign lasted from 1 July to 18 November (141 days), yet only the first day with its unprecedented British casualties (before or since) of 57, 740 officers and men, is popularly remembered. Indeed, as Ian Beckett has remarked, "there can be few historical episodes where the gap between professional and public comprehension is greater."[1]

It is doubtful if this obsession with the Somme campaign in general and 1 July in particular as the symbol of unnecessary losses and incompetent generalship has been consistent and continuous since the end of the First World War. Indeed, it seems more appropriate that doubtful distinction be bestowed on Flanders, and especially Ypres, where the British effort on the Western Front was concentrated from October 1914 and culminated in the Third Ypres offensive of 1917, which was more costly in casualties, fought in more hellish conditions, and harder to justify politically than the Somme. This offensive produced the perfect name for pacifists, satirists, and critics of the Western Front—Passchendaele (Passion Dale): a terrible valley in which thousands of British soldiers met their sacrificial deaths.

My contention in this chapter is that historians' attitudes toward the Somme have never achieved a consensus since 1918, with regard to either the campaign's significance in British strategy and tactics or its contribution to the outcome of the war. Much of what is now assumed, by all but a handful of specialists, to be the objective "truth" established during or soon after the war probably dates only from the rediscovery of the Western Front in the radical 1960s. Although the controversialists of the 1960s did not, of course, invent the catastrophe of the first day, 1 July, they gave it lurid prominence in books, articles, and radio and television programs. "By implication they seem to ask, in effect, why on the 133rd day of the Battle of Verdun, the British Command did not tell the French (and Russians) that one day of this sort of thing was quite enough."[2] Without in any way wishing to excuse military miscalculations or ignore the resultant heavy casualties, it may be suggested that this ritual invocation of "the first day of the Somme" tells us more about the assumptions and attitudes of the 1960s and later generations than about the actual history of the war and its impact on the British public at the time.[3]

Even if the course and outcome of the battle had been more obviously successful, the Somme would still occupy a central and controversial place in British interpretations of the war, due to its timing and the exaggerated expectations placed on it. Though it did not mark the debut of the Kitchener volunteer armies, nor, as modern myth implies, was it fought solely by them, it was nevertheless by far the biggest British offensive to date, assisted by the greatest concentration of artillery (more than fifteen hundred guns firing 1.7 million shells in the preliminary bombardment). No fewer than 97 of the 143 infantry battalions involved in the battle consisted of New Army volunteers, some of whom annoyed the Territorials and regulars by their supreme confidence in their ability to finish off the war.[4] On the home front, overoptimistic press reporting, magnified by propaganda, created a dangerous delusion that this was the "Big Push," which was bound to succeed. In particular, as Stuart Sillars has shown, overwhelming attention was given to the supposedly irresistible power of the preliminary bombardment, which, it was assumed, would so shatter and demoralize the enemy that victory would be gained with minimal casualties. It would literally be a walkover. The reporters, magazine illustrators, and the mass of the public alike were willing the attack to succeed and could not conceive of failure.[5]

The public's notion of war was founded on preindustrial romantic and patriotic imagery (exemplified by Caton Woodville's illustrations), and in 1916 these illusions were still being nourished by the handful of war reporters in France sanctioned by the British High Command. Some reporters were irked by the strict censorship, but others, such as W. Beach Thomas, were to become notorious because of the deceitful picture they continued to convey. Popular notions of war as a glorified game were encapsulated by the famous incident of Captain Neville of the Eighth East Surreys leading his men by kicking a football into no-man's-land. Most popular papers stressed the heroic courage of the British attackers, but Beach Thomas went to nauseating lengths: "The very attitudes of the dead, fallen eagerly forward, have a look of exuberant hope. You would say that they died with the light of victory in their eyes."[6] This and much more in similar vein would later provoke a furious reaction, when the truth about conditions, casualties, and very limited progress eventually filtered through. Insofar as later "antiwar" writing, by Siegfried Sassoon, Robert Graves, and others, was directed at this sort of deception, it was entirely justified. It would be a great mistake, however, to assume that the horrendous losses on the first day were immediately known at general headquarters (GHQ), much less at home. Press censorship, willingly abetted by patriotic editors, muffled and delayed the impact of the heavy losses, but these could not be long concealed from towns such as Accrington, whose localized "Pals Battalion" suffered particularly severely. Taking this as an example, the first terrible indication of the tragedy occurred about ten days after the battle, when a train full of wounded soldiers arrived in Accrington, causing anguished rumors of a massacre. This was substantiated when the local newspaper published half a page of photographs of the killed and wounded. Even two weeks after the attack, the War Office still refused to release the full statistics of losses to the mayor of Accrington. But it eventually emerged that on 1 July, in just twenty minutes, the Accrington Pals had suffered 585 casualties out of about 700 men; 235 were killed, and of these it is grimly significant that 135 have no known grave.[7] Given the prebattle euphoria and the subsequent official evasiveness about the scale of casualties, it is easy to comprehend how an extreme reaction of bitterness and disillusionment might set in. Who, for example, except professional military historians, now remembers that on 1 July, the southernmost sector of the British offensive was successful?

Neither the early books about the battle nor the celebrated film *The Battle of the Somme,* first shown to mass audiences in Britain in August 1916, conveyed a realistic sense of the conditions or any idea of the casualties. The film, shot by G. Malins and J. D. MacDowell, contains some genuine combat footage, notably the explosion of the enormous mine under the Hawthorn Redoubt, but much of it was faked behind the lines. Its emphasis on the artillery preparations and bombardment, plus the absence of sound, conveyed little impression of the realities of battle. The film's captions and the accompanying publicity were reassuring, exciting, and patriotic.[8]

After the war General Sir Douglas Haig, who had served as

commander in chief of the British Expeditionary Force (BEF), and his supporters were quick to establish a defense of British strategy in the Somme campaign. The essential points were as follows. The campaign had been forced on an unwilling Haig by General Joseph Joffre, the French commander in chief, to take pressure off the French at Verdun, in which it had succeeded. Haig's persisting belief that the BEF could achieve a breakthrough with "decisive" results was played down in favor of the need to wear down the enemy in an attritional battle. In this, so the argument ran, it had been the real turning point in the war: the Germans suffered irreplaceable losses and were steadily forced to give ground, culminating in their retreat to the Hindenburg Line in the early months of 1917.

This is not the place for an extended discussion of this enormous and still controversial topic; especially as one of the key issues—comparative casualty statistics—can never satisfactorily be resolved.[9] Suffice it to say that a positive interpretation of the purpose, course, and outcome of the campaign was established early on and has continued to have reputable supporters, even if for the general public it has now become an article of faith that the operation was a complete and unmitigated failure.

A brief sampling of the British press after 1918 shows that surprisingly little attention was paid to the battle on the anniversaries of 1 July.[10] In 1925, for example, *The Times* recorded (on p. 16) that it was the anniversary of the opening of the battle: it included ninety-eight *in memoriam* notices of individuals and nine collective (regimental) notices. On that day, the *Daily Express* contained no references whatsoever to the battle. Five years later, *The Times* published sixty-seven individual notices and seven collective entries; the *Daily Express,* nothing. By 1935, *The Times's* individual entries had declined to thirty-eight (three not for Somme casualties), while collective entries remained at seven; the *Daily Express* again ignored the anniversary. On 1 July 1966, however, the modern preoccupation with the Somme is evident. In addition to individual (fourteen) and collective (seven) entries, *The Times* included a long editorial prefaced by Siegfried Sassoon's poem "When the Barrage Lifted."

There had, it stated, been many previous bloody British battles: "But the Somme has become a national symbol for the waste of human life." The editorial went on to assert that some 410,000 British casualties had been suffered for a strip of land thirty miles long by, at most, seven miles deep, of no strategic or tactical value. Both armies (British and German) had been pretty well destroyed, but whereas the German losses had been irreplaceable, Kitchener's army to some extent was. The post-Somme conscript army could meet the Germans on level terms. "Yet, when all the defense arguments have been heard, the fact remains that 'the flower of Britain's generous manhood' was lost at the Somme." The editorial ended by stressing the pacifist influence of the war poets, including Sassoon, Edmund Blunden, Graves, Wilfred Owen, and David Jones, that had previously been prevalent.

Significantly, on 1 July 1966, the *Daily Express* published a long article by Henry Williamson about the first day of the battle, and on 2 July there was an account of six Somme winners of the Victoria Cross revisiting the battlefield. Recollections by survivors, such as the distinguished economic historian R. H. Tawney, became almost obligatory on succeeding anniversaries. The fact that nearly all of these "old soldiers" have "faded away" forces journalists now to make the annual pilgrimage to contrast the present tranquillity of the Picardy landscape with the scarcely imaginable noise and desecration of that sunny morning in 1916. ("Could so vast a slaughter—in all history none was ever greater—so enormous a destruction, have been accomplished in so arcadian a landscape, so small a compas?" wrote Nigel Buxton in 1986.)

It is a shock to retrace our steps from these post-1960s commemorations of ruination and losses to a second film of *The Somme,* directed by M. A. Wetherell, produced by Gordon Craig, and scripted by Geoffrey Barkas in 1927.[11] This eighty-minute silent film is positive, even celebratory and patriotic, in the tone of its captions. It describes the main episodes (1 and 14 July, 15 September, and the final successful offensive) in detail, not ignoring the depiction of casualties and dreadful conditions but stressing that the enemy (characteristically called Jerry) was outfought. The battle as a whole is termed "a glorious tragedy" but only near the end of the film are appalling losses mentioned. The winning of four V.C.s (only one winner an officer) is reenacted; a few battalions are named; and prominence is given to the contributions of Scots, Anzacs, Canadians, and South Africans. It is the assertion of the film that "our victorious forces" were deprived of a crowning achievement only by the onset of torrential rain, ruining any chance for significant advance. The German withdrawal to the Hindenburg Line in 1917 signified a type of victory, though, as a Tommy puts it, "Jerry's 'opped it"—he pulled back to avoid a worse disaster thereby abandoning nine hundred square miles. Consequently, "the sacrifice has not been in vain." Further research might reveal how audiences received the film. One reviewer criticized it as "unreal" for taking "a romantic boy adventure book angle," but elsewhere, along with the director's other war films, it was said to have earned a great deal of prestige for the British film industry.

It is hard to imagine such an unapologetic patriotic film being made a few years after the remarkable impact—particularly on the middle classes—of the spate of memoirs, novels, poetry, and plays about the Western Front that appeared between 1928 and 1932. There is no need to list them here or discuss them at length.[12] By no means all of them were concerned with the Somme; there is no consensus about what they were recalling in bitterness or saying goodbye to; and a minority could not be construed as "antiwar" in a polemical or any other sense (e.g., Charles Edmonds's *A Subaltern's War*). Nevertheless, for all their differences in form, tone, and attitude, Correlli Barnett is surely correct to argue that a broad message was conveyed to the readers: of idealism turning to disenchantment, the incompetence of the higher command and staff and the futility of the fighting, the obscenity and terrors of the modern battlefield. Their popularity is undeniable. R. C. Sherriff's play *Journey's End* was reprinted thirteen times in 1929, and several of the memoirs and novels were immediate best-sellers. Barnett has argued convincingly that this outpouring of "antiwar" literature was unrepresentative of the officers and enlisted men alike, and Douglas Jerrold pointed out in a celebrated pamphlet that, by focusing on the anguish of individuals, these works ignored both the collective activity of armed forces and the wider purposes for which they fought.[13] Others have noted the ambiguity of the targets in the memoirs of ex–junior officers such as Graves and Sassoon, who had fought bravely and were proud of their regiment.

Nicholas Hiley has formulated the most original interpretation of the "war books boom" as a more or less deliberate reworking of wartime experience to meet the perceived public need of the late 1920s. This need was to prevent the recurrence of war by demonstrating that the democratic opinion was naturally pacifistic. The message made a profound impact, at least among the intelligentsia, but had the unfortunate effect of obscuring for a time the unpalatable truth that in certain circumstances war might again be unavoidable—even for a pacifistic democracy.[14]

The cumulative and enduring effect of this antiwar phase, with its savage criticisms of Haig and of the conduct of the war on the Western Front, specifically the Somme, has made it almost obligatory for later writers and commentators to deal with the battle on terms established by the critics. This was vividly exemplified in the Second World War when the American chief of staff George Marshall was urging the speedy opening of a second front in Europe. "It's no use," Churchill's adviser Lord Cherwell replied, "you are arguing against the casualties on the Somme."[15]

The British chief official historian, Sir James Edmonds, unintentionally provided ammunition for critics of the Somme campaign by concluding the first volume on the subject, published in 1931, with the end of the first day of operations—1 July 1916. It would be some years before the publication of the second volume, leaving the public quite some time to digest the failure of the first day of the campaign. Edmonds's conception of his role and his ambivalent attitude toward Haig continue to puzzle historians,[16] but it seems clear that under pressure from senior participants in the campaign, notably General Sir Archibald Montgomery-Massingberd (who became adjutant-general in 1931 and chief of the Imperial General Staff in 1933), he was prepared to tone down criticism, suppress evidence, and put his own idiosyncratic interpretation on comparative casualty statistics. The result is a dense and generally bland narrative that reaches no firm conclusion about Haig's intentions on the first day and contrives to suggest that he consistently pursued a strategy of attrition, which paid off: total Anglo-French casualties being reckoned at 623,907 and German at a round figure of 600,000. In the second volume, however, Edmonds contended that German casualties should be increased by 30 percent to allow for lightly wounded men not previously included. Later research, which is rarely used by popular writers, proved Edmonds's calculations to be demonstrably false.[17] For our purposes, the main point about the official history volumes on the Somme is that their irregular appearance served spasmodically to awaken controversy over Haig's generalship and over particular episodes in the campaign, such as the decision to employ tanks for the first time on 15 September and to persist with the offensive into November despite dreadful conditions.

The very influential historian Captain B. H. (later Sir Basil) Liddell Hart's changing views on Haig and the Somme were greatly influenced by the "credibility gap" between what he read in the official history and the (generally) more critical information that Edmonds leaked to him confidentially. As a convalescent junior officer in 1916, Liddell Hart had compiled a remarkably eulogistic account of the battle then still raging, and in *Reputations* (1928) he was still taking a moderate and balanced view of Haig. In *The Real War* (1930) his tone was more critical, but he at least covered the whole campaign in outline; allowed that it had had some positive, albeit indirect, effects at Verdun and on other fronts; and ended by noting the severe strain suffered by the Germans—even if their heavy casualties were largely due to their commanders' faulty tactics. Eight years later, in *Through the Fog of War,* Liddell Hart based his short critique entirely on the official history's account of the first day, devoting only one sentence to the following 140 days. He concluded rhetorically that the mutual destruction in the battle was not a necessary contribution to the final issue, "while even as an actual contribution it was not comparable to the blockade." It was, in sum, "the summit of strategic senselessness—until 1917."[18] As a result of his prolific output of books and articles, readable style, and confident, critical tone, Liddell Hart probably had more influence than any other historical writer in the 1920s and 1930s in shaping the reading public's notions about the British army's inept performance on the Western Front and the unwisdom of its heavy commitment there.

There were, however, other, contrary interpretations that were widely read, most notably C. R. M. F. Cruttwell's *A History of the Great War 1914–1918,* first published in 1934 but remaining in print in several later editions. Cruttwell, an Oxford don who had been seriously wounded in the war, took a more balanced and judicious view of the war as a whole than did Liddell Hart, as was appropriate both to his profession and his publisher—Oxford University Press. Though perfectly capable of acidulous criticisms of British generalship and tactics, Cruttwell, took a remarkably positive view of the Somme campaign. While allowing that the ground gained was not impressive, he argued that the British advantage had put the enemy under intolerable strain. The retreat to the Hindenburg Line proved that the Germans could not spare the troops to man the extended line of November 1916. He concluded that the campaign, "in spite of the grave tactical mistakes which opened it," had rendered a serious service to the Allies. "In spite of the slaughter, the British army gained experience rather than discouragement, as the attack before Arras was to show in the coming spring."[19]

In similar vein was Cyril Falls's *The First World War,* which was not published until 1960 but owes its genesis to the 1930s, when the author was a member of the team of British official historians. Falls admitted that British tactics in 1916 had been clumsy but said this was mainly due to the amateur character of the army, whose instructors were too stiff and conventional. "Yet for determination and devotion the army that fought on the Somme has never been surpassed." While making due allowance for other contributing factors, such as the initial success of the Brusilov Offensive launched by the Russians on the

Eastern Front, Falls concluded that Haig and Joffre had succeeded in wearing down the German army, as General Erich von Ludendorff and others had admitted after the war.[20]

As saturation coverage of the Second World War declined and the approaching fiftieth anniversary of the war of 1914–1918 caused a revival of publishers' interest in the earlier conflict, a new generation of historians rediscovered the Great War in which their fathers or even grandfathers had fought. Several were directly, and nearly all indirectly, influenced by Liddell Hart—by 1960, the doyen of British military historians—with a firmly established and highly critical view of Haig and the conduct of war on the Western Front. In addition to his generous help and advice, Liddell Hart's enormous archive was invaluable to writers on the First World War.[21]

Hence, for these and other reasons springing from the more radical and antiauthority zeitgeist of the decade, most of the new historians were inclined to debunk rather than "bunk" (in Alex Danchev's term) the conduct of the earlier war.[22] Though concerned with 1914–1915 rather than the Somme, Alan Clark's title *The Donkeys* (1961) epitomized the fashionable (and highly publishable) approach to British generalship in the war.

There were, of course, historians who resisted what might be called the Liddell Hart orthodoxy. Most notable for his courage, industry, and sheer perseverance was John Terraine, who began his personal counteroffensive with *Douglas Haig: The Educated Soldier* in 1963.

These opposing viewpoints clashed in the making of what was then the biggest ever British television documentary series—*The Great War* (1964). Terraine, who scripted the Somme and Passchendaele programs, sought to put over a positive view of British generalship and the primacy of the Western Front, with the corollaries that heavy casualties had been unavoidable and had contributed significantly to eventual victory. The outcome—depressing or salutary, depending on one's views on the war, on the relative importance of the printed word or visual image, and on the value of television in shaping public opinion—was a clear victory for the medium over the message. Audience research reports showed that the terrible battlescapes and the sufferings of all participants (soldiers, civilians, and animals) had reinforced earlier (pre-1939) views about the horrors of war and the waste of young manhood. A new generation was unable to assimilate the implicit revisionism embodied in Terraine's scripts. Thus, in complete contradiction of his intentions, this enormously successful series reinforced the notion of the utter futility of the First World War.[23]

Of course, the *Great War* series and the astonished flurry of publications in the 1960s did not end the controversies over the Somme and Haig's generalship. Like his hero, Terraine never acknowledged defeat and continued an attritional barrage, which, for all its duds, shortfalls, and misses, gradually ground down at least some of the opposition's defenses.

He deplored the obsessive concern of so many writers with the disastrous first day of the Somme, preferring to stress the army's endurance over the remaining 140 days of the campaign at what he considered to be an acceptable casualty rate (just under three thousand per day), given that Britain was now beginning to bear the main burden of fighting the German army on a strongly defense sector. The "payoff" for Terraine lay in the Germans' propensity to counterattack rather than yield ground (some 330 incidents are listed) and the fact that the great majority of their 125 divisions on the Western Front were exhausted in the attritional battle on the Somme, obliging Ludendorff eventually to retreat to the Hindenburg Line. Terraine has also made determined attempts, particularly in *The Smoke and the Fire,* to demolish Somme "myths" such as the primacy of the machine gun when, in fact, artillery was more effective; the alleged misuse of tanks on 15 September; and, above all, the mistaken attribution of failure on 1 July to the infantry having to carry a sixty-six-pound pack.[24] In sum, while not going so far as to claim that the Somme was a victory in itself, Terraine argues that it was a necessary battle in which an attritional strategy was deliberately pursued, resulting in a fatal drain on German manpower that contributed significantly to the eventual collapse in 1918.

Two books from the 1960s specifically about the Somme deserve a brief discussion. Brian Gardner's *The Big Push* (1961) was an unambitious popular account relying on mostly familiar published sources. Gardner was sure that a historical verdict would be reached on the battle during the 1960s but hedged his bets as to whether it would be seen as a victory or a disaster. His illustrations and sketch maps clearly suggested the latter, as did his concluding chapter heading, "Napoo!"—soldiers' slang for "useless" or worse. Anthony Farrar-Hockley's *The Somme* (1964) was much more thoroughly researched, including interviews with survivors, and benefited from the author's firsthand knowledge of the army's command structure and his distinguished combat experience. He provided excellent coverage of the long gestation of the battle, the first day, and "the long struggle" up to mid-September, but then rather skimped on the final stages. In his summing up, Farrar-Hockley appreciated that casualty statistics were of crucial significance but accepted Edmonds's calculations as reliable. He felt that Haig and other national commanders had been made scapegoats but concluded unconvincingly that the Somme battle (and the war as a whole) "was ultimately the responsibility of the peoples of Europe and the United States, who permitted conditions to come to such a pass."[25]

The 1970s witnessed the emergence of a new style of popular military history, the best of it exploiting the oral and written testimony of veterans but also reinforcing the growing public obsessions with the Somme as the archetypal "futile" battle of the First World War. There is no need here to rehearse the many merits of Martin Middlebrook's *The First Day on the Somme* (1971),[26] which makes admirable use of veterans' recollections to convey a vivid sense of what the battle was like without polemical intent. Although his coverage is wider than the title suggests, Middlebrook's technique was better suited to the detailed study of an episode rather than a broader account of a long battle, let alone the whole war. Similarly, John Keegan confined himself to the first day of the Somme to delineate *The Face of Battle* (1976). Subsequently there have appeared numerous detailed studies of particular episodes in the Somme campaign or the experience of particular units, including the Ulster Division on the first day, the Thirty-eighth (Welsh) Division at Mametz Wood, the South Africans at Delville Wood, and the Australians at Pozieres. Many of these studies by enthusiasts (amateurs in the best sense) are critical of strategy, tactics, and the higher command while evincing admiration for the courage and endurance of the troops—one recent study of the fighting at Arras in 1917 is titled *Cheerful Sacrifice*,[27] an uncomfortable notion at the turn of the new century. This genre, and especially a current series about the Pals Battalions, is above all concerned to commemorate and celebrate the achievements of individuals and units, based on personal recollections and close study of the ground. This is a very different approach from the popular debunking mode of the 1960s.

One unexpected consequence of the treatment of the First World War in the 1960s, as Alex Danchev has pointed out,[28] was to resurrect the ghosts of the 1920s and early 1930s for a new generation that was antiestablishment, antiauthority, and profoundly skeptical of the values of the "officer class." Since then, there have been encouraging signs (for professional historians) that the Somme campaign is at last being approached as history—as distinct from a national trauma—and so properly placed in the full context of technological development, politics, and strategy. Studies such as Shelford Bidwell and Dominck Graham's *Fire Power* (1978), Tim Travers's *The Killing Ground* (1987), and Robin Prior and Trevor Wilson's *Command on the Western Front* (1992) prove that it is possible to bring the full range of scholarly qualities successfully to bear even on such a controversial and emotion-encrusted subject as the Somme campaign. Unfortunately, a vast gulf still lies between the influence of these historians and the pervasive popular stereotype of 'Butchers and Bunglers.'[29] In a sense, as Ian Beckett remarked, "the early writers of the 1920's and 1930's took the Somme out of history" and converted it into a national myth.[30] This tendency was reinforced by the debunkers of the 1960s—most notoriously by the film *Oh! What a Lovely War* in 1969. Though still experiencing the occasional counterattack, historians are now "digging in" and grappling with the task of restoring the Somme to operational and strategic history. Its place in cultural and social history is also being firmly established.

NOTES

1. I am most grateful to Dr. Ian Beckett for allowing me to make use of his unpublished lecture on the historiography of the Somme, delivered at the Imperial War Museum, London, in November 1986.

2. John Terraine, *The Smoke and the Fire* (London: Sidgwick and Jackson, 1980), 108, 205, 221–22.

3. Brain Bond, ed., *The First World War and British Military History* (Oxford: Oxford University Press, 1991). See especially the chapter by Alex Danchev—"Bunking and Debunking: The Controversies of the 1960's."

4. Beckett, lecture, 2.

5. Stuart Sillars, *Art and Survival in First World War Britain* (New York: St. Martin's, 1987), 56–57.

6. Quoted in ibid., 71.

7. Peter Crookston "Death of the Pals," *Sunday Times* colour supplement, 30 October 1988. See Martin Middlebrook *The First Day on the Somme* (London: Allen Lane, 1971) app. 5, 330, for a list of battalions suffering more than five hundred casualties in addition to the Accrington Pals.

8. Sillars, *Art and Survival,* 49–51; Beckett, lecture, 3. See also Roger Smither, "A Wonderful Idea of the Fighting: The Question of Fakes in the *Battle of Somme,*" *Imperial War Museum Review,* no. 3 (1988): 4–16.

9. M. J. Williams, "The Treatment of the German Losses on the Somme in the British Official History," *Journal of the Royal United Services Institute (JRUSI)* (February 1966): 69–74. See also the chapters by David French "Sir James Edmonds and the Official History: France and Belgium" and Keith Simpson "The Reputation of Sir Douglas Haig" in Bond, ed., *First World War.*

10. I am indebted to Robert Stevens for the following information on the Somme anniversary, as reported in *The Times* and the *Daily Express.*

11. I thank the British Film Institute for allowing me to see this film, and Simon Baker for sending me a detailed synopsis, credits, and other information. Dr. Nicholas Hiley called my attention to the existence of this film.

12. See Correlli Barnett, *The Collapse of British Power* (1972), 428–435;

Strand Alan Sutton and Noel Annan, *Our Age: Portrait of a Generation* (1990), 66–76.

13. Douglas Jerrold *The Lie about the War,* Criterion Miscellany—No. 9 (London: Faber and Faber, 1930), 22–23, 46–47.

14. Nicholas Hiley "The New Media and British Propaganda 1914–18" (unpublished paper); and idem review of Stuart Sillars's *Art and Survival in First World War Britain, Times Literary Supplement,* 16–22 September 1988.

15. Quoted by Beckett, lecture, 7.

16. See David French's chapter in Bond, ed., *First World War.*

17. See Williams, "Treatment of the German Losses"; and his earlier article "Thirty Per Cent: A Study in Casualty Statistics," *JRUSI* (February 1964): 51–55.

18. B. H. Liddell Hart, *The Real War* (Boston: Little, Brown, 1930), 221–23, 267; and idem, *Through the Fog of War* (London: Faber and Faber, 1938), 257–8. See also Hew Strachan's chapter, "'The Real War': Liddell Hart, Cruttwell and Falls," in Bond, ed., *First World War,* for the contrasting careers and comparative merits of these historians.

19. C. R. M. F. Cruttwell, *A History of the Great War 1914–1918,* 2d ed. (Oxford: Oxford University Press, 1964), 277.

20. C. Falls, *The First World War* (London: Longmans 1960), 177–78. Hew Strachan rates Falls's study higher than those of Cruttwell and Liddell Hart, particularly for its balance and objective judgment.

21. See my introduction in Bond, ed., *First World War,* 6–7.

22. See Danchev's "Bunking and Debunking," 280–81.

23. Beckett's lecture, 8, provides an admirable summary of Terraine's main concerns and arguments.

24. See Terraine, *Smoke and Fire,* 23.

25. A. H. Farrar-Hockley, *The Somme* (London: Pan, 1964), 242–53.

26. See Peter Simkins's discussion of Middlebrook's work and influence in Bond, ed., *The First World War,* 302–3.

27. Jonathan Nicholls, *Cheerful Sacrifice: The Battle of Arras 1917* (London: Leo Cooper, 1990). An excellent study despite its tendentious title.

28. In Bond, ed., *First World War,* 287.

29. John Laffin, *British Butchers and Bunglers of World War One* (Surrey: Beamley Books, 1988).

30. Beckett, lecture, 11: "In a sense, the early writers of the 1920's and 1930's took the Somme out of history and slowly the historians are coming round to the problem of giving it back to history." In adapting Ian Beckett's conclusion (with permission), I can only resort to the hackneyed phrase "Great minds think alike!"

CHAPTER 8

The Elusive Victory

The BEF and the Operational Level of War, September 1918

Niall J. A. Barr

The Great War on the Western Front was different in its conduct and character from any conflict that preceded it. Its officers struggled to make sense of a seemingly horrible conflict, which did not correspond to their training or to their preconceptions about the nature of war. Neither the Napoleonic conception of decisive battle, as greatly desired by French commanders, nor Count von Schleiffen's strategy of an enormous enveloping maneuver bore fruit in the maelstrom of 1914. With the onset of trench deadlock, the battles fought on the Western Front became bloody attritional contests in which the manpower and resources of entire nations were put to the test. Michael Howard has noted that in the Great War, battle "acquired a frightful, Moloch-like existence of its own unrelated to strategic or political objectives."[1] Developments in technology had outstripped levels of tactical innovation.

Yet the armies of the West slowly adjusted to the challenges of the Great War and in so doing began to lay the strategic and tactical foundations of modern war. Although the Great War gave birth to the main technological elements of modern maneuver warfare, such as sophisticated artillery techniques, section-level infantry tactics, and the development of tanks and air forces, it is difficult to find clear-cut examples in the conduct of Great War that point toward the ideas of modern maneuver warfare. This can make analysis of the Great War beyond the level of "siege warfare" difficult and makes it no easier to place the war within the continuum of strategy, operations, and tactics that leads from Helmuth von Moltke to today's idea of AirLand Battle.

This chapter examines the planning, preparation, and execution of just one of the British Expeditionary Force (BEF) attacks on the German army on the Western Front in the autumn of 1918. The British First Army's assault on the Drocourt-Queant Switch Line on 2 September 1918 was only one of the many engagements fought by the BEF during the final Allied offensives, but this battle in particular highlights the difficulties in mounting a successful offensive operation during 1918 and reveals a great deal about the extent to which the BEF—and particularly its commander Field Marshal Douglas Haig—understood the new problems of modern warfare and the means that could be used to overcome those difficulties.

In his survey of the Great War, the eminent military historian John Keegan describes attempts to study the learning process of the BEF from 1914 to 1918 as "a pointless waste."[2] At the same time, the battles fought by the BEF in the autumn of 1918 have remained relatively unexplored next to the traumas of 1916 and 1917. In British popular culture, 1918 remains the forgotten year of the Great War, and to a large extent the same is true in military historiography. Much remains to be researched and written concerning operations in 1918. In fact, it is in the last dramatic phases of the war—particularly the period known to the B.E.F as the "Hundred Days" from 8 August to 11 November 1918—that we can learn much about the problems of battle in the Great War and come to a deeper understanding not only of this first great technical-industrial conflict but also of its place in military history and the development of military theory and doctrine.

General Sir Frederick Maurice wrote about the British successes during the "Hundred Days" soon after the war:

> I am convinced that the achievement of the National Army of Great Britain transcends even that of her old Regular Army. . . . Starting on August 8, it fought uninterruptedly and victoriously for three months, driving the enemy back 120 miles, taking more than twice as many prisoners and more than three times as many guns as it had lost, and completely routing the German armies by which it was opposed. This is . . . an achievement to which no words can do justice.[3]

British soldiers in November 1918 understood that they had just won a great victory, but they did not tend to analyze this success in any great detail or profound way. While the soldiers took it for granted that they had won, the British people began to question the cost necessary for such a victory. By the 1930s, the whole idea that the British army had won a great victory had been called into question. British military writers such as Basil Liddell Hart and J. F. C. Fuller rejected the style of fighting on the Western Front and attempted to provide alternatives to such fighting. While Fuller suggested that the greater use of technology and, in particular, more imaginative use of the tank might have broken the deadlock more quickly, Liddell Hart also suggested that the use of his strategy of the "indirect approach" might have avoided major British commitment to the fighting on the Western Front.[4] Within Germany in the interwar years, the technological solution of fast-moving armored formations was seen as the way to avoid a future war of attrition and deadlock.[5]

Yet it was in the Soviet Union that military thinkers were first able to place the fighting on the Western Front within its proper context. They were scathing about the Allied victories of 1918. They argued that "even after the German front fell because the stability of the defense could not hold, the Allies still wasted four months in 1918 to push back deployed German forces a total of only 100 kilometers."[6] Harsh criticism indeed. Soviet military theorists such as V. K. Triandafilov and Aleksandr A. Svechin were among the first to understand the enormous changes wrought in warfare and to foresee developments that could solve some of the problems of modern war. In their analysis of the changing nature of war and battle, Soviet writers came to realize that the quest for a single, decisive battle was no longer possible. Carl von Clausewitz had argued that the decisive battle—a single point in time and space—was the "struggle for real victory, waged with all available strength."[7] Napoleon had certainly been an exponent of the classic decisive battle, but no battle in the Great War had resembled his victories. G. Isserson noted that the "World War did not provide a single operation which could rightly be considered an operational solution to the problem of achieving victory."[8] The enormous strength of the defensive when matched with a positional front and combined with the enormous complexity in organizing and conducting offensive action meant that the war ended "under the omen of insurmountable difficulties in organizing and conducting an offensive operation."[9]

Soviet military thinkers proposed their theory of "operational art" to replace what they saw as the "dead end" of the imperialist military system. This rejected the idea of a linear front and the search for victory through one decisive battle. Instead, it was suggested that a linked series of successive engagements fought throughout the enemy's breadth and depth would lead to operational paralysis, which could then be exploited by a second echelon that would wreak havoc in the enemy's rear. The cumulative effect of all these operations would lead to victory.[10] In fact, the battles of 1918 informed much of the study of the Soviet theorists in the interwar years, and the battle for the Drocourt-Queant Switch Line on 2 September demonstrates just how close Haig and his armies came to implementing this higher level of war.

Much had changed in tactics and technology since 1914. While much of the equipment utilized in 1918 was exactly the same as in 1914, tactics and techniques had become much more sophisticated. Artillerymen had developed new methods for ensuring accuracy even with indirect fire. The use of precise mapping of the Western Front, calibration of guns, and acknowledgment of the importance of meteorology had all increased accuracy, while new techniques such as "silent registration" and the use of new and more sensitive fuses had all increased the effectiveness of the artillery. Flash spotting and sound ranging had vastly improved counterbattery techniques. Less dramatic but just as important was the development of the Royal Artillery into a mass arm. The British had famously introduced the tank as a means of trench crossing and wire crushing and by 1918 had developed tactics and improved versions of their tanks. The tank of 1918 was still not impressive in terms of reliability, speed, or maneuverability but was a useful weapon for infantry support. Just as important, the infantry of 1918 bore little resemblance to the battalions of 1914. A British infantry battalion in 1918 had an establishment of just 550 men (often much fewer in reality) but could produce greater volumes of firepower due to its allocation of trench mortars, Lewis guns, and rifle grenades. The introduction of section tactics, which stressed the importance of fire and movement, gave much-needed flexibility to the infantry. Infantry now had the capability to fight through the tactical problems and obstacles that were encountered on the Western Front. Meanwhile, airpower had really come of age during the Great War. Improved aircraft types such as the Sopwith Camel, SE5a, and DH4 bomber meant that the new Royal Air Force could perform all the important air missions, from air superiority, interdiction, and reconnaissance to ground support. Aircraft in 1918 were still flimsy, slow, and vulnerable, but they made an important contribution to the conduct of operations. The glue that held many of these improvements together was the increased use of wireless communications. While wireless sets were still heavy and difficult to transport, the use of wireless communications in aircraft and tanks provided much better control and information to commanders in 1918 than had been available in 1914. The importance in all these technological improvements and tactical developments was that the BEF learned to work and fight together. It was the combination of all these developments, in the use of combined-arms warfare, that made the BEF such a formidable instrument in the autumn of 1918.[11]

After the Battle of Amiens on 8 August 1918, it was clear that the conditions on the Western Front were changing. During the battle, the British Fourth Army, utilizing surprise, 450 tanks, and an extremely heavy but well-timed bombardment, managed to penetrate up to eight miles into the German defenses and capture over four hundred guns and twenty thousand prisoners at the cost of only four thousand casualties.[12] Such results were unprecedented for the BEF, which was more used to counting gains in yards and casualties in tens of thousands. Although German resistance began to stiffen on the 10 and 11 August, the British were able to continue their advance, and when the Third Army began a new offensive to the north around Albert on 23 August, the pressure on the Germans increased still further.

It was in an atmosphere of swelling prisoner-of-war cages and glowing reports of further advances that General Douglas Haig, the commander in chief of the BEF, began to scent victory. His views on battle were well formed after many years of experience and study—but they also tended to be stereotyped and methodical. Haig had started his career in the Great War believing that there was a set pattern to battles that enabled a commander to discern the course of the fighting. First there would be reconnaissance and a meeting engagement in which strong advanced guards could turn the balance of advantage. This would be followed with a test of strength and firepower—known as a "wearing out fight"—in which the side that maintained its morale and willpower would begin to gain the upper hand. At the end of this "wearing-out fight," the enemy would begin to crumble, and the moment for pursuit was at hand. This was the moment to release the exploitation forces—the horsed cavalry—to clinch the victory. Haig did not change these basic ideas throughout his command on the Western Front, although the length of the wearing-out fight certainly altered from a few hours or days to weeks, months, and even years.[13]

On 22 August 1918, Haig issued his army commanders with an important set of instructions that demanded greater boldness and audacity in pursuit:

> The methods which we have followed, hitherto, in our battles with limited objectives when the enemy was strong, are no longer suited to his present condition. . . . To turn the present situation to account the most resolute offensive is everywhere desirable. Risks which a month ago would have been criminal to incur, ought now to be incurred as a duty.
>
> The situation is most favorable; let each one of us act energetically, and without hesitation push forward to our objective.[14]

Haig clearly believed that the "wearing-out fight," which had lasted through three long years of warfare, was almost over, and that the pursuit phase of a defeated enemy was about to begin. Since Haig believed that the time for the decisive battle had arrived, he ensured that his Cavalry Corps was concentrated for the coup de grâce. When, on 23 August, General Byng gave an order that would have separated the First and Second Cavalry Divisions widely, Haig insisted, "I wished the cavalry kept together as a corps for strategical objectives in view of the possibly rapid disintegration of the enemy."[15]

Haig had made optimistic assessments of the military situation before, both in 1916 and 1917, and as the events of late August and early September 1918 were to show, he was premature in ordering the abandonment of methodical progress for all-out pursuit. Even the success experienced by First Army from 26 August onward did not precipitate the breakup of the German army, and Haig's Cavalry Corps remained unused. Nonetheless, while the detail of this order was premature, the spirit was correct. Indeed, the entire BEF began to scent victory as it pushed through the old Somme battlefields in a matter of days—progress that in 1916 had been measured in months and thousands of casualties. The BEF became, for the first time in years, an army on the move.

The German army, with its reserves exhausted, was unable to mount any major counterstroke, and on the night of 24–25 August, the Germans began a phased withdrawal of three miles across the old Somme battlefields to the "Winter Line," a series of defenses stretching from the River Somme at Peronne to the Drocourt-Queant Line in the north. The German High Command hoped that an orderly withdrawal to the strong defensive positions of the Winter Line would gain time in which to rest and reorganize its battered divisions. Even as the German army withdrew, its troops were being treated to a barrage of morale-boosting orders of the day. A German Second Army order written by General von der Marwitz on 25 August was typical of such statements. He warned:

> People with anxious temperaments saw everywhere squadrons of tanks, masses of cavalry, thick lines of infantry. It is high time that our old battle-experienced soldiers spoke seriously to these cowards and weaklings and told them of the deeds that are done in the front line. Tanks are no bogey for the front line troops who have the artillery close behind them. . . . There are no reasons for any panic. On the contrary, the troops in the front line never considered themselves victors in the way that they have at present.[16]

For the few "old battle-experienced soldiers" left in the German ranks, von der Marwitz's words must have possessed a terrible irony. The truth was that the German army had been forced into retreat by "squadrons of tanks" and "thick lines of infantry," backed up by tremendous weights of artillery. The real situation in the front line was described by the letters of ordinary soldiers, as in this extract from a letter by a man in the 217th Reserve Infantry Regiment in the 225th Division:

> In the last few days our company has lost over 70 men and I am amongst those who remain—10 NCOs and 26 men, and still we have to hold on. Our Battalion is now equal to a company and company to a platoon. Our three regiments are now only equal to one.[17]

German divisions were being exhausted and fought to destruction due to lack of reserves, and the constant pressure produced by the British attacks was beginning to create real demoralization. The large number of bullish statements produced by German senior officers during August and September simply reinforced the impression that the German High Command was not in full touch with military reality. Even Ludendorff's naming of the Winter Line as the position the German army would hold throughout the winter months, until the spring campaign of 1919, was simply another example of this misguided propaganda.

Yet the linear strategy of the Great War, as noted by Soviet theorists, created enormous problems for the attacker. The German army held a continuous front line from Switzerland to the sea that gave it considerable defensive advantages. During August 1918, the Allies had managed to put the German forces under enormous pressure. Rather than see their front crack open, however, which would restore mobile warfare but on distinctly unfavorable terms, the Germans quite naturally preferred to withdraw to new defenses, shorten their line, rest their reserves, and allow the entire process to begin again.

This meant that this period of unexpected success was also a great test for the British armies. Without numerical superiority, the same troops and formations had to be committed to battle day after day, in a series of more or less continuous operations. Just as the fighting troops became weary from constant activity, so the engineers and supply troops had to struggle to keep up with a constantly moving front. British success depended not just on the increasingly evident confidence and skill of the fighting troops but on the ability of their logistic chain to match the rate of advance. The BEF could only follow up the German withdrawal, not exploit the German weakness, break through, and pursue. Much of the reason for this lay in the state of technology available to the armies on the Western Front. In the advance, the British did not have tactical mobility beyond the boots of its soldiers and the hooves of its horses. The Mark IV tank could not make more than four miles per hour, while the Whippet, which could make eight miles per hour, was not reliable enough for pursuit operations. An additional problem was that the artillery found it very difficult to bring up sufficient guns through the devastated battlefields to support the advance. Thus, the limits of technology in 1918 made it almost impossible to engage in high-tempo operations that would develop sufficient speed and momentum to break into the German rear.

Significantly, as the British armies followed up the German retreat, they began to shake out into mobile warfare. Haig had been convinced of the need for strong advanced guards in mobile warfare ever since his Staff College days, and he now ordered their use against the retreating German forces. These advanced guards, consisting of one brigade of infantry with an attached squadron of cavalry, actually worked well, allowing the British forces to follow up the German retreat and also to close up for a more formal attack if necessary.

Nonetheless, British divisions continued to watch their flanks carefully and move in phased lines. When Byng set far-reaching objectives for his Third Army on 25 August, these were found to be hopelessly overoptimistic. This was due to the nature of the German defense, which, although it now approximated a rearguard, remained tough and difficult to overcome. The German rearguards were composed of their most reliable troops—and generally consisted of a web of machine-gun posts sited for all-around defense, backed up by field guns and mortars. It was still perilous to incur risks against such a determined foe.

Nonetheless, the continued British advance caused the Germans much concern. The German troops on the Third Army front formed the link between the withdrawal further south and the defenses of the Hindenburg and Drocourt-Queant Lines further north, and the eight German divisions in the line were reinforced by a further eleven divisions in an effort to stabilize the position. These reserves had mounted numerous counterattacks on 25 August that were beaten back and dispersed mainly by British artillery fire, but this effort had limited the gains made by the Third Army. More important, however, these reserves had bought only a temporary respite from the Third Army's pressure, just one day before the First Army began its offensive south of the Scarpe and widened the scope of the British offensive still further. The fact that German reserves had been drawn onto the Third Army meant that the First Army's attack was to have much greater impact.

The actions of the First Army need to be examined in greater detail, as it is quite clear that Haig envisioned this attack as the decisive action that would break open the front and restore mobile warfare—and allow the pursuit to begin. General Sir Henry Horne of First Army had been warned on 15 August to make ready for an offensive. On 24 August, the objective of the attack was fixed: the First Army was to advance and "pierce the DROCOURT-QUEANT line and subsequently to operate in a south-easterly direction against the right flank of the enemy's troops opposed to the Third Army."[18] Haig explained in a conference to his commanders that

> I wished the First Army to attack by surprise and advance as rapidly as possible astride the Cambrai road with the left secured on the river Scarpe. First the Monchy le Preux position will be taken, then the position of the Drocourt-Queant Line will be taken, and next the Marquion—Canal du Nord Line. If all went well, and the advance was accomplished rapidly, the enemy might be still holding his positions in front of Third Army. It would then be the role of First Army to operate against the right flank of this part of the enemy's forces.

While the Fourth and Third Armies made steady progress against the German troops opposing them, Haig hoped that the First Army would penetrate the German Winter Line and then lever the German troops off their position. Haig clearly saw this coming battle as the decisive blow.

To be ready for the pursuit phase, he augmented the Cavalry Corps with an infantry brigade in buses and

> extra machine-gun batteries in motors, with the object of exploiting the situation which I hope will arise after we get the Marquion–Canal du Nord line. We might possibly get the chance of pushing on the cavalry corps to the Bois de Bourlon to intercept the enemy's retreat.[19]

While it is often argued that the idea of using horsed cavalry in 1918 was hopelessly obsolete, the Cavalry Corps organization displayed some remarkably forward thinking. It consisted of three weak cavalry divisions, one infantry brigade mounted in trucks, and a motorized machine-gun battalion (actually the Household Cavalry). With the exception of the reliance on horses, such an organization would not have looked out of place in 1939. Although only fourteen thousand men strong, the Cavalry Corps was Haig's only exploitation force, his only weapon to exploit a victory—if it appeared.[20]

Horne, after breaking through the Drocourt-Queant Line, was to press on "as fast as possible against Marquion where the enemy's depots are which supply the whole of von Below's Army."[21] Haig believed that it might be possible for the First Army to rupture the Drocourt-Queant Line, outflank the Hindenburg Line, *and* destroy von Below's supply lines. By achieving a deep penetration of the German lines in front of Cambrai and then outflanking the German positions to the south, the First Army would strike a shattering blow and force the disintegration of the German army. While many historians have argued that Haig's conception of battle was archaic, this plan does display elements of what the Soviets later termed *operational art.* The advance of the Third and Fourth Armies had drawn the German reserves away from the point of the attack, enabling the First Army to attack a weakened sector of the front. At the same time, the First Army's offensive was to be mounted on a narrow frontage supported by heavy concentrations of artillery—a technique that bore a striking similarity to German and Soviet techniques of the Second World War.

Nonetheless, Horne's First Army faced a very difficult task. The northeastern slopes of the Artois hills rose to three hundred feet above the Scarpe and the Sensee at Monchy le Preux, while the rivers cut deep valleys into the hills. The Germans had fortified all the commanding high ground with four defensive lines, and there was a jumble of switch lines, strong points, and debris left from the earlier battles of the war. At Monchy le Preux, the old British and German trenches from March 1918 had been repaired to form the German outpost line. Two miles to the east, the Fresnes-Rouvroy Line guarded the main positions of the Drocourt-Queant Line that lay one mile farther east. The Drocout-Queant Line was the southernmost section of the Wotan I–Stellung that extended the main Hindenburg Line (the Siegfried Stellung) northward from Queant. The defensive positions of the Drocourt-Queant Line were formidable, if not quite as powerful as the main Hindenburg Line. The line consisted of a front system of two trenches positioned on the forward slope and a support system of two trenches on the reverse slope. All the trenches were well equipped with concrete shelters and machine-gun posts, and the entire position was heavily wired. Further east was the unfinished Canal du Nord, which still acted as a considerable barrier to forward movement.

Another surprisingly modern aspect to the British planning was the use of sophisticated security and deception measures. These measures were employed before the offensive to persuade the Germans that the attack would be made north of the Scarpe. Heavy artillery bombardments, combined with tank-infantry training, dummy ammo dumps, casualty clearing stations, and false radio traffic, were all designed to draw German attention away from the Arras-Cambrai road. In fact, such efforts were the clear precursor of Montgomery's famous Operation Bertram before Second Alamein and the Soviet use of *maskirovka* during the Second World War. The First Army's deception measures proved very successful, as a report on 30 August acknowledged:

> The activity on the Army front north of the SCARPE, combined with the open employment of tanks for training, undoubtedly had the effect of leading the enemy to suppose that our attack was coming north of the SCARPE River, and by inducing the enemy to locate his reserves so as to meet attack on that front, assisted materially in facilitating the progress of the operations which have been carried out so successfully by the Canadian Corps.[22]

Not only had the Germans prepared to meet an offensive north of the Scarpe, but they also had thrown precious reserves into the fight against the Third Army on 25 August. The success of these deception measures was vital; without them, the First Army's attack might have stalled among such powerful defense works.

While Horne's northern flank was mainly protected by the flooded Scarpe and Sensee Rivers, elements of Eighth Corps operated as a flank guard north of the river. Horne, as had become habitual for him, gave the main assault to General Currie and his Canadian Corps of four divisions, which would be supported by the British First and Fourth Divisions. Artillery support was generous, with fourteen brigades of field artillery and nine brigades of heavy artillery available, and nine tanks were also attached to each division. The offensive was also granted considerable air support. One Royal Air Force (RAF) brigade, consisting of five scout squadrons, one day bomber and one night bomber squadron, with two squadrons for tank-air cooperation and two squadrons for tactical ground support, was assigned to the First Army.[23] During late August, low clouds, mist, and rain hampered flying, but on every day except 28 August, patrols of fighters, cooperation aircraft, and ground-support aircraft worked hard to support the advance. The British artillery needed constant air spotting for its heavy bombardments, but due to the low cloud and frequent smoke screens, the cooperation aircraft had to fly at under two hundred feet, within the arc of the British shells and exposed to intense AA machine-gun fire, at the cost of heavy casualties.[24]

On 26 August, the Second and Third Divisions of the Canadian Corps attacked along a four-mile frontage aligned along the main Arras-Cambrai road, from Neuville Vitasse up to the river Scarpe. The numerous valleys leading to the Scarpe and Sensee, however, made the going difficult, and there was heavy fighting even before the first objective was reached. Even in the face of heavy machine-gun fire, the Canadians broke through the Monchy le Preux defenses and established themselves in the 1917 German trench system. With this attack, the First Army had successfully extended the scope of the British offensive. Now three armies were attacking along a forty-mile front, which increased the pressure on the German defenses. Furthermore, this attack had taken the Germans by surprise, leaving their reserves maldeployed to meet this new threat.

Meanwhile, the Third and Fourth Armies continued to make ground against the German rearguard defense, and on 27 and 28 August the Australian Corps, operating south of the Somme, was able to advance over four miles, pushing through numerous abandoned villages. Even with the use of skilled artillery and machine-gun tactics designed to delay the British advance as much as possible, the Germans had to abandon vast quantities of supplies and equipment, and aggressive British patrolling managed to bag a considerable number of prisoners caught in the act of retreat.[25]

On 27 August, the First Army ordered the Canadian Corps to attack in two stages to punch through the Fresnes-Rouvroy line. With heavy rainfall during the night, the ground was difficult, and the Germans, through determined resistance, were able to hold the majority of their positions. Indeed, the Germans mounted a series of counterattacks that day, which were held off only with difficulty.

Given the tough resistance encountered by the First Army, a note of caution began to intrude into Haig's thoughts concerning the forthcoming operation:

> I directed Horne not to attack the Drocourt-Queant line until he had all the means at hand to follow up success, in the hope of passing the Canal du Nord line and pushing forward the Cavalry Corps at once.[26]

The only problem with caution in the face of fierce German resistance was that if Horne was forced to revert to methodical tactics, the Germans would be able to reinforce the front, and the chance for a rapid advance would disappear. Only one day after the opening of the First Army's offensive, the opportunity for Haig's decisive stroke was beginning to fade away.

The First Army continued its attack the next day in order to close up to the Drocourt-Queant Line, but the Germans, realizing the importance of their positions, were not prepared to relinquish them without a tough fight. At 11 A.M., the Third Canadian Division pierced the Fresnes-Rouvroy Line by storming the high ground between the Cojeul and the Sensee, but the Second Canadian Division were driven out of this line by a counterattack.[27]

By the end of this day, both the Second and Third Canadian Divisions had been fought out. Over the past three days of action, they had advanced nearly five miles and captured 3,380 prisoners, but in the process they had lost 254 officers and 5,547 men as casualties.

These were very heavy casualties and gave a measure of just how fierce German resistance was. During the night, both were relieved by the First and Fourth Canadian Divisions.[28]

On the night of 28 August, the Germans crossed over to the east bank of the Somme, giving up nearly four miles of ground, allowing the Fourth Army to push up to the river while, farther north, the Germans evacuated Bapaume and withdrew between two and four thousand yards. North of Bapaume, however, the Germans held their ground.[29]

On 29 August, the First Army was engaged in minor operations to complete the capture of the Fresnes-Rouvroy Line and come within assaulting distance of the Drocourt-Queant Line.[30] Just as his divisions mopped up and readied themselves for the main attack, however, Currie was beginning to have a less sanguine view of the imminent operation. At a conference held at Canadian Corps Headquarters, he stated that

> there had been hard fighting yesterday (28th), and there was every prospect of heavy fighting against the D-Q line. He was operating so as to get into position for attack on this line and did not want to hustle his divisions forward onto this line without preparation. He considered 20 brigades, R.F.A., and 10 brigades, R.G.A. necessary to carry out this attack.[31]

Currie's desire to mount a methodical attack certainly contradicted Haig's original idea for a brisk offensive that would bounce the Germans out of their positions, but such ideas agreed with Haig's more cautious orders of 27 August. Currie's request for lavish artillery support (more artillery than he began the offensive with) also meant delay and made it less likely that his troops would meet an unorganized defense. While Currie now tended toward caution, Horne continued to support Haig's earlier and more optimistic assessment and believed that the Cavalry Corps might be able to "gain the high ground about BOURLON WOOD and to deny the crossings of the CANAL de l'ESCAUT, south of CAMRAI, to the enemy."[32]

Meanwhile, further south, Haig issued orders to the Fourth Army on 30 August that were designed to change the tempo of Rawlinson's advance. While the Fourth Army would press the German rearguards across the whole front, the principal pressure was to be exerted on the northern flank of the army. It was hoped that an attack here might turn the Germans off the line of the Somme without having to mount a hazardous river crossing. South of Peronne, the Australian Corps was not to "commit itself for the present to a general advance."[33]

These Fourth Army orders reveal Haig's intentions. The Germans, secure behind the barrier of the Somme, would be pinned by the limited advance of the Fourth Army so that when the First Army broke through the Drocourt-Queant Line and pushed southwest to operate against the German communications, the German troops on the Somme might be trapped against the river line. "Vigorous action" was needed by the Fourth Army to keep the Germans opposite them alert and in position, but major attacks were far from desirable, as this might precipitate a further German retreat, which would reduce the effectiveness of the First Army's operational turning movement. This shows the relative sophistication of the BEF's operations in 1918. Far from operating in a linear fashion, Haig was attempting to puncture the German lines at one critical point and doom the rest of the German forces to envelopment and capture. However, the problem remained whether the BEF could actually move fast and far enough to achieve these ambitious goals.

Just as Henry Rawlinson chafed at being given a supporting role to the First Army's operations further north, so John Monash, the commander of the Australian Corps, was unwilling to take a back seat to the operation on the northern flank of the Fourth Army. Monash persuaded Rawlinson to allow him to mount an attack to seize Mont St. Quentin, an eminence that commanded the line of the Somme and Peronne, even though Rawlinson "considered it over-bold to attempt with so weak a force." The hill was covered with thick belts of wire and old trench systems, and the western slopes could be swept with converging machine-gun fire. Given the opportunity to rest his corps and minimize casualties, Monash chose instead to push his troops yet further in offensive action.[34] While Monash believed that the Germans must not be given any chance to strengthen their defenses around Mont St. Quentin and Peronne, this attack, by driving the Germans away from the safety of the Somme, was actually counterproductive to Haig's overall plan.

While the Third and Fourth Armies were held up, the First Army made considerable progress on 29 August. With a well-organized barrage rolling in front of the Canadian troops, the Germans were taken by surprise and did not mount serious resistance. The German defenses of the Vis en Artois Switch Line were rapidly mopped up and held against numerous German counterattacks throughout the day.[35] Notwithstanding the tough German resistance over the previous days, the First Army was inexorably advancing toward the Drocourt-Queant Line, but not at the rapid pace desired by Haig.

Indeed, although the First Army had made good progress on 30 August, Haig's growing sense of caution meant that he now altered the entire purpose of the assault on the Drocourt-Queant Line. Given the stiffening of German resistance and the delays imposed on the British advance, he recognized that "the opportunity for the employment of a large body of cavalry to move forward and seize the CANAL DU NORD crossings as soon as the QUEANT-DROCOURT line is forced appears . . . to have passed." Instead, exploitation would be restricted to one cavalry regiment and one battalion of armored cars that would cause "all possible disorganization to the enemy's forces."[36] Haig had been forced to admit that the German army had not reached its breaking point. The assault on the Drocourt-Queant Line was no longer to be the decisive act that would see the final disintegration of the German army, but rather a conventional attack that might drive the Germans back but could not destroy them.

On 31 August, Monash's plan to seize Mont St. Quentin went ahead and became one of the great feats of arms of the Great War. The hill was stormed, catching its defenders by surprise, and Mont St. Quentin village was initially captured; but a strong German counterattack drove the Australians back. Heavy fighting continued throughout the day, and Monash ordered up more troops to complete the capture of Mont St. Quentin village and Peronne. The next day, the Australian Corps capitalized on their previous success with the capture of Peronne and Mont St. Quentin. The Fifth and Second Australian Divisions advanced at 6 A.M. through misty, drizzling rain and with heavy artillery support firing from both banks of the Somme. The Second Australian Division recaptured Mont St. Quentin village, and the troops consolidated on the east edge of Mont St. Quentin wood. Meanwhile, after heavy fighting, troops from the Fifth Australian Division crossed over Peronne's wide moat using two precarious footbridges, and by 8:45 A.M., most of the town was in Australian hands.[37] Monash's daring assault had captured a key German position and had rendered the whole of the German line along the Somme untenable. This made another German retreat inevitable even before the First Army launched its blow against the Drocourt-Queant Line.

Meanwhile, unaware of such developments, a Canadian Corps Operation Order was issued on 31 August that gave detailed plans for the Drocourt-Queant attack. The frontage for the attack was very narrow and aligned along the Arras-Cambrai road. The First Canadian Division on the right, with the Fourth Division and one brigade of the Fourth Canadian Division on the left, reinforced with "3rd Bde. Tank Corps, one regiment of Cavalry and [the] 17th Armored Car Battalion,"[38] were to punch through the defenses, after which the reserves, First Division, and the Independent Force would leapfrog through and then fan out to the north and south to roll up the line. The four objective lines would take the First Army through the Drocourt-Queant Line and onto the high ground overlooking the Canal du Nord. Further exploitation might reach beyond the Canal du Nord. Brutinell's Independent Force of Cavalry, Cyclists, and Armored Cars was "if successful to exploit success towards CAMBRAI and North and East of that town, moving forward as soon as the bridgehead has been occupied by infantry of the 1st and 4th Canadian Divisions." Since the attack frontage was narrow, great care was given to protect the flanks of the operation.[39] The plan sketched out what was to be a strikingly modern operation. The idea of penetrating defenses on a narrow frontage with heavy concentrations of artillery and infantry and then using reserves to fan out and cause maximum destruction would not look out of place in a plan of attack from the Second World War.

The importance placed on the attack can be judged from the fact that Currie received the lavish artillery support he had asked for, consisting of twenty brigades of field artillery and eleven of heavy artillery. The air support was equally impressive. Two squadrons were assigned to support the attacking divisions, while other squadrons were tasked specifically to deal with tank cooperation, enemy antitank guns, and the Cavalry Corps, and for the first time, all of the cooperation squadrons were coordinated from one headquarters. Detailed instructions were also given concerning the use of tanks in the operation. The night before the attack, FE2b aircraft of the night bombing squadron were to fly over the front lines on their way to and from their objectives to drown out the noise of the assembling tanks. Two companies of tanks were to be allotted to each attacking division, but although tanks would assist in the break-in and the mopping up of the Drocourt-Queant Line, the tanks were not to be used beyond this, except on the flanks near the Buissy Switch and Etaing.[40] This was a conscious decision to preserve tank strength for organized attacks against strong, fixed defensive lines, rather than expend them ineffectually in pursuit operations. Commanders were prepared to use lavish amounts of tank support for an important attack but, given the speed of the advance and the mechanical unreliability of the tanks, were not prepared to fritter this important resource away.

The attack on the Drocourt-Queant Line was meticulously planned, but as the First Army made its thorough and detailed preparations for the assault, Haig, with his earlier optimism now gone, actually had second thoughts about the attack. He noted in his diary on 31 August, "I sent Lawrence to see Horne and Currie and tell them that I have no wish to attack the Queant-Drocourt line, if they have any doubts about taking it."[41] This provides us with a very different picture not only of Haig but of the entire BEF Gone were the days of 1916, when massive casualties could be taken and simply replaced. We have already noted that two Canadian divisions had been pulled out of the line. Infantry—and particularly the high-quality Dominion divisions—were a wasting asset, and high casualties for no result simply could not be afforded. Not only were British preparations for an attack highly professional and thorough, but Haig was managing the manpower of the BEF much more carefully than he had done two years previously.

The next day, 1 September, just hours before the First Army launched its attack, Haig showed that his views on the strategic situation had changed dramatically. In a conversation with Byng, he mentioned that "owing to our shortage in men, I was opposed to doing more attacking than was absolutely necessary." This was a complete reversal of his previous arguments with Byng and of his 22 August "pursuit" order. Yet, when Lawrence mentioned the great success around Peronne and suggested sending a cavalry division to exploit the situation, Haig disagreed because he wanted

> to *wear out* the enemy by continually attacking him, and so to prevent him from settling into a strong position. The decisive moment will arrive when the Americans attack in force. The British Army must still be able to attack then, and to have the means of exploiting the victory and making it decisive. I therefore wished the Cavalry Corps to be kept as strong as possible. . . . By this procedure I hoped to have an efficient cavalry corps ready to act vigorously when the decisive moment comes and reap the fruits of victory.[42]

Even though the fighting around Peronne demonstrated the kind of weakness that, on 22 August, Haig had seen as proof that the Germans were about to crumble, he now returned to older ideas. Haig readopted his view that the British army was engaged in a wearing-out fight (even though this contradicted his advice to Byng), and now the decisive moment had shifted from the attack on the Drocourt-Queant Line to the opening of the forthcoming American offensive.

Haig's optimism and pessimism had fluctuated throughout the war as he waited for "the decisive moment." Of course, as the Soviet theorists of the interwar years later realized, this decisive moment—as one point fixed in time and space—would never come. Warfare had changed beyond recognition from Napoleon's day, and Haig had not entirely grasped this problem. Yet, at the same time, Haig and his commanders were groping toward an operational solution to their problems. By using three armies that launched offensives one after the other—as one offensive died down, another would begin—Haig and his army commanders were "continually attacking" the Germans and creating an intolerable pressure on German reserves and high command. But the decisive moment when he would release his cavalry to clinch a glorious victory would never come. Each time the German High Command feared that it could not hold its present positions any longer, it simply ordered another withdrawal to preserve the linear front and deny Haig and his armies the decisive moment they sought.

That night, Lawrence returned from his meeting with Horne and was able to tell Haig that "all were quite confident and his attack will go in at 3am to-morrow as planned"; but with his own fears about the attack calmed, Haig reacted badly to a telegram from Henry Wilson, the Chief of the Imperial General Staff (CIGS), marked "Personal," that expressed concern about the coming attack. Wilson warned Haig that the War Cabinet "would become anxious if we received heavy punishment in attacking the Hindenburg Line [meaning the Drocourt-Queant Line] WITHOUT SUCCESS." These were exactly the same concerns that had made Haig hesitate, but receiving such doubts from the CIGS made him angry:

> It is impossible for a C.I.G.S. to send a telegram of this nature to a C-in-C in the Field as a "personal" one. The Cabinet are ready to meddle and interfere in my plans in an underhand way, but do not dare openly say that they mean me to take the responsibility for any failure though ready to take credit for every success.[43]

It is hard not to feel at least some sympathy for Haig over this exchange. Haig could hardly have changed his plans or postponed this major attack just a few hours before it began. The plan had been discussed in great detail and already modified many times: it was far too late to change it further. Like any Great War commander, with the plan fixed and all preparations made, Haig could only wait for the outcome.

On 2 September 1918, at 5 A.M., just as dawn broke, the First Army began its assault on the Drocourt-Queant Line. On the right flank, the First Canadian Division attacked on a sixteen-hundred-yard frontage with eight tanks in support. Even though there was still fighting for the start line on the left flank, and the open right flank caused difficulties throughout the day, in the center the advancing troops encountered only light hostile artillery fire and little resistance. Indeed, once the Canadians reached the front German trenches, large numbers of the German defenders simply surrendered, and by 7:30 A.M. the entire front and support system had been captured. After consolidating, the Canadians surprised and captured a full battalion's worth of prisoners in Cagnicourt and, in a dash across two thousand yards of open country, managed to seize the Buissy Switch Line. German resistance here stiffened considerably, and many of the supporting tanks were knocked out by German artillery fire and concealed field guns. By 6 P.M., the open right flank was finally sealed by supporting British troops, and by 11 P.M., the entire Buissy Switch Line was in Canadian hands.

The Fourth Canadian Division met with pockets of determined resistance in capturing the front system but was still able to reach its initial objectives on time. In continuing the advance, however, the troops crossing the long crest near Mont Dury were exposed to intense machine-gun fire, and it was only with grim determination that the area was cleared. On the division's left, intact barbed-wire entanglements caused considerable difficulty, and it was only with bitter fighting that the division was able to wrest control of Dury from the Germans. In the second-phase attack, made at 8 A.M., very little headway was made against Villers-les-Cagnicourt, and by mid-afternoon all the brigades in the front line were held up by strong

German resistance. On the left, the Fourth Division captured its section of the Drocourt-Queant Line but was unable to go farther.

Meanwhile, the Independent Force had made numerous attempts to push along the Cambrai road, without success. Much of the difficulty was caused by a one-thousand-yard lane astride the Arras-Cambrai road that had been left clear of artillery fire to allow the Independent Force to advance unhindered in the exploitation phase. Unfortunately, this also meant that the German defenders were unsuppressed and able to frustrate any further attempts to advance. Although the First Army had experienced considerable early success, exploitation of these gains had proved difficult in the face of increasingly determined German opposition.

The British effort in the air over the First Army also met with difficulties. From the start of the attack, the RAF made a major effort, and the ground-attack squadrons flew flights in rotation throughout the morning, with as many as ninety aircraft over the battle area at any one time, causing considerable disruption to the German defenders. The German air force reacted vigorously, however, and large fighter formations inflicted heavy RAF losses during the morning. There were two intense periods of dogfighting between the German and British fighter patrols around Marquion, and German fighters managed to attack the slow two-seater aircraft of the cooperation squadrons, which lost fifteen aircraft from four squadrons. Total British air losses were heavy, with thirty-six aircraft lost for the loss of only one German aircraft.[44] During the late morning and afternoon, as Haig noted, the "enemy had local superiority in the air, and prevented our airplanes from observing over Ecourt St Quentin,"[45] and German fighters were able to harass the advancing British troops. This limited—and temporary—control of the air, however, could not reverse the serious defeat that the First Army had inflicted on the Germans.

By the end of 2 September, the First Army had punched clean through the very strong defenses of the Drocourt-Queant Line over a frontage of seven thousand yards and had taken more than eight thousand prisoners. Meanwhile, the Fourth and Third Armies had continued their crumbling of the Winter Line. Since 24 August, the British armies had, with hard fighting, pushed through the old Somme battlefields and then come up against the prepared positions of the Winter Line. But these fortifications, meant to hold the British for months until 1919, had held only for a matter of days. By midday on 2 September, the German High Command was forced to order the abandonment of the Winter Line, and begin the fifteen-mile withdrawal to the Hindenburg Line proper—the line they had held before the start of their March offensive. The bankruptcy of Ludendorff's strategy was becoming apparent, and little was left to the Germans but continued retreat, in the vain hope that the British would not follow up rapidly.

As the events of 2 September proved, Haig's decision not to commit the Cavalry Corps had been correct; the German army had not disintegrated and had still put up tough resistance. Yet Haig in his diary concentrated on the example of the German surrenders early in the day: "Discipline in the German Army seemed to have gone—if this is true, then the end cannot now be far off, I think. To-day's battle has truly been a great and glorious success."[46] Here was an echo of his optimism of mid-August. The battle on 2 September *was* a "great and glorious success," but Haig's grandiose scheme for his First Army that would result in a decisive victory had come to naught.

Even after the British pursued aggressively and punched through the main German defenses of the Hindenburg Line on 27 September, the pattern of events repeated itself. The Germans withdrew once again, leaving the British unable to exploit their success. This highlights the difficulties inherent in the BEF's attempt to restore mobile warfare and seek a "decisive battle." The problems in getting this huge and complex machine to move fast enough and far enough to achieve a breakout were too great. In essence, the BEF was straining against the limits of technology as they existed in 1918. The BEF's operations, in the use of its armies in cooperation; in its use of surprise and deception to deceive the Germans of the main blow; in its use of aircraft, tanks, artillery, and infantry in combined-arms attacks—and indeed in the sophisticated and professional planning of offensives—all pointed toward the future. The BEF could now force the Germans back through its skilled use of combined arms and improved technology. But no matter how many tactical battles the BEF could win, given the existing technology and no reliable mechanized force that could penetrate deeply behind German lines, the British were incapable of winning an operational victory. Haig's only exploitation force—the Cavalry Corps—could not be used lightly, and its effectiveness was drastically limited through its continued use of horsed cavalry. With only fourteen thousand men in its ranks, it was a one-shot weapon, and Haig waited anxiously for the moment to launch it in a decisive action that never came. The lack of what the Soviets later termed a *second echelon*—and the reliable mechanized and armored vehicles that formed them—meant that the full exploitation of a victory was beyond the BEF in 1918.[47]

But far from the operations of the BEF demonstrating the bankruptcy of the "bourgeois military system"—or the futility of the fighting in 1918--these last offensives clearly show Haig and his army at the height of their powers. The foundations of modern maneuver warfare and of operational art had been soundly laid, but it took another generation of generals to put them into practice.

NOTES

1. Michael Howard, *Studies in War and Peace* (London: Temple Smith, 1970), 36.

2. John Keegan, *The First World War* (London: Hutchinson, 1998), 315.

3. Major General Sir F. Maurice, *The Last Four Months: The End of the War in the West* (London: Cassell and Company, 1919), 249–50.

4. See J. F. C. Fuller, *On Future Warfare* (London: Sifton Praed, 1928); B. H. Liddell Hart, *The Decisive Wars of History: A Study in Strategy* (London: G. Bell, 1929) 159–231.

5. H. Guderian, *Achtung—Panzer!* (London: Arms and Armour, 1992).

6. G. Isserson, "The Evolution of Operational Art," trans. Harold S. Orenstein, in *The Evolution of Soviet Operational Art, 1927–1991: The*

Documentary Basis, vol. 1: Operational Art, 1927–1964 (London: Frank Cass, 1995), 76.

7. C. von Clausewitz, *On War* ed. and trans. Michael Howard and Peter Paret (Princeton: Princeton University Press, 1976), 248.

8. Isserson, "Evolution of Operational Art," 46.

9. Ibid.

10. David M. Glantz, *Soviet Military Operational Art: In Pursuit of Deep Battle* (London: Frank Cass, 1991), 18–27.

11. J. P. Harris, with Niall Barr, *Amiens to the Armistice: The BEF in the Hundred Days Campaign 8 August—11 November 1918* (London: Brassey's, 1998), 23–57.

12. Ibid., 103–4.

13. Tim Travers, *The Killing Ground: The British Army, the Western Front and the Emergence of Modern Warfare 1900–1918* (London: Allen and Unwin, 1987), 85–100.

14. Douglas Haig to Army Commanders, OAD 911, 22 August 1918, War Office (WO) 158/2415225, Public Record Office (PRO), Kew.

15. Field Marshal Earl Haig's Diary, 23 August 1918, WO 256/35, PRO.

16. Annexe to Third Army Intelligence Summary no.1153, 2 (e), WO 157/165, PRO.

17. Third Army Intelligence Summary, 27–28 August, WO 157/164, PRO.

18. 24 August 1918, Sir Herbert Lawrence to Sir Henry Horne, OAD 907/13, WO 158/241, PRO.

19. Haig's Diary, 25 August 1918, WO 256/35, PRO.

20. S. Badsey, "Cavalry and the Development of Breakthrough Doctrine," in P. Griffith, ed., *British Fighting Methods* (London: Frank Cass, 1996), 138–74.

21. Haig's Diary, 25 August 1918, WO 256/35, PRO.

22. "First Army Report to All Corps Commanders," 31 August 1918, from a summary of information on 30 August, WO 158/191, PRO.

23. H. A. Jones, *The War in the Air* (Oxford: Clarendon Press, 1937), 485.

24. S. F. Wise, *Canadian Airmen and the First World War* (Toronto: University of Toronto Press, 1980), 552–53.

25. A. A. Montgomery, *The Story of the Fourth Army in the Battles of the Hundred Days August 8th to November 1918* (London: Hodder and Stoughton, 1919), 86.

26. Haig's Diary, 27 August 1918, WO 256/35, PRO.

27. G. W. L. Nicholson, *Canadian Expeditionary Force 1914–19* (Ottawa: Queens Printer, 1962), 431–32.

28. Ibid., 432.

29. Montgomery, *Story of the Fourth Army,* 93–94.

30. Nicholson, *Canadian Expeditionary Force,* 432–36.

31. Notes on Conference at Canadian Corps Headquarters, 29 August 1918, WO 95/178, PRO.

32. Ibid.

33. HQ Fourth Army to Corps, 30 August 1918, WO 158/241, PRO.

34. Robin Prior and Trevor Wilson, *Command on the Western Front: The Military Career of Sir Henry Rawlinson 1914–18* (Oxford: Blackwell, 1992), 342; Montgomery, *Story of the Fourth Army,* 97–98.

35. Nicholson, *Canadian Expeditionary Force,* 432–36.

36. Notes on Conference at Canadian Corps Headquarters, 30 August 1918, WO 95/178, PRO.

37. Montgomery, *Story of the Fourth Army,* 99–103.

38. First Army Order no. 262, 31 August 1918, WO 158/191, PRO.

39. Ibid.

40. Canadian Corps Operation Order no. 234, 31 August 1918, WO 95/179, PRO.

41. Haig's Diary, 31 August 1918, WO 256/35, PRO.

42. Haig's Diary, 1 September 1918, WO 256/36, PRO.

43. Ibid.

44. Wise, *Canadian Armies,* 553–54.

45. Haig's Diary, 2 September 1918, WO 256/36, PRO.

46. Ibid.

47. C. J. Dick, "The Operational Employment of Soviet Armour in the Great Patriotic War," in *Armoured Warfare,* ed. J. P. Harris and F. Toase, (London: Batsford, 1990), 88–123.

CHAPTER 9

Scientists at War

The Development of Radar and Jet Propulsion in Britain

Eric Bobo

> The Story of the human race is War. Except for brief and precarious interludes, there has never been peace in the world; and before history began, murderous strife was universal and unending.[1]

This remark, taken from Winston Churchill's 1924 essay "Shall We All Commit Suicide?" contains his bleak vision of the future of warfare. Continuing, Churchill remarked that the only reason man has survived thus far was that "up to the present time the means of destruction at the disposal of man have not kept pace with his ferocity."[2] But now, thanks mostly to "Science on the Side of War," humanity confronts for the first time the possibility of total annihilation: "It was not until the dawn of the twentieth century of the Christian era that War really began to enter into its kingdom as the potential destroyer of the human race."[3]

Churchill certainly understood the role science had already played on the battlefield. Tanks, chemical warfare, airplanes, submarines, and machine guns were all introduced during the First World War, and all were products of the application of science to essentially offensive warfare. Churchill expected this trend to continue, which would result in deadlier weapons that, as he mentioned, held the potential to destroy humanity. Consider his predictions concerning explosives:

> Might not a bomb no bigger than an orange be found to possess a secret power to destroy a whole block of buildings—nay to concentrate the force of a thousand tons of cordite and blast a township at a stroke? Could not explosives even of the existing type be guided automatically in flying machines by wireless or other rays, without a human pilot, in ceaseless procession on a hostile city, arsenal, camp, or dockyard?[4]

These statements concerning the future application of science to offensive warfare certainly make Churchill appear to have had remarkable prescience. Unfortunately, he would see most of his fears take shape during the course of the Second World War. Amid the darkest days of that conflict, Churchill became Britain's prime minister and found himself faced with the threat of total warfare prosecuted against his own homeland. For the survival of his own nation, then, Churchill turned to science, the force he realized might destroy the world, for its own defense.

Thus, as Great Britain mobilized for war, so did her scientists. These men provided key advancements in many areas that had a significant impact on the war effort. This is not to say that these advancements came without a heavy price. As Carl von Clausewitz noted, all warfare depends on a trinity of support: people, army, and government. Any "theory that ignores any one of them . . . would conflict with reality to such an extent that for this reason alone it would be totally useless."[5] Despite their necessary interrelationship, the stress of war often sets these segments of society against one another, thus harming the overall goal of winning the conflict. Nowhere is this failure more evident than in the relationship between British scientists and their own government during World War II. It is understood today that modern warfare demands technological advancement. Without it, any military force deteriorates and crumbles in the face of a new, modern, and technologically superior enemy. Therefore, states have a duty to encourage advancement and technological innovation of their armed services, but when a state is faced with a crisis, technological innovation can often be misapplied by the leg of the trinity least able to do the job correctly: the government.

> One of the most bizarre features of any advanced industrial society in our time is that the cardinal choices have to be made by a handful of men: in secret: and, at least in legal form, by men who cannot have a first-hand knowledge of what those choices depend upon or what their results may be.[6]

Charles Percy Snow certainly understood the political ramifications of the wartime necessity of mobilizing science as part of the total war equation. He worked with some of the brightest scientific minds in Great Britain during the Second World War and understood that science and technological innovation were integral parts of the war effort. He also understood, however, that the fog of war could strain the relationship between those who produce advancement and those who evaluate and apply those advancements.

Scientific research and application, whether for offensive or defensive purposes, faced tremendous obstacles as Great Britain moved steadily toward war. Any advancement required the government, the military, and scientists to work in harmony. Examples of such cooperation and success, however, are few in the history of warfare. In the main, scientific and technological development has followed two paths during wartime. The first path sees the correct application of science to warfare, usually through the efforts of a dogged scientist or an enlightened governmental official who nurtures an invention from its inception, provides it with the necessary funding, and sees it to fruition. Other times, advancement becomes lost along a dark path that leads into the bureaucratic abyss of mismanagement, miscommunication, and misapplication.

This model is best illustrated by two examples of British scientific achievement developed before and during the Second World War, and in fact, both were located within the confines of air defense preparations. The first example took a more enlightened path of development and proved to be of extreme importance in the defense of the island. Without the development of what the British termed RDF, or RADAR, air defense against Luftwaffe bombers may have been impossible.[7] The second, one of the most revolutionary technological achievements of World War II, struggled for acceptance, funding, and manpower. The development of jet propulsion, and its subsequent delayed application to air defense, can be traced to the efforts of one man: Frank Whittle.

The advent of airpower during the Great War and the strategic bombing doctrines of the interwar period caused many in Britain great worry. Airpower could easily overfly the English Channel, robbing Britain of her traditional defense. As war clouds loomed in Europe during the 1930s, the mood of the British government was decidedly gloomy; to the point that ranking ministers questioned whether or not Britain could be defended against aerial attack at all. Stanley Baldwin, a former prime minister who ironically would be returned to that post from 1935 to 1937, stated before the House of Commons in November 1932 what he perceived to be the bleak future for his homeland should a conflict materialize:

> I think it is well for the man in the street to realize that there is no power on earth that can protect him from being bombed. Whatever people may tell him, the bomber will always get through.[8]

The mood only worsened after the 1934 summer air exercises. The Royal Air Force carried out mock raids against British air defenses, and the final evaluations of the exercises revealed serious flaws in a system designed to keep enemy planes out of British industrial and governmental centers. Had this been an actual attack, the Air Ministry would have been destroyed in the first wave, followed closely by the Houses of Parliament.[9] Air Chief Marshall Sir Arthur Longmore, who commanded one of the defense zones in the air exercises, wrote that "successful interception by fighters of raiding bombers required more accurate information from the ground as to movements of hostile formations than was at the time available."[10] At the time, then, Baldwin's statements concerning the invincibility of the bomber appeared true; but without knowing it, Longmore had stumbled onto the answer to this problem. At the time of the exercises, RAF leadership had no way of knowing the locations and numbers of attacking planes. If accurate information could be obtained early in an enemy attack, RAF fighters could launch a coordinated response. No device existed that could provide this information at the time, but dedicated scientists and government officials investigating the problem of air defense soon discovered the secret to early warning.

The poor defense performance prompted two major reactions. The most vocal reaction came from members of the government who protested openly against this intolerable situation. Winston Churchill, while leading the movement to rearm, denounced the "cursed, hellish invention and development of war from the air" which had placed his homeland in this vulnerable position.[11] More important, Churchill's closest adviser on scientific matters, Frederick Lindemann,[12] publicly railed against the gloomy forecast provided by the government in an August 8, 1934, letter to *The Times*:

> In a debate in the House of Commons on Monday on the proposed expansion of our Air Forces, it seemed to be taken for granted on all sides that there is, and can be, no defense against bombing aeroplanes and that we must rely entirely on counter-attack and reprisals. That there is at present no means of preventing hostile bombers from depositing their loads of explosives, incendiary materials, gases, or bacteria upon their objectives I believe to be true; that no method can be devised to safeguard great centres of population from such a fate appears to me profoundly improbable.

Continuing, Lindemann outlined a plan of action to counter the threat from above:

> If no protective contrivance can be found and we are reduced to a policy of reprisals, the temptation to be the "quickest on the draw" will be tremendous. It seems not too much to say that bombing aeroplanes in the hands of gangster Governments might jeopardise the whole future of our Western civilisation. To adopt a defeatist attitude in the face of such a threat is inexcusable until it has been shown that

> all the resources of science and invention have been exhausted. The problem is far too important and too urgent to be left to the casual endeavours of individuals or departments. The whole weight and influence of the Government should be thrown into the scale to endeavour to find a solution."[13]

Lindemann had succinctly outlined this high-stakes game. Being a man of science, Lindemann understood that not all avenues of defense had been explored. In his mind, exploring scientific solutions was favorable to an acceptance of the inevitable.

The second reaction to the air defense problem came from within the British Air Ministry itself. A young civil servant, A. P. Rowe, labored on the headquarters staff of the ministry's director of scientific research. Rowe decided to explore the problem of strategic bombing, and he examined the fifty-three files on air defense available at the ministry. From his research, Rowe made a simple but startling conclusion:

> It was clear that the Air Staff had given conscious thought and effort to the design of fighter aircraft, to methods of using them without early warning or balloon defenses. It was also clear however that little or no effort had been made to call on science to find a way out. I therefore wrote a memorandum summarizing the unhappy position and proposing that the Director of Scientific research should tell the Secretary of State for Air of the dangers ahead. The memorandum said that unless science evolved some new method of aiding air defense, we were likely to lose the next war if it started within ten years.[14]

Simply put, scientific research and development had been concentrated on offensive weapons, while little or no effort was spent on making certain that defensive capabilities kept pace with offensive innovation. Rowe forwarded his observations to Henry Wimperis, the director of scientific research for the Air Ministry, who quickly took action and set into motion a chain of events that would lead to the formation of the Committee for the Scientific Study of Air Defense. This group's mandate was to "consider how far advances in scientific and technical knowledge can be used to strengthen the present methods of defense against hostile aircraft."[15]

The selection of Henry Thomas Tizard to chair this group of scientists was instrumental in laying the groundwork for victory in the Battle of Britain. Born in 1885 at Gillingham in Kent to a navy officer, Tizard was an excellent scientist with an Oxford education in chemistry. During the Great War, Tizard learned to fly airplanes, not to engage in dogfights with the enemy but rather to investigate how to make bombing more effective.[16] Tizard's remained active in the defense research after the war. He served in the Department of Scientific and Industrial Research during the German military buildup of the 1930s and later as chairman of the Aeronautical Research Committee. From his new position as chair of the Committee for the Scientific Study of Air Defense, Tizard gathered together the greatest scientific minds in Britain for the sole purpose of developing a new policy for the defense of the home islands.

The group, which Tizard deliberately kept small, first met on 28 January 1935. He asked both Wimperis and Rowe to serve, as well as A. V. Hill and P. M. S. Blackett, both imminent scientists in their own right. Hill was one of the most distinguished physiologists in the world and had won a Nobel Prize in 1922. Blackett, a noted physicist, later went on to win his own Nobel Prize. Tizard had chosen the members of his committee primarily due to their scientific knowledge. However, Tizard was a realist and made certain that these men also had important political[17] and military connections. Hill had served in the First World War and edited a classic work on antiaircraft gunnery, while Blackett had been a professional naval officer before he turned to science.[18]

The men Tizard gathered together now began a tremendous undertaking. Their responsibility departed radically from the previous roles given to scientists working within the government. Instead of researching new ways to attack or defend, they concerned themselves with learning whether or not Britain could be defended from air attack at all.[19] Initially, the committee was interested in ways to use radio waves to produce a "death ray," a pulse of energy to destroy either an airplane's electronics or the pilot's flying capabilities. In an effort to ascertain whether such a weapon was feasible, Wimperis contacted Sir Robert Watson-Watt.

Robert Watson-Watt graduated from University College, Dundee, in 1912. In 1915, he worked as a meteorologist at the Royal Aircraft Factory at Farnborough. While there, he believed radio waves could be used to locate thunderstorms and provide pilots with an advanced warning of inclement weather. In 1927, Watson-Watt became superintendent of a small station of the National Physical Laboratory at Slough, and in 1933, he became superintendent of a new radio department at the National Physical Laboratory, Teddington.

Watson-Watt, intrigued by Wimperis's query, quickly realized that the energy requirements for such a task would be astronomical. Watson-Watt and his aides then adjusted their calculations to determine whether or not a plane would *reflect* radio waves. If they did, then these reflected waves could be measured and used to determine the position of planes in the air, even though they might be beyond the range of human vision. In February 1935, Watson-Watt and his assistant Arnold Wilkins hurriedly formulated a paper, "Detection and Location of Aircraft by Radio Methods," outlining his experimental results that indicated airplanes could be located with the use of radio waves. On February 14, 1935, Wimperis and Tizard discussed the paper, and the next day Wimperis suggested to Air Chief Marshall Hugh Dowding that £10,000 should be given to Watson-Watt to fund further experiments. He outlined his proposal in the following manner:

> Mr. Tizard and I have sufficient confidence in Mr. Watson-Watt's work not to regard this as a necessary preliminary, but should you prefer such a check made I can arrange for a metal aeroplane to fly from the Royal Aircraft Establishment, Farnborough, to Ditton Park (Slough) in about ten days, by which time Mr. Watt would have sufficient time to modify the Ditton Park transmitter so as to make it better suited for this variation of its normal duties.[20]

On 26 February 1935 the test was carried out, and its success marked the beginning stage for what was to become, in British terms, RDF.

The project developed quickly and efficiently in the early months of 1935. As a result, shortly after its inception in January 1935, Tizard's committee found itself in charge of the "raw material from which the weapon of radar was to be hewn."[21] In its early stages, very few people knew or understood the importance of the success of the first trial, and Tizard's committee had managed to avoid governmental interference. The Air Ministry itself had taken the lead, appointed Tizard to one of its committees, and had so far shouldered the expenses. Therefore, the project remained sheltered and out of the bureaucratic eye. All that would change in July 1935.

Politics invaded what had been an apolitical situation when Professor Frederick Lindemann managed to secure an appointment to Tizard's committee in July. Born in Baden-Baden, Germany, in 1886, Lindemann made his name as a physicist. He crossed paths with Tizard in Berlin, where they both were studying at the University of Berlin under Professor Hermann Nernst. The men became friends, but as Tizard noted, "There was always something about him which prevented intimacy. . . . Still, we remained close friends for over twenty-five years, but after 1936 he became a bitter enemy."[22] After finishing his education, Lindemann worked at the Royal Air Force Laboratory at Farnborough from 1915 to 1918, where he learned to fly in order to investigate the effects of aircraft spin.

After World War I closed, Lindemann served as director of the Clarendon Laboratory, Oxford, from 1919 to 1956. During this time, he befriended Churchill and often drove from Oxford to Chartwell to visit him on the weekends. From these visits and discussions, Lindemann became a trusted adviser. Once Churchill rose to power, the professor found himself advising him on scientific matters and tied to the political world through his association with Churchill.[23]

Once a part of Tizard's committee, Lindemann tried to push several of his ideas into discussion by sending a memorandum early in July 1935 to the committee chair, outlining his position on air defense. The bizarre nature of Lindemann's suggestions threatened to sidetrack the committee and prompted Tizard to remark that his counterpart's ideas "depend largely on two pre-conceived notions, both wrong."[24] First, Lindemann felt that artillery barrage was useless, estimating that it would take ninety thousand shells to bring down one airplane. Since airplanes were fragile machines, as they needed to be light to fly, he posited that dropping small mines of high explosives attached to parachutes would be more effective. His second suggestion involved the production of a cloud of some sort that could be placed in the path of an airplane that would foul the engine, thus rendering it ineffective. Most important, Lindemann felt that early detection of attacking planes could never be accomplished with any efficiency.[25]

Tizard, determined to develop radar, clashed with Lindemann's desire for other, more tangible defenses. The feud lasted for an entire year. At each meeting, Tizard pushed for radar while Lindemann demanded other alternatives. Finally, in July 1936, Blackett and Hill resigned in disgust and blamed Lindemann for the division in the committee. The committee was reconstituted shortly thereafter, with Lindemann noticeably absent.[26] Tizard had won the battle for now but perhaps had lost the political war. Tizard was a scientist involved in a political world. Lindemann by contrast, was a canny politician. By attaching himself to Churchill, Lindemann assured himself of an important role in the scientific war.

Despite the governmental intrigue surrounding the Tizard Committee, radar managed to develop under the watchful eye of scientists, not bureaucrats. By September 1935, Watson-Watt had managed to boost power, successfully find and locate planes, and gauge the height of an incoming plane to within one thousand feet. The project had evolved with such speed that the British government approved funding for the construction of a chain of five RDF stations twenty-five miles apart along Britain's east coast. The success of the project generated more governmental attention, and it soon began to fall under the direction of the government. Watson-Watt became director of communications development for the Air Ministry in 1937, which assured governmental control over the emerging technology.[27] By 1940, the project had developed the cavity magnetron, a small and efficient source of microwaves. Resistant to jamming and much more powerful than regular radio waves, microwaves generated in a small device such as the magnetron proved to be the ultimate breakthrough in British radio detection.[28]

Necessity had driven the British to prepare for their aerial defense, and it proved to be one of the decisive moves of the preparedness struggles of the 1930s. Despite political intrigue that threatened the overall success of the RDF project, dedicated scientists and governmental officials shepherded a key technological advancement through its early days. The radar watch permitted a coherent, organized defense of the island without standing patrols that placed strains on pilots, ground crews, and machines. The radar barrier provided the British with some security against surprise attacks and allowed the RAF to mass a defense without guesswork. Overall, radar provided the RAF with the all the additional information concerning the movements of hostile aircraft that Air Chief Marshall Longmore had clamored for in 1934. Without the advent of radar, and more important, its correct application to the defense of Great Britain, the eventual outcome of the war may have been different.[29]

While the development of radar gave the British an edge in the coming air war, the defense of the home islands still fell to the men and machines of the Royal Air Force. Scientific development and industrial might worked together to provide RAF pilots with some of the finest aircraft of World War II. Much has been written about the Supermarine Spitfire and its contributions to the war effort and the defense of British air space, but its success may have doomed other projects that held even higher promise. Mired in technical and financial problems, another fighter powered by a revolutionary propulsion system struggled for acceptance, manpower, and testing. Jet-propulsion technology promised faster planes with higher operating ceilings and climb rates. For pilots, that meant a survivability edge. For some in the government enamored by these possibilities, it meant a war-winning machine, a weapon of exploitation to drive the enemy from the skies. For others, it was a monumental waste of time, effort, and funds. Unlike radar, the development of jet propulsion faced three tremendous difficulties: governmental acceptance, financial woes, and engineering problems.

Born in 1907, Frank Whittle's boyhood coincided with the development of aviation, including the use of aircraft in the First World War and the formation of the RAF in 1918. Whittle grew up in an engineering background (his father owned a general engineering business in Leamington Spa), but he was particularly fascinated by aircraft. He joined the RAF in 1923 as an apprentice aircraft fitter. Later, he was selected for pilot training at the RAF Staff College, Cranwell, where he was soon flying solo.[30] While at Cranwell, still only twenty-one years of age, Whittle began to consider the possibilities of jet propulsion as applied to aircraft and deduced that "if very high speeds were to be combined with long range, it would be necessary to fly at very great heights where the low air density would greatly reduce resistance in proportion to speed."[31] Whittle set to work on the problem, and by 1929 he had made much progress, both theoretical and practical, on the development of the turbojet engine.

Unfortunately for Whittle, his thoughts and ideas proved almost too revolutionary for the time. The aircraft industry knew only piston engines, the performance of which was improving steadily as the war approached. Whittle's proposals for an alternate form of propulsion, while scientifically sound, forced the establishment to choose between developing the already-established piston engine further or abandoning working machines in favor of Whittle's radical proposals.[32] Nevertheless, Whittle took his invention to W. L. Tweedie, a technical officer in the directorate of scientific research at the Air Ministry, who then referred him to A. A. Griffith. The Air Ministry subsequently declined to give developmental funding for Whittle's proposals, citing limitations in materials.[33] The government, especially the Air Ministry, had missed a chance to make a significant technological breakthrough. There was little doubt that Whittle's engine was not practicable in 1929, yet some form of research program should have been established to explore further its developmental possibilities.[34]

The failure of the government to provide any research-and-development funding forced Whittle to approach various industrial firms, including the British Thomson-Houston Company, Armstrong-Siddely, and Bristol Aircraft, in an attempt to finance the engine. All were put off by the practical difficulties and high costs involved in making the engine. Without any backing, Whittle applied for and received a patent for his engine in January 1930.[35]

By 1935, Whittle had almost given up hope that his ideas would ever be developed. He was then at Cambridge, where the RAF had sent him to work for a degree in mechanical engineering. While there, he excelled in his studies but met with financial despair. His original patent came due in May 1935, but he was unable to pay the required £5 and his patent lapsed. His luck changed, however, when two ex-RAF officers, eager to secure funding for the development of the engine, approached Whittle. With the aid of the investing firm of Falk and Partners, the group secured funding and formed Power Jets, Ltd.[36]

Whittle could now proceed with limited funding, but the formation of Power Jets by no means solved all the problems facing the designer. The limited assistance from investors meant that testing and design work could begin, but Whittle had to proceed with caution. At this point, two major problems faced Whittle and his partners at Power Jets. The original patent had lapsed and could not be renewed, so Power Jets had to be content in patenting improvements to the original design. Whittle, concerned by this, ordered that "in no circumstances should anybody connected with the aircraft industry be approached . . . if one of the big aircraft firms became involved, then the only thing which could stop them from proceeding independently was whatever patent protection he and the others happened to have."[37] In other words, Whittle now concerned himself not only with production but also with the protection of his and the investors' financial gain. Financial and patent issues would continue to play a detrimental role in the development of the engine.

Second, Whittle faced a technological and engineering nightmare. While his plans seemed feasible on paper, in reality he faced tremendous difficulties. Jet propulsion as Whittle imagined it required that compressed air be heated in a combustion chamber by the burning of fuel. This heated air would then turn a turbine, providing power for the compressor. The remaining air, forced out of the back of the engine, would then provide the thrust to propel the airplane. Problems arose, though, because the high temperatures coupled with extreme revolutions per minute tended to destroy the turbine assembly. Whittle was forced to solve these engineering problems without significant governmental aid or funding, and this cost precious time in the engine's developmental phases.

Work continued throughout 1936 on the manufacture of the required engine parts. The British Thomson-Houston Company (BTH) did most of the engineering work, and on 12 April 1937 the first engine was tested. According to Whittle, the engine made a "shriek like an air raid siren" so that all the onlookers ran for cover.[38] While the tests were not deemed a success, the engine did produce power and the compressors and turbines worked at limited efficiency. Whittle's ideas worked, but they still had to be refined. Acceptance of his breakthrough, however, proved to be an even greater challenge.

Power Jets still needed governmental involvement in the project, especially for funding and testing facilities. The Air Ministry asked Dr. A. A. Griffith to evaluate Whittle's results, and he issued his report in February 1937:

> In its present form the proposed jet propulsion system cannot compete with the conventional power plant in any case where economical flight is demanded. It is of value only for special purposes, such as the attainment of high speed or high altitude for a short time, in cases where take-off requirements are not stringent. In order that the pro-

> posed system may become a competitor in the field of economical flight, a large improvement, of the order of at least 50–100 percent, must be made in the ratio of take-off thrust to power plant weight.[39]

Ironically, Griffith submitted his report to an Air Ministry subcommittee chaired by Henry Thomas Tizard. As noted earlier, Tizard had worked tirelessly to see through the development of radar despite beliefs in the government that defense was somewhat beside the point, for the "bomber will always get through." In the main, the government relied on the construction of its own strategic bombing force to act as a deterrent to enemy attack. Such a policy required the development of bombers with heavy load capabilities and longer ranges. Tizard, somewhat more defensively minded, realized that a high-altitude, fast jet fighter would give the RAF a tremendous advantage over the lumbering bombers of any attacking force. If radar could pinpoint an incoming attack, swift, short-range planes could be sent into the skies to destroy the heavier and slower enemy bombers. Though persistent in his advocacy of the jet engine, Tizard could not yet sway governmental opinion, and the Air Ministry chose to rely on conventional reciprocating engines rather than the experimental Whittle engine. The ministry granted only limited funding to Power Jets for results of a first series of tests and a limited second series run of tests. In total, ministry support came to less than £2,000.[40]

Power Jets continued to labor under a financial cloud throughout 1937, culminating with a near financial collapse in the summer of 1938. Whittle and the design team moved their facilities to an abandoned BTH foundry at Ladyworth and continued to test the engine there until its main parts, including the turbine, disintegrated in June 1938. Whittle had the engine rebuilt because they could not afford to replace all the parts, and testing of this new version began in October 1938, achieving higher levels of success.[41]

During this period of semi-independent testing, Whittle and the Power Jets design team had managed to produce a working jet engine, albeit a temperamental one. There is little doubt that much more progress could have been made, and more disasters averted, had the Air Ministry provided the effort with any type of major funding. Even the limited successes achieved by Power Jets during this period, however, began to win Whittle more advocates within the scientific and political communities. Tizard was by now more convinced than ever that jet propulsion was worthy of further study, and he redoubled his efforts toward securing additional funding for the project.[42]

Others soon joined Tizard in his campaign. In June 1939, David R. Pye, director of scientific research for the Air Ministry, visited Whittle and observed the engine at work. Pye, who had been very skeptical, immediately voiced his enthusiasm for Whittle's achievement. Two weeks later, Whittle received a contract for a flight engine. In January 1940, Air Vice Marshall A. W. Tedder and Henry Tizard visited Whittle to witness a demonstration. The engine performed well, and Tizard wryly remarked, "A demonstration which does not break down in my presence is a production job."[43] Whittle had achieved great success despite a lack of support and funding from the government, and he had managed to secure funding and some semblance of acceptance within the governmental community. Power Jets, however, soon faced another challenge as a result of its success.

Once the decision was made to enter production, the fate of Power Jets proved very uncertain. After all, the company was really a think tank dedicated to the production of *one* working engine for test purposes. Whittle, concerned that Power Jets would fall by the wayside, attempted to dominate the engineering side of the project by clinging to its patent rights, which inevitably caused friction. In a phone conversation in April 1940 with Mr. Whyte, a Power Jets official, Air Vice Marshall Sir Arthur Tedder threatened to remove ministerial support from Power Jets unless patent issues were solved: "You see . . . if there is going to be a hold up on this, we shall have to reconsider our whole attitude toward Power Jets—as to whether Power Jets will perform a useful function in the future."[44] Tedder, though sometimes difficult, was not an enemy of the project. In an Air Ministry minute sheet from

March 1940, concerning jet propulsion, Tedder wrote that "although it is 'futuristic' and a gamble its development should be pushed on as fast as possible with a view to its use in this war."[45] Patent issues, however, continued to cause problems with any attempt to spread out production to other firms. Continuing, Tedder outlined the situation in March 1940: "Up to the present the bulk of the manufacturing work has been done by B.T.H. There has been intense friction between Power-Jets and B.T.H. . . . It has also been pointed out that although B.T.H. have experience of turbine work . . . , once it comes to the making of complete units for installation in aircraft is much finer and smaller work than what they are accustomed to."[46] In other words, BTH was the company most familiar with the work but perhaps the least able to turn their work into true mass production.

The government did not accept delay over patent issues for long, and in April 1940, the Air Ministry chose the Rover Company as the manufacturers of the Whittle Engine. Power Jets found itself isolated from the project and restricted to a purely research capacity. The Ministry of Aircraft Production next awarded the Gloster Aircraft Company a contract to build an airframe suitable for flight using Whittle's engine. The result, the Gloster E.28/39, took off from Cranwell on 15 May 1941, powered by a Whittle engine. The project was a success in that the engine worked. The government never managed to work out production difficulties, however, and the fighter, the Gloster Meteor, did not enter squadron service until July 1944.

One interesting footnote to this comparison between the development of radar and the jet engine ties both together in a new light. In the summer of 1940, a small, nondescript group of scientists headed by Tizard left Britain on a goodwill exchange mission to the United States. This mission, more than any other act, cemented the Anglo-American alliance, when the British shared their technical and scientific military secrets with officials from the United States and Canada. Among these items was the cavity magnetron, the heart of British radar research. In return, the British sought a technological secret from the Americans: the Norden bomb sight. The desire to obtain the bomb sight demonstrates Britain's continued emphasis on strategic bombing and its role in the war. In the hopes of a quid pro quo exchange, the British chose to use the magnetron as a bargaining chip. While they were unsuccessful in obtaining the bomb sight, James Phinney Baxter III, official historian of the U.S. Office of Scientific Research and Development, remarked, "When the members of the Tizard Mission brought one [a magnetron] to America in 1940, they carried the most valuable cargo ever brought to our shores."[47] Jet propulsion items were never discussed at these meetings, though certainly Tizard could have used this information as another bargaining chip in the effort to acquire the bomb sight. While the overall goodwill mission was considered a success, failing to disclose jet propulsion secrets to the United States caused further delays in bringing the technology into service.

Finally, in March 1941, the British disclosed much of their information concerning jet propulsion to General Henry "Hap" Arnold, chief of staff of the U.S. Army Air Corps. Arnold, in Britain on a fact-finding mission, returned to America with the knowledge that the British were in possession of a potentially revolutionary weapon. He made arrangements with the British to gather information, and by July 1941 the United States had enough information regarding the jet engine to pass on to the General Electric (GE) Company. GE oversaw the production of the engines while Bell Aircraft Company built the airframe. The first GE production engine, the I-14, ran in February 1943. It was replaced by the I-16, which powered a fully armed P-59A to a height of 46,700 feet in July 1943. At the same time, the British were still having trouble reaching heights above twenty-five thousand feet. By the end of the war, most ministers familiar with jet propulsion realized the Americans had not only caught the British but had overtaken them in terms of engine power.[48]

The Tizard Mission's failure to bring U.S. manufacturing and manpower into the jet propulsion effort in 1940 caused another year's delay. This neglect of the project may have been due to the unsolved patent issues or a lack of acceptance by the greater scientific community. Many within that community, including Tizard, understood the importance of this type of propulsion, especially after the first flight. Others, within the government, refused to accept the project, even by 1943. In a letter to Air Chief Marshall Sir Charles F. A. Portal, chief of the Air Staff, one member of the Ministry of Aircraft Production voiced his feelings concerning jet propulsion work:

> If the war with Germany ended in 1944 and you had a talk with Goering immediately afterwards and asked him how he accounted for Germany's defeat, he might reply as follows:—
>
> We always kept on thinking that one or other of our secret weapons was a war winner—magnetic mines, high flying bombers, navigation and bombing by beams, rocket and jet propelled aircraft, and so forth—all of which had limited success. The amount of scientific and technical effort which we diverted to achieve these comparatively small and short-lived advantages was tremendous. Above all, jet/rocket propulsion swallowed up an enormous part of our energies so that the normal development of reciprocating engines and the improvement of conventional aircraft were neglected.

Continuing, the minister warned of possible future problems:

> I am seriously concerned about the possibility that we may squander a disproportionate amount of effort on projects which can have no bearing on the course of the present war. . . . We have got to keep technically ahead of the enemy, but it is *one* jump ahead that we want to keep. We should not divert our energies from the pursuit of the next logical improvement of current types in the hope of producing a completely revolutionary aircraft. . . .
>
> This is not an argument against pressing on vigorously with the already tried types of jet propulsion machines. It is merely a caution against shelving them, together with reciprocating engine development, in order to chase after other schemes of enormous but unverified potentialities. In other words, whilst this war is on let us have quick returns even if they are comparatively small.[49]

British science produced two technologically sound inventions in the days preceding World War II. Radar proved to be one of the keys to the successful defense of English skies from the Luftwaffe during the Battle of Britain. Ironically, this technological victory slowed the development of the jet-propelled fighter because it limited the need since for an extremely fast but short-range airplane. By 1943, and especially after the Normandy landings in June 1944, airplanes needed either to carry bombs into Germany or to have enough range to protect the bombers. The Air Ministry felt that Whittle's concept did not match those criteria. This opinion quickly changed when the Germans introduced the ME-262 jet-propelled fighter, which caused near panic in the government. Both the ME-262 and the Gloster Meteor, however, were too late for this war, and it would not be until the Korean conflict that jet-propelled fighters took center stage.

The successes and failures regarding the development of radar and the jet engine in Britain are instructive regarding the relationship between government and science in time of war. Radar was a success, in spite of governmental involvement, due to the efforts of a group of scientists determined to see its technology applied correctly. At the same time, Frank Whittle struggled to gain acceptance for a new technology so radical that it is perhaps surpassed only by the atomic bomb in its effects on modern society. He was unable to overcome his problems with the government, and for good or ill, the development of the jet suffered accordingly. The relationship between science and government during wartime in Britain was, then, a hodgepodge of success and failure. Simply put, the British government did not make full, effective use of its scientific and technological resources. Britain "muddled through," and it can be argued that placing too much emphasis on the jet engine would have indeed been a mistake. As the world enters an era of unprecedented technological advance, however, such guesswork in the relationship between government and science regarding warfare is unacceptable. Modern states cannot afford to follow the British example during World War II. As we enter an age where technology allows a "see deep–strike deep" capability and computers can launch devastating cyberattacks at the speed of light, the work of individual scientists operating on their own, no matter how well meaning, is not likely to produce needed results. The relationship between government, science, and the battlefield must be seamless. The state that does not pay heed to this advice will face almost certain defeat.

NOTES

1. Winston S. Churchill, "Shall We All Commit Suicide?" (1924; reprinted by International Churchill Societies, Washington D.C. 1994). Much has been made of Churchill's governmental and administrative abilities. Paul K. Alkon, in his essay by the same name, observed that Churchill possessed a scientific mind as well, evidenced by his prophetic statements concerning the future of warfare. See Paul K. Alkon, "Shall We All Commit Suicide?" *Finest Hour* (Journal of the International Churchill Societies) 94 (Spring 1997): 18–23.

2. Ibid.

3. Ibid., 4. Churchill used this essay in several of his later works. The original essay was published in the September 1924 issue of *Nash's Pall Mall* magazine. It was later reprinted in Churchill's 1932 collection of essays *Thoughts and Adventures / Amid These Storms* and in *The Gathering Storm,* printed in 1948. See Alkon, "Shall We All Commit Suicide?" 18.

4. Ibid., 8–9.

5. Col. Harry G. Summers Jr., "What Is War?" *Harper's* 268 (May 1984): 75–78; Carl von Clausewitz, *On War,* ed. Michael Howard and Peter Paret (Princeton, N.J.: Princeton University Press, 1976).

6. C. P. Snow, *Science and Government,* The Godkin Lectures at Harvard University, 1960 (Cambridge, Mass: Harvard University Press, 1961), 1. Snow opened his lectures with this remarkable insight into the relationship between politics, technology, and warfare that this paper attempts to illustrate. Snow defined "cardinal choices" as those which determine whether we live or die.

7. The British referred to their invention as RDF, presumably meaning "Radio Direction Finding." RADAR, an acronym standing for Radio Detection And Ranging, was a U.S. Navy code word proposed by Commander (later Captain) S. M. Tucker and adopted in November 1940. The U.S.

Army adopted the term in 1942, and in 1943 the British adopted it as a replacement for RDF See Henry E. Guerlac, *RADAR in World War II,* The History of Modern Physics 1800–1950 Series, vol. 8 (New York: Tomash Publishers, 1987).

8. Max Hastings, *Bomber Command* (New York: Dial, 1979), 38–39; B. Bruce-Briggs, *The Shield of Faith: A Chronicle of Strategic Defense from Zeppelins to Star Wars* (New York: Simon and Schuster, 1988), 21.

9. Ronald W. Clark, *Tizard* (London: Methuen, 1965), 106.

10. *The New Cambridge Modern History,* 12 (Cambridge: Cambridge University Press, 1968): 283.

11. Winston Churchill, *The Gathering Storm* (Boston: Houghton Mifflin, 1948), 84.

12. Prof. Lindemann is discussed in greater detail later in this piece. For more biographical information, see the Earl of Birkenhead, *The Prof. in Two Worlds: The Official Life of Professor F. A. Lindemann* (London: Collins, 1961) R. F. Harrod, *The Prof: A Personal Memoir of Lord Cherwell* (London: Macmillan, 1959); Thomas Wilson, *Churchill and the Prof* (London: Cassell, 1995).

13. *The Times* of London, August 8, 1934.

14. A. P. Rowe, *One Story of Radar* (Cambridge: Cambridge University Press, 1948), 4–6; Clark, Tizard, 108–13.

15. Clark, *Tizard,* 111; C. P. Snow, *Science and Government,* 24–25.

16. J. G. Crowther, *Statesmen of Science* (London: Dufour, 1966), 318.

17. For official invitations, see Public Record Office (PRO) Air Ministry, Kew 2/4481. Descriptions of the scientists themselves can be found in Snow, *Science and Government,* 25–26.

18. Snow, *Science and Government,* 27.

19. Clark, *Tizard,* 115–16.

20. Sir Robert Watson-Watt, *Three Steps to Victory* (London: Odhams, 1957), 109.

21. Clark, *Tizard,* 118.

22. Clark, *Tizard,* 16. Quoted from Tizard's first draft of his autobiography in 1957.

23. Martin Gilbert, *Churchill: A Life* (New York: Henry Holt and Co., 1991), 463–464.

24. Clark, *Tizard,* 125.

25. Ibid.

26. Snow, *Science and Government,* 34–35; R. W. Clark, *The Rise of the Boffins* (London: Phoenix House, 1962), 46–47.

27. Robert Buderi, *The Invention That Changed the World: How a Small Group of Radio Pioneers Won the Second World War and Launched a Technological Revolution* (New York: Simon and Schuster, 1996), 67–75.

28. Ibid., 82–88.

29. Guerlac, *RADAR in World War II,* 123.

30. John Golley, *Whittle: The True Story* (Washington, D.C.: Smithsonian Institution Press, 1987).

31. Frank Whittle, *Jet: The Story of a Pioneer* (London: Frederick Muller, 1953). While at Cranwell, the curriculum required him to write a term thesis, which he titled "Future Developments in Aircraft Design." His thesis was later published in the *RAF Cadet College Magazine* as "Speculation." These "speculations" are the origins of his thinking concerning the application of jets, rockets, and high-altitude flight. See Frank Whittle, "Speculation," *RAF Cadet College Magazine* (Fall 1928): 106–10.

32. Edward W. Constant II, *The Origins of the Turbojet Revolution* (Baltimore: Johns Hopkins University Press, 1980), 15–20. The turbojet developed side by side with existing piston-engine technology. Both competed for research and funding.

33. Constant, *Origins of the Turbojet Revolution,* 184.

34. Most scholars who have studied this period in turbojet development believe that serious progress could have been made with the application of significant funding at this point. See Constant, *Origins of the Turbojet Revolution*; M. M. Postan, D. Hay, and J. D. Scott, *Design and Development of Weapons* (London: HMSO, 1964); and Robert Schlaifer, *The Development of Aircraft Engines* (Cambridge, Mass.: Harvard University Press, 1950).

35. Golley, *Whittle,* 36.

36. Ibid., 65–72; Constant, *Origins of the Turbojet Revolution,* 186–87.

37. Golley, *Whittle,* 66.

38. Ibid., 87.

39. A. A. Griffith, "Report on the Whittle Jet Propulsion System," *PRO* ARC 2897.

40. Constant, *Origins of the Turbojet Revolution,* 190–91; Schlaifer, *Development of Aircraft Engines,* 433.

41. Constant, *Origins of the Turbojet Revolution,* 191–92. By this time

Whittle had introduced "vortex blading," which increased efficiency of the turbine by twisting the turbine blades at an angle. This improved combustion and fuel efficiency.

42. Clark, *Tizard,* 95–99.

43. Whittle, *Jet,* 96.

44. Phone Conversation between Air Vice Marshall A. W. Tedder and Mr. Whyte of Power Jets, Ltd., 18 April 1940, *PRO* Ministry of Aviation 15/2312.

45. A. W. Tedder, Minute Sheet, 28 March 1940, *PRO* AVIA 15/421.

46. Ibid.

47. James Phinney Baxter III, *Scientists against Time* (Boston: Little, Brown, 1947), 2. For more information on the significance of the exchange, see David Zimmerman, *Top Secret Exchange: The Tizard Mission and the Scientific War* (Montreal and Kingston: McGill-Queen's University Press, 1996).

48. Schlaifer, *Development of Aircraft Engines,* 426.

49. Letter from Minister of Aircraft Production to Air Chief Marshall Sir Charles F. A. Portal, Chief of the Air Staff, 6 September 1943, *PRO* Air Ministry.

CHAPTER 10

✦ ✦ ✦ ✦ ✦ ✦ ✦ ✦ ✦ ✦ ✦

War and Black Memory

World War II and the Origins of the Civil Rights Movement

Neil R. McMillen

In 1993, some fifty years after the Allies defeated the Axis powers in Europe and Asia, I began interviewing some of the eighty-five thousand black Mississippians who served in uniform during World War II.[1] These conversations with American GIs did not neglect what might narrowly be defined as traditional "military history," but they focused more fully on the social meaning of the war. The emphasis was always on matters of race: the more painful aspects of black service in a Jim Crow military, the connections between soldiering and citizenship, the relationship between wartime patriotic sacrifice and postwar racial struggle. My interests and my questions generally centered on the ultimate American wartime irony: the conscription of blacks to fight abroad for the very liberties they were denied at home. By interviewing survivors in a thinning population of World War II veterans, I hoped to document both how ordinary black men and women from one southern state remembered their service in the last Jim Crow war and how that service may have influenced their understanding of the black place in what white Mississippians once called "a white man's country." By gathering life experiences from the historically nameless, I hoped the better to distinguish between the wartime aspirations and agendas of the nation's black leadership elite and those of rank-and-file black conscripts from the lower South.

What I found was more complicated and more interesting than what I had expected. These oral accounts, I believe, illuminate aspects of wartime experience that might otherwise escape notice. No writer could fabricate stories quite so compelling as these; no academic narrative could speak so eloquently of the American dilemma as the plain words chosen by these Mississippians.

The civil rights movement, it is now widely believed, was borne on the wings of war, World War II.[2] The remote origins of this modern freedom struggle, of course, are remote indeed, traceable through more than three centuries of black protest and striving, back to the early period of the slave trade. Its immediate catalyst, moreover, may have been any one or a combination of specific postwar developments: the southern bus boycott movement that began, almost unnoticed, in Baton Rouge, Louisiana, in 1953, and then reached critical mass two years later in Montgomery, Alabama; the Supreme Court's landmark school desegregation ruling of 1954; or the black rage that overflowed nationwide after the unpunished murder in 1955 of teenager Emmett Till in Money, Mississippi.

For fundamental first causes, however, many scholars now look to the Second World War, to home-front dislocations and overseas sacrifices, to the moral contradictions of a war for "four freedoms" fought by a democratic nation with a segregated workforce and a Jim Crow military. They look to a series of interrelated economic, demographic, political, cultural, and other changes in the structure of American society that emanated from (or were accelerated by) total war and that collectively created a context favoring racial progress. Not least, they look to the returning black troops who, as Woman's Army Corps (WAC) veteran Luella Newsome explained, "wouldn't take it any more."

Transformed by wartime experience, African Americans of the World War II generation, soldier and civilian, often could not immediately act on their emergent convictions. Yet they entered the postwar period more determined than ever to exercise their constitutional rights. Not a few of them, even many who lived in the darker reaches of the Deep South, announced for all to hear that the United States must now live by its own ideals. Not all black southerners literally "returned fighting," but many of them shared as never before the conviction that the humiliations of Jim Crow were intolerable. "It had to change," the Mississippi WAC explained, "because we're not going to have it this way anymore."[3]

Stated so baldly, the argument that the postwar black awakening flowed more or less directly from wartime black militancy may seem problematic on its face. There is, after all, the problem of that apparently quiescent ten-year interlude between war's end and the rise of anything that can reasonably be described as an effective, mass-based black movement capable of stirring a nation's conscience and changing a region's laws. Moreover, as the historian Harvard Sitkoff has recently argued, the reductionism of the "Double-V" argument may well have led historians to overstate the case for an unconditional black two-front war for freedom abroad and freedom at home.[4] We should be wary of a determinism that reads every squall on the wartime color front as a storm of racial militancy. It does seem at least arguable that historians, by and large as committed a group of "continuitarians" as can be found anywhere, could have been led into teleological error and, by reading history backward, could have found more continuity between wartime yearning and postwar confrontation than there may in fact be.

It is arguable. But the war narratives collected for this project suggest that while the Second World War in the short term left the structure of white control in Mississippi very much intact, it nevertheless touched the lives of Mississippi's black servicemen and women in ways their white oppressors both feared and underestimated. Most veterans in this sample believed that though the war may not have changed their hometowns, it nevertheless changed them. It gave them new perspectives and new aspirations they could not always act on immediately but that influenced the subsequent course of their lives.

"Oh, it changed a great deal of things," Wilson Ashford said of his two years in the army. "It changed the individual." Shouting to be heard over the din of the small automotive garage he had operated since returning to Starkville in 1945, this seventy year-old Oktibbeha County NAACP (National Association for the Advancement of Colored People) leader told me that the black soldier learned from wartime military service "that he was able to compete":

> That's where the change come in. That was the first step in change, where you would feel that you could do it. You were always taught that you can't do, you can't do. [Whites] they'd always tell the negative side. Nothing positive. From that standpoint it gave you an opportunity to see things in a different light.

Then, looking squarely at his interviewer—the nearest emblem of racial oppression—Ashford closed the subject: "I believed you was wrong all of the time, the way blacks were being treated. And after I got out [of the army] I *knew* you was wrong."

His entire postwar life seemed testimony to that conviction. Soon after his discharge, Ashford became a founding member of the local NAACP; he paid his poll taxes unfailingly and in 1950 became one of the first black registered voters in his county. Thereafter, as the freedom struggle unfolded, he became active in county voter registration drives and a persistent advocate of school integration; eventually he was appointed to the local school board and to the state School Board Association, became a lay minister in his church, served on a citywide biracial committee and on the board of trustees of a local community college, and became president of the local Habitat for Humanity. In 1993, the Starkville Area Chamber of Commerce honored this veteran with its Community Service Award, in recognition of an exemplary lifetime of civic responsibility.[5]

For this white interviewer, old enough barely to remember the war but fully a generation younger than these interviewees, one of the more striking aspects of the stories I collected is the shock of *non-*

recognition. I grew up listening to the war stories of white men, older friends and family who served in Europe and the Pacific. I began this oral history project, moreover, as a historian with some knowledge of wartime racial discrimination. Yet the memories I recorded moved and disturbed me in unexpected ways. So different were those white stories of my youth from these black stories that they might well have been from a different war. And in fact, in important respects, they were.

War, as those who study it often say, is not experienced by everyone in the same way. For the disempowered and the marginalized, for segregated black men and women, the shape of war is dictated by their secondary status. The American defense establishment of World War II was not, as these veterans frequently explained, Colin Powell's military. As the war began, black participation in either civilian war work or military training was rigidly circumscribed by race. Shut out of the better defense jobs and restricted largely to noncombat military assignments, African Americans were initially excluded from both the Marine Corps and the U.S. Army Air Corps; they were accepted by the navy, but only as messmen, and by the army only in segregated units. The Red Cross at first refused "black blood," and the War Department planned to train few African American military officers. Under the pressure of black outrage and extreme national emergency, some adjustments were made, and blacks were more fully utilized in the defense of their country. But from first to last, Uncle Sam was not, during World War II, an equal opportunity employer, and black troops were often subject to the embarrassment and humiliation of separate and unequal military service.[6]

Some black veterans, however, clearly had the time of their lives in the military. "I enjoyed every minute of it," a Laurel women said of her two years in the Women's Army Corps. "We were given opportunities to do all the things I never would have done had I not gone to the army."[7] Others believed that military service gave them pride and purpose they might otherwise never have experienced. "I had low self-esteem," said former supply sergeant Nathan Harris of his youth in the Mississippi Delta. "The army built me up and made me proud of myself. It sounds stupid. But that's the way it was. And I've still got it in me."[8] In a third example, Dabney Hamner, a veteran combat soldier, remembered action during the German counteroffensive of December 1944 as a defining moment in his life. "I'm about the only black that I know was in the infantry," he said. "The biggest pride I ever had in my life [was] when the guy put that combat infantry badge on me."[9]

Yet black memories of the war years are rarely star-spangled memories. The war these veterans recall bears little resemblance to the so-called Good War of romanticized American white popular memory.

Although these narratives document the diversity of black wartime experience and perception, one thread running through all of them speaks to the pain of a pariah race—to the humiliations and the burdens that were all too often visited on African Americans in a Jim Crow military. In nearly every case, the veterans described how the white soldiers who preceded them abroad had tried to export American racial values to civilian populations in England, the European continent, or the South Pacific. In one version of a story that reflected common black experience, Charles H. Jones Sr., a soft-spoken retired Clarksdale school teacher who had served in the European theater of operation as a medical corpsman, remembered that the first obstacle he confronted overseas was the idea, deeply implanted in the English mind by white GIs, that blacks were so low on the evolutionary scale that they had tails that came out at night. The English, as he said, had never known black people:

> I met a very interesting family. This man and his wife befriended me and I was very appreciative because I didn't know anybody. The first time they invited me to dinner his daughter put a large pillow in my chair. I just figured what the heck, to make me nice and comfortable. So one day I asked her [the wife] and she said, "Well they had told us you had tails and I didn't want you to sit on it. The whites had told us that you had this tail and you were monkeys."

"Oh that embarrassed me," the veteran recalled. "This is what we had to face." In time, Jones said, the British learned otherwise: "they found out that we were good people too."[10]

Slanderous white accounts of black character were perhaps still more wounding. Henry Murphy, a veteran from a Piney Woods county who, as a supply sergeant, had been wounded in Germany in 1945, forced a thin smile as he related his personal encounters with the tail myth. "After we got on through," he said of the black men in his company, "we laughed about it. It just didn't catch hold to us at all." Murphy pretended no amusement, however, when he recalled, "They [American whites] also said that we was rapists, murderers, and thieves. That bothered me. I don't know of anybody in my outfit [who] raped anybody. I imagine there are some people who probably was raped by black folks and by some white folks. I don't know. But I know in my outfit we didn't go around raping or rob[bing] anybody."[11]

Some veterans remembered that just wearing their nation's uniform could cause problems in a Deep South town. WAC volunteer Brunetta Garner, who served in the medical corps, recalled that what she wore mattered little to whites in the village of Ellisville. They regarded her, she thought, "the same way they did before I left: 'There's a nigger with a uniform on.'"[12] A retired handyman and sometime custodian from Scott County, Mississippi, insisted that no white ever dared challenge his right to appear in khaki: "You're looking at one black man here never been scared of white folks. Eugene Russell is his own man. Don't nobody tell me what to do. Wore my uniform till it got too small for me."[13] Still others, however, thought that white resentment was palpable. Whites often feared, they recalled, that service in the armed forces, especially service overseas, might elevate black status, might make blacks "uppity." Clemon Jones, an elderly Jackson native who, as a postwar emigré, had once been commander of the American Legion in New Jersey, still remembered half a century later the hostile stares his uniform attracted when he was on leave in Mississippi: "You could tell they resented it because if you met one or two of them on the streets they'd look at you like you was an alligator or something. You could tell they was talking about you."[14]

The case of Rieves Bell, the soldier who went home to Starkville on furlough in 1943, only to be imprisoned after an encounter with whites who objected violently to his military attire, is well documented in military archives and still remembered by local blacks. Outnumbered on a city street by three young white men who taunted, pushed, and then tried to strip off his uniform, he fought back, wounding one of his tormentors with a knife. Despite the efforts of the military to handle, through military channels, what seemed to be a case of self-defense, civilian authorities tried and sentenced Bell to three and a half years in the state penitentiary at Parchman.[15]

Other examples, perhaps less dramatic, are remembered only by the soldiers themselves. In one of the most affecting of these, Henry Murphy, a Purple Heart winner who served in two wars, recalled that when he returned to Hattiesburg in 1946, his father, a local minister, met him at nearby Camp Shelby with a change of civilian clothes, lest he invite trouble from white residents and police: "He told me not to wear my uniform home. Because the police was beating up [black] GIs and searching them. If they had a white woman's picture in his pocket, they'd kill him." Although Murphy had no pictures, he did as his father wished. His was no hero's welcome; as the older man drove the family Chevrolet to Hattiesburg, the returning soldier slipped out of his olive drabs and put on overalls and a jumper, the uniform of a field hand.[16]

Incidents of open racial conflict, including deadly firefights between black and white American personnel, recur in these narratives. The hidden, or perhaps forgotten, history of the American soldier during World War II includes a disturbing number of bloody encounters, some of them full-fledged race riots, on or near military training centers, particularly at southern camps, posts, and stations where roughly 80 percent of all black enlistees were trained. Mississippi, as one might expect, had its full measure of such conflicts. Fifty years after the fact, one can hardly imagine the horror many northern black soldiers felt when they learned they had been assigned to a Mississippi base, to their minds not unlike being sold south into slavery.[17]

The worst examples of GI racial conflict, however, seem to have occurred overseas where, as some interviewees explained, the black soldier had two enemies: the Wehrmacht and white American troops. One veteran actually remembered that he feared the Germans rather less than the men of what he called "the Bloody 1," the First American Division he encountered near Nuremberg:

> They was the worst of all. The most prejudiced. Most of them came out of—you're going to be surprised—like California, New York, Ohio, [as well as] Arkansas, Mississippi and they were hell-double-breasted. If you faced that "Bloody 1," you'd better be ready, you hear me, or you would not survive. They would kill you and throw you in the middle of the street. They didn't care anything about blacks at all. Do you hear me?

He described fistfights, knife fights, firefights—open warfare between white and black Americans in the army of occupation that in some cases resulted in death. Fearing he would be killed or mutilated, Sergeant Murphy learned to avoid any contact with the men of the First Division: "I wasn't going to let them castrate me because you understand they were castrating—hit your seeds with a razor. I didn't want no part of that, so I decided to stay in camp. That wasn't no rumor, that was true."[18]

It must be said that many black veterans remembered warm friendships with white soldiers, and some of these friendships survive to the present day. Napoleon Evans, a retired Laurel police officer who, as a young draftee, had been attached to the Ninety-sixth Engineers Regiment in the Philippines and New Guinea, believed the war fostered interracial understanding in ways civilian life had not. "You might not have had all of the rights you wanted," Evans remembered, "but some of them did treat you like you was a human being. A lot of blacks and whites in service together, I think they learned to have respect for each other. The whites I know learned to have more respect for blacks than they did at first."[19]

Not a few other veterans, by contrast, recalled virtually no contact with white Americans, apart from with their officers, who were nearly always white and very often southern. As these men and women put it, growing up in color-conscious Mississippi was good preparation for military life, for being what some called "half a soldier": "Everything was segregated then, everything—different water fountains, different bathrooms, different camps."[20] "Black folks on this side of the track, and white people on their side of the track. That's the same thing within the Army: All your blacks over here, and all your whites over there."[21] The two races were fully as isolated in northern camps or overseas as they had been at home. They did not train together. They generally did not play together. They rarely served together.

Inevitably, however, other veterans told deeply moving stories of interracial friendships made and interracial friendships shattered as wartime emergency gave way to peacetime normalcy. In one version of this all-too-common memory, Ben Fielder, a retired hotel bell captain who served in both Europe and the Far East, recalled a last, long train ride across the country, from California to Mississippi, at war's end. This staff sergeant and his white traveling companion had for days exchanged stories of family and sweethearts; they had shared their meals, their dreams, their hopes for the future. They had gotten, Fielder believed, beyond "this race nonsense" and "become tight"—until the train crossed into Dixie. Then everything changed, as the white southerner assumed his traditional role of dominance and, without a trace of subtlety, informed the black soldier that the war was over, that they were back home, that—as the black soldier remembered—"I was still just a nigger. Not an American soldier anymore. Just a nigger."[22]

Other interviewees related similar experiences amid the ebb and flow of war. "When the bombs were falling," Dabney Hamner said of Hitler's desperate Ardennes offensive of December 1944, there "wasn't no black or white—we was just the same, black soldiers and white soldiers, all Americans." Napoleon Coney of Pike County, another infantry veteran who proudly displayed his five bronze stars, remembered his years in the European theater as "the only time in my life I felt like a man. It's perfect. When you're in combat those folks forget that you're black. When the shells start, incoming mail—BOOM—your whole self is shaking. [Color] don't make no difference."[23]

Another man, a twelve-year veteran with a remarkable gift for re-creating the historical moment, first described a racial "hornet's nest," a bloody standoff between himself, another black soldier, and two white GIs in Germany, a conflict that resulted in serious injury and perhaps death to one of the whites. Threatened by the armed white Americans and fearing the worst, this soldier's companion lashed out with the butt of his M-1 rifle: "We had no choice because they was fixing to do us in. I mean blood flew everywhere. I don't know whether he's living or dead. We left." Then, as if to put the story in some larger context, the veteran explained that during the thick of things, race rarely divided American fighting men: "Yes, we supported them during the war. We gave our hearts to them during the war, and they gave theirs to us." A noncommissioned officer assigned to a trucking unit, he remembered close quarters and easy relations with whites in the early aftermath of the Normandy invasion, amid the fog of battle in France. "We picked up their dead. We even hauled them on spearheads. We slept in the same areas they slept in. We shared the rations." Although some whites initially resisted an integrated mess, troops of both races quickly discovered that racial convention was impractical at the front: "They decided that oh, what the hell, everybody eat together, work together, and get along." "When we was supplying them, they needed gasoline, food, ammunition. Everything was lovely. They welcomed us until after the war." During the German breakthrough, he remembered, his unit got out of their trucks "to fight the Germans for them and with them." But once the war was over, he recalled, "everybody went back into their little old world. They decided we was the scum of the earth after the war. They never did decide we was equal, you know that. But they needed us when the Germans [were] giving them hell." Asked how such could be, he thought a moment and said: "Well, it was just a mirror of our civilian life, that's all. The way of life that we've been living all them many years."[24]

Perhaps the deepest meaning of such memories is to be found in the very singularity of the interracial moment. Combat allowed a shared American experience, one not circumscribed by race; the Jim Crow army was never so color-blind as when it was in a foxhole under fire. Wilson Evans of Harrison County, for example, explained that he never felt so accepted, so free from the burden of color, as he did during the Battle of the Bulge: "That was the first time during twenty-seven months in the service that I was an American soldier. For those, what, six, maybe eight, ten days there was no black or white soldiers. We was all soldiers. White was afraid of dying as blacks. And there was no color. During the breakthrough I did see that Americans could become Americans for about eight or nine days."[25]

Among the questions of most interest to this project were those that centered on the American wartime dilemma, on the moral paradox of a democratic nation deploying a Jim Crow military to liberate Europe and Asia. Few of the interviewees came from families that voted during this period. Most scoffed at the very idea that their parents enjoyed the protection of the Fifteenth and Nineteenth Amendments. "When we went in World War II, we didn't have no rights. We couldn't vote," remembered Sam Jackson, a retired custodian who had served in the South Pacific. "Man, you'd get killed [just] talking about voting."[26] Most recalled brutal treatment by white police: "If you were a black violator you got a beating first. After they got you handcuffed you got a whipping before they'd take you in."[27] All, when asked, recounted the humiliations of Jim Crow public transportation, of being forced to sit at the back of the bus, behind a screen—what William Nicholson of Wayne County called "an iron curtain," "a Jim Crow curtain," positioned to shield white eyes from black passengers.[28]

"Conditions here was ridiculous," Jonestown native Nathan Harris said of the Mississippi Delta. The example this veteran noncommissioned officer used was one that had been painfully seared into the memory of virtually all African Americans of the World War II generation: the status of black American citizens relative to the nearly 250,000 white German prisoners of war (POWs) interned in the South. "I'll tell you what," he said, "if you was a black boy here in Mississippi, when they brought those Germans over here as prisoners they got more privilege than you did as a citizen—German prisoners! Right here around this country." "Blacks didn't have nothing," and the worst part, Harris thought, was that "at the time we didn't know any better. We thought that's the way it was supposed to be."[29]

Veteran after veteran used virtually the same words—"we thought it was supposed to be that way"—explaining, often, that even in the North they encountered in more subtle forms the racism they were accustomed to in Dixie,[30] and that not until they served overseas did they fully understand the depths of their own degraded citizenship. Growing up separate and unequal, Clarksdale navy veteran Haywood Stephney recalled, was like "living in a closed circle":

> Because you grow up in this situation you don't see but one side of the coin. Having not tasted the freedom or the liberty of being or doing like other folks then you didn't know what it was like over across the street. So we accepted it. When you're not exposed to much you don't get much. But after seeing what some of the other world was doing then I realized how far behind I was. As we began to move and stir around and learn other ways then we had a choice—a comparison. I could contrast this with that.

With a new sense of himself and his race, Stephney remembered thinking: "Now it's going to be difficult to get me back in total darkness. That was my attitude. Deep within me I made up my mind I've got to move out of the situation that I'm in to something better."[31]

Eugene Russell of Scott County, a self-described "jailhouse preacher," detailed a similar process of growth and enlightenment:

> We were young men. We all knew the treatment we already had been getting here in Mississippi and everywhere else. So we don't think

> nothing about that. We thought that was just the way it was supposed to be. We was dumb to the facts and didn't know.

Thirty-three months and nineteen days in the Pacific theater, however, opened Russell's eyes. "It started coming to me after we got to Australia," he said. "I noticed the difference right then. The Australians called us the 'tan Yanks,' not niggers and monkeys." "There wasn't nothing but white people [in Australia]," he remembered, "but you was welcome. And the farther you go up [finger raised, tracing an imaginary map]—the Dutch Indies, I don't care how white you was or how black you was, you were still treated right. You was just a man. When you went on to the Philippines, you were just a man there." "I wouldn't take nothing for it," he said of such experiences. "I got a better understanding in life and my destiny and where I'm headed to. I saw how folks can live. It opened up a new world to me, opened up [my eyes] to the racial problems."[32]

These memories of changing personal perspectives and transforming encounters overseas, of new and revealing light shed on old injustices, become more poignant still when paired with memories of return and reunion in the immediate postwar period, as GI became civilian. Consider but a single story, this one told by Dabney Hamner, a veteran of the 125th Infantry Division who had been wounded in Germany and who believed that "the only time in my life I [had] felt like a man was in Europe." He returned to Clarksdale, Hamner remembered, with a chest full of decorations and not a little pride that, as he put it, "I'd been hopping in and out of foxholes in five battles for American democracy." One of the first whites he met stopped to admire his medals: "Ow-w-w look at the spangles on your chest. Glad you back," the white man said. "Let me tell you one thing though, don't you forget." Forget what? the veteran inquired. "That you're still a nigger."[33]

The message was not always so blunt, but time and again one finds evidence, in these narratives of black homecomings spoiled by pointed reminders that this was still Mississippi, that the old rules still applied.[34]

Why, then, did these Mississippians fight? What were they fighting for? Did they think this might not be their war? Or did they expect that the uniform might be a ticket to full citizenship? At a conscious level, did they then embrace the ideology of the "Double V"? Do they now believe—did they then suspect—that for blacks, World War II was a "war on two fronts" and thus a staging point for the modern freedom struggle?

The answers found in these narratives are so diverse, the expressions of personal motivations often so complicated and ambiguous, as to defy confident analysis. Human diversity is the oral historian's most vexing challenge; the men and women represented in this sample did not speak with one voice on these or any other issues. Some few veterans admitted that they did not know why they went to war or, until much later, what they were fighting for.[35] Others reported that they were not eager to fight in Europe or Asia for what their own people did not have at home, that they put on a uniform only because they were drafted or because military pay was better than that of any other job then open to black Mississippians.[36] "I wasn't thinking about volunteering," one conscript remembered. "Most of my people always said that it was a white war and blacks didn't have no business going."[37] Another declared that, after weighing the alternatives, he had concluded he would fight for the lesser of two evils: as bad as Theodore Bilbo was, Adolph Hitler was worse.[38]

From a somewhat different perspective, an infantry veteran from the Gulf Coast said that his prewar experiences with Jim Crow had stripped him of all patriotic feeling for his country. He was unmoved by the Japanese attack on Pearl Harbor: "I figured that if the Japs took America that I would fare better." Even after he was drafted, even as he slogged across Europe from Normandy to Berlin, he saw little difference between German and American racism. He rather hoped, he said, that the Nazis did not want his mother's Harrison County home, but if they wanted downtown Gulfport, "hell, they could have it."[39]

That bitter memory aside, most of these men and women insisted that, however resentful of social injustice, they served in uniform because this was their country too. "We were second-class citizens," said James Briwder Jones, a veteran of the 761st Tank Battalion said. "We were called niggers, and we were deprived of certain privileges. We lived as a nation within a nation." Still indignant that his unit never got the recognition it earned for cutting off a German spearhead-for as he believed "saving Bastogne"—this Jones County NAACP branch president nevertheless remembered the black tankers' pride "in just being Americans. We always felt that someday that shackle that held us down would be broken. We had faith in this country. This is the only country we had."[40]

Such hopes and sentiments were widely shared. One finds in these narratives rich layers of counterpoint: shocking accounts of interracial bloodlettings and wounded spirits, balanced by the uncomplicated love of flag and country that seems so natural to the World War II generation. Twenty-year veteran flyer Alva Temple, one of the celebrated Tuskegee Airmen, was perhaps more forceful than most when he declared that his wartime patriotism was unconditional. By war's end he had flown 120 combat missions over Italy and East-Central Europe and had been awarded the Distinguished Flying Cross. Acknowledging that some African Americans "said that blacks should not participate in that war because we weren't getting a fair deal," this retired lieutenant colonel nevertheless insisted, "Well, I never felt like that. I felt it was my job, and I did it the best that I knew how. I was going to make my contribution. I did think that the contribution would have some effect on things being equalized, blacks getting a better deal in the United States." "On the other hand," Lt. Col. Temple said, "that was not a condition why I went. I would have gone regardless of what happened. I felt it was my duty to do that."[41] None said it more eloquently, but the thought recurs in interview after interview: I went because I loved my country.

Expressions such as these might invite derision in some circles today, but these interviewees did not seem naive. Few admitted to any illusions; none truly expected a hasty change of racial heart back home.[42] None of them—not even those who reported weekly wartime exposure to the *Pittsburgh Courier,* the most militantly antisegregation black newspaper published in the United States—remembered hearing the protest slogans of the "Double V" and the "War on Two Fronts." Not one professed to have had what Dr. David W. White, Mississippi's first black optometrist and a veteran of the 1895th Engineers, called a "crystal ball," a clairvoyant understanding when they answered the call to arms that the war marked the beginning of Jim Crow's end.[43] In fact, some reported that they expected nothing when they went to war, and so well schooled were they then in Jim Crow's ways that they were neither surprised nor even particularly disappointed to learn on their return that nothing had changed; that, as German Levy of Brookhaven expressed it, "the pancake hadn't turned over," that "you come back home right into the same world you left."[44]

Three years in the Army Air Corps had made Delta postal worker William McDougal less patient with white injustice but not a fool. "All the white folks they was worried about the reaction from [black] soldiers," he remembered of his return to Coahoma County; but "we didn't push nothing in that time because they was running everything. We didn't have no say so about nothing then because we weren't voting." Understanding that "one man couldn't do too much," McDougal didn't "think too much about the segregation. I didn't entertain the idea that I was going to change it. No, but I had the idea, look I'm trying to better myself."[45]

Employing the same blend of social realism and personal ambition, retired university Professor Matthew Burks said of his return from three years in the Philippines, "I knew the way conditions were, and I really did not expect them to have changed within that period of time." Discharged in December 1946, he resumed an interrupted education in a segregated Mississippi high school. "It really didn't bother me that much at the time," Burks remembered. "I just went on and fitted into the situation as I had before." In fact, this Neshoba County sharecroppers' son did return fighting, a long-time friend remembered, "but fighting to overcome his background." That fight eventually carried Burks to a career as teacher and administrator at Mississippi Valley State University, after degree programs at Tougaloo College (B.A.), Boston University (M.A.), and the University of Virginia (Ph.D.).[46]

Without exception, these returning veterans found more continuity than change in their Mississippi hometowns; there would be no real progress, they patiently explained, until after "the Freedom Riders," until the advent of the "civil rights people"—common metaphors for the upsurge of black activism in the 1960s.[47] In fact, many interviewees affirmed that what they would later call civil rights mattered less to them in 1946 than a decent job. "I didn't know that much about the right to vote," a Brookhaven draftee admitted. "But I did know that I deserved the right to have a job." The moral contradictions of the war had largely escaped his notice, he said, until he returned to a job market that was monopolized almost entirely by whites: "That's when it began to dawn upon me, that's when I began to realize that it was just something wrong in this country."[48] A retired Clarksdale educator, by contrast, remembered that his father had managed to "sneak in and register" to vote during the war, and that he himself had left the military wanting "the same rights as anybody else." His first priority, however, was his family's welfare: "When I came back here the only thing I was looking out to do was to try to survive. I didn't bother about that other [civil rights]. I wanted to make a living for me and my wife and children and that's all."[49]

Others explained that job and wage discrimination in Mississippi, more than any other postwar disappointment, pushed them into either outmigration or reenlistment. "Many veterans in the South came back determined not so much to change things but to get the hell out of here," explained former state senator Henry Kirksey, who had been a wartime artillery officer. "That was one of the most often expressed intents for those veterans that I talked to.

You've got to get out of the South because they can't stand it anymore." After the war Kirksey—who as a major had been Mississippi's highest-ranking black soldier during World War II—finished his college education, married, and in 1948 reentered the army as an enlisted man when he concluded that he could support his family in no other way.[50]

Yet, if these Mississippians generally had few short-term expectations, their narratives nevertheless document a widely shared hope that black wartime sacrifices would lead eventually, in ways they could not then fully imagine, to a better day. Alva Temple, the Tuskegee airman who attached no strings to his patriotism, also affirmed that "when you defend your country, I think you were due full citizenship privileges."[51] And, like that airman, these interviewees more often than not—and very often in the same breath—placed their military experience in a broader context of racial uplift. Asked what they were fighting for, all agreed with a retired shoe repairman from Hattiesburg who said, "We were fighting for what we didn't have." Yet veteran after veteran also explained that they fought for the future, for their children and their grandchildren. They went to war hoping that Mississippi might eventually be a better place for all of its people—not immediately but eventually.[52]

Their words were as varied as the men and women who spoke them. But if a single veteran could be said to have somehow captured their collective understanding of the social meaning of the war, it may well have been Douglas L. Conner, a Starkville physician who acknowledged that, though a college graduate when drafted, he never encountered "the expression 'Double V for victory'" until he studied the history of black wartime experience many years after his discharge. Conner thought, moreover, that the men in his unit, the Thirty-first Quartermaster Battalion stationed in Okinawa, did not in any conscious way understand that the war for the four freedoms was for blacks a "war on two fronts," a war for democracy over there and back home: "We talked about our hatred for segregation and for the Japanese, but we never tied the two together that way." He also remembered the ambivalence many black Mississippians shared as the war began. After Pearl Harbor, he sensed among his fellow students at Alcorn College no "gung-ho spirit of 'Come on, let's get it over with.' Most of us said that we would not volunteer for a segregated service. If we're called, we'll go. We'll fight this war. We'll do what we can to win, but at the same time we need to do something to make sure that we don't go back to where we were before the war." Once in uniform, he came to think that he was "fighting for future generations because I had a sense that things would get better for African Americans in time." Fifty years after the fact, he remained persuaded that "if there had not been a war, I think it would have been harder to fight segregation and to change Jim Crow laws." "The important thing about World War II," he believed,

> is that there was proof that blacks could enter the service and fight as heroically as anybody else. The air people in Tuskegee, Dorie Miller, and the others gave the blacks a sense that they could succeed and compete in a world that had been saying that "you're nothing." And at the same time because of the world war, I think many people, especially blacks, got the idea that "We're going back, but we're not going back to business as usual. Somehow we're going to change this nation so that there's more equality than there is now."

At war's end, Douglas Conner had no better idea than did any other returning soldier just how that change might be realized. He recognized that "things would not get better on their own," and he now counted himself among those servicemen and women who were "determined that somehow Mississippi had to do better than it had been doing for black Americans." Yet nothing suggested to him that real change was in the postwar air. When he was mustered out of the army at Camp Shelby in June, 1946, his first act as a civilian was not to storm the county courthouse but to catch a northbound train to a summer job in a Chicago steel mill. That fall he took his savings and his GI benefits to Washington, where he entered Howard University Medical School. In 1951, after an internship in St. Louis, Dr. Conner returned to Mississippi to open a medical practice in Starkville. The next year, at the age of thirty-two, he became one of a tiny fraction of adult black Mississippians (perhaps 4 percent) who were registered to vote. Thereafter he was a force for gradual change in his community, and he never forgot the conviction he carried home from the Pacific: "my black skin was not a valid excuse for anyone holding me back—or, for that matter, for holding myself back."[53]

How then shall we understand these narratives? What do these voices say about war as the analogue of reform? What do the narrators tell us about the connections between black participation in the last Jim Crow war and the modern black freedom struggle that engulfed the South in the decade after the *Brown* decision?

However one might choose to read them, these interviews speak directly to the humiliations of separate-and-unequal wartime experience and far more ambiguously about that experience as the genesis of postwar black struggle. One could not persuasively argue from the oral testimony I have collected that fifty years ago the generality of black soldiers, much less the generality of black soldiers from Mississippi, returned from the war with radical change on their minds, determined to vote at any cost, determined immediately to challenge the discrimination that for more than three centuries had kept their people down. Although historians understand that wars and their aftershocks very often ultimately unsettle traditional social patterns, these black men and women returned to hometown racial environments that were depressingly familiar. Much as they hated segregation and disfranchisement, they had known little else and they were not then prepared to fling body and soul upon racial ramparts that were at least as impregnable in 1946 as in 1941. Until the 1960s, few of them found reason to believe that genuine racial progress was imminent—or even possible—in Mississippi.

Some veterans did situate the antecedents of the modern freedom struggle in the World War II years. "We paved the way, we broke the ice, we opened the door," Nathan Harris insisted.[54] Many more, however, saw little relationship between black sensibilities in the 1940s and black protest in the 1960s. In these memories, the war waged for the liberation of Europe and Asia is often described as a defining moment and sometimes a personal turning point, but it does not emerge as America's liberating social divide, the watershed from which flowed a torrent of subsequent social change. Indeed, what seems most remarkable about these narratives is the fact that so many of the aging black men and women who constructed them declined to invest wartime experience with postwar meaning. The narrators, by and large, did not read back into the period before 1946 the attitudes and expectations that in the period after 1954 would be so common to the black community.

That fact is of no small significance, of course. But if the black veterans themselves do not remember World War II as a black war on two fronts, if they seem less prepared than historians to link wartime aspiration to postwar confrontation, does it mean that historians must now radically revise their histories of recent American race relations, severing all connections between the 1940s and the 1960s? The answer is surely no. I do confess that what I expected to hear at the outset of this oral history project was rather different from what I in fact heard. While I did not assume that history makers and history writers would interpret the past in precisely the same way, I expected, nevertheless, greater convergence of perspective than I found between participant observers and scholars. Yet, when read and understood in the context of a larger body of documentary evidence, these fascinating conversations persuade that because of the war and the events and circumstances that eventually flowed from it, the Jim Crow system was doomed. In the short term, the war did not massively reshape race relations. In the war's immediate aftermath, fearful white southerners were if anything more unyielding, and the color line held for nearly two decades. Despite isolated acts of immense heroism and racial militancy, sustained black protest on a mass scale did not erupt for half a generation.

Nevertheless, though delayed, the war's impact was decisive. As these narratives suggest, patriotic service at home and abroad provided new perspectives on ancient white wrongs and ultimate black possibilities. The war helped shape an emerging racial consciousness; it paved the way for an emerging black awakening. It underscored the moral contradictions of a nation that professed human rights and practiced white supremacy. It illuminated some of the darker places in the American social landscape. It made plain, for all who would see, that in a modern democratic society the citizen soldier, of whatever race, must be both a full soldier and a full citizen. And the war also fostered the development of a larger societal framework within which successful struggle for human rights could be waged. Though never an independent variable in the equation of sweeping social change, World War II nevertheless accelerated political, economic, legal, and intellectual tendencies already in motion. It set in train a combination of international, national, and regional forces that ultimately undermined the legal foundations of segregation, discrimination, and disfranchisement.

And it was precisely this convergence of war-born developments, personal and impersonal—this combination of human agents pushing for social justice and of societal forces conducive to racial change—that brought a Second Reconstruction to the American South. As these interviewees understood only too well, the courage and conviction they attributed in substantial part to their service experiences were in themselves not enough. What they perhaps understood less fully was how the war—*their* war—contributed to the emergent structural changes in the nation's political and economic life, in its laws and social values, and in its public policy that ultimately opened the way for fundamental racial progress. In this favoring context, black courage and conviction could help transform a system that had once sent these Mississippians to fight abroad for what they did not have at home.

NOTES

A version of this essay was presented at a symposium at the University of Southern Mississippi in October 1995. In revised form, it was first published as "Fighting for What We Didn't Have: How Mississippi's Black Veterans Remember World War II," in Neil R. McMillen, ed., *Remaking Dixie: The Impact of World War II on the American South* (Jackson: University Press of Mississippi, 1997). By permission, it appears here with minor revision.

1. I have audiotaped more than 120 hours of conversation with nearly fifty black Mississippians of the World War II generation. In all but two cases they were veterans, although the spouses of some veterans did participate in some phases of the interviews. Most of these subjects were men. The sample was not scientifically selected. Most sections of the state were represented, through either the interviewees' places of current residence or their places of upbringing. Although postwar employment histories ranged from custodian and garbage collector to educator to health professional—and education levels varied from the barely literate to the highly trained—the sample probably included somewhat more white-collar workers and professional people than the state's black population in general. Despite the sample's technical limitations, I believe that the views expressed are reasonably representative.

The transcriptions rendered here are literal in every sense, save one: in the interest of more readable narratives, I have not littered these pages with ellipsis points to indicate the omission of habitual filler words, frequent false rhetorical starts, or even material extraneous to a particular line of thought or story. Because unrehearsed speech is rarely seamless, and interviewees often revisit a single topic repeatedly over the course of an interview, I have in some cases merged these related and recurrent strands of conversation into a single, coherent discourse. These few liberties can be identified by the page numbers cited in the endnotes. Otherwise, of course, the respondents' words, syntax, and meaning are reproduced exactly as given. The narratives are temporarily closed, pending completion of transcript processing, but interested scholars will ultimately enjoy unrestricted access through the Mississippi Oral History Program, (hereafter cited as MOHP), University of Southern Mississippi, Hattiesburg.

2. See, for example, Richard Dalfiume, "The 'Forgotten Years' of the Negro Revolution," *Journal of American History* 55 (June 1968): 90–106; Richard Polenberg, *War and Society: The United States, 1941–1945* (Philadelphia: Lippincott, 1972), 99–130; Harvard Sitkoff, "Racial Militancy and Interracial Violence in the Second World War," *Journal of Ameri-*

can *History* 58 (December 1971): 661–81; and Neil A. Wynn, *The Afro-American and the Second World War* (New York: Aolmes and Meier, 1975).

3. Luella Newsome, interviewed by the author, December 14, 1993, vol. 486, p. 57, MOHP. (Except where otherwise indicated, all citations are to MOHP interviews, and all interviews are by the author. The absence of numbers for volume and page indicates an interview not fully processed at the time of this writing.)

4. Harvard Sitkoff, "African American Militancy in the World War II South," in McMillen, ed., *Remaking Dixie,* 70–92.

5. Wilson Ashford, December 3, 1993, vol. 477, pp. 13, 25. See also Sears Ward, April 19, 1994.

6. Ulysses Lee, *The Employment of Negro Troops,* United States Army in World War II, Special Studies (Washington, D.C.: Center of Military History, United States Army, 1966); Richard Dalfiume, *Desegregation of the United States Armed Forces: Fighting on Two Fronts, 1939–1953* (Columbia, MO.: University of Missouri, 1969); Neil A. Wynn, *The Afro-American and the Second World War,* rev. ed. (New York and London: Homes and Meier, 1993), 21–38.

7. Luella Newsome, December 14, 1993, vol. 486, p. 57.

8. Nathan Harris, December 2, 1994, vol. 598, p. 8. In another example, an infantry sergeant remembered the happiness of his soldiers while on assignment in Western Europe: "their only question was, 'Why can't it be this nice in America?'" (Dabney Hamner, December 2, 1994, vol. 606, p. 22).

9. Dabney Hamner, December 2, 1994, vol. 606, p. 56.

10. Charles H. Jones, Sr., December 2, 1994. But see also John C. Berry, November 16, 1993, vol. 471, p. 6; Roscoe Simmons Pickett, November 5, 1993, vol. 468, p. 12; Nathan Harris, December 2, 1994, vol. 598, p. 12; Napoleon Coney, November 28, 1994; James B. Jones, December 14, 1993, vol. 499, p. 25; Claude Montgomery, Jr., December 2, 1994, vol. 597, p. 9.

11. Henry Murphy, October 24, 1994, vol. 595, p. 20.

12. Brunetta Garner, interviewed by James Nix [December 1993], vol. 495, p. 10.

13. Eugene Russell Sr., March 3, 1994, vol. 498, p. 37. See also Clell Qualls, March 2, 1994, vol. 609, p. 36; Sears Ward, April 19, 1994; Charles H. Jones Sr., December 2, 1994; Haywood Stephney, December 2, 1994; James Boykins, November 23, 1993.

14. Clemon F. Jones, September 21, 1994. Similar experiences are described in Nathan Harris, December 2, 1994, vol. 598, p. 16; Ollie Jackson, February 3, 1994, vol. 529, pp. 18–20; Douglas Conner, December 2, 1993, vol. 479, p. 11–13; Alva Temple, December 3, 1993, vol. 481, p. 18; Edward L. Thompkins, April 16, 1994, vol. 592, p. 22.

15. Vertie Bell to Elliott Freeman, September 14, 1943; S. D. Redmond to Truman Gibson, September 27 and December 12, 1943; Leslie Perry to Henry Stimson, November 23, 1943; T. Hughes to Truman Gibson, December 12, 1943, all in Box 198, Records of the Office of the Secretary of War (Hastie Files), Record Group 107, National Archives, Washington, D.C.; miscellaneous Rieves Bell materials, Soldier Troubles, Legal File, 1940–1955, Box B156, Papers of the National Association for the Advancement of Colored People (NAACP), Library of Congress, Washington, D.C.; *Washington Tribune,* December 11, 1943; Wilson Ashford, December 3, 1993, vol. 477, pp. 30–31.

16. Henry Murphy, October 24, 1994, vol. 595, p. 25.

17. See, for example, letters from soldiers, spouses, and parents in Soldiers Complaints, Legal Files, 1940–1955, Boxes B148–B151, NAACP Papers.

18. Henry Murphy, October 24, 1994, vol. 595, p. 21. Although military records document many interracial conflicts, some veterans believed that the War Department was engaged in a continuing coverup. See, for example, Edward L. Thompkins, April 16, 1994, vol. 592, pp. 12, 17, 19. For a more problematic exposition of this argument, see Carroll Case, *The Slaughter: An American Atrocity* (n.p., FBC, Inc. 1998), an imaginative and unsubstantiated retelling of a much-rumored military conspiracy to coverup the murder of more than a thousand mutinous black troops, allegedly by white army personnel acting under orders, at Camp Van Dorn, near Centreville, Mississippi, in 1943. Mr. Case is generous to credit me for supplying what he thinks to be the documentary record essential for his "fact-based fictional account" of the "story the army has not told." Although I did supply copies of the authentic documents so meticulously reproduced in his book, he has interpreted them in ways that seem improbable to me. I share his keen interest in local hearsay about mass murder at Camp Van Dorn, but the evidence I have seen does not lead me to share his conviction that the United States Army directed the secret summary execution of roughly one-third of the black 364th Infantry Division after a well-documented race riot at that World War II facility.

19. Napoleon B. Evans, interviewed by James Nix, November 27, 1993, vol. 489, p. 14. See also Edward L. Thompkins, April 16, 1994, vol. 592, pp. 77–83.

20. Wilson Ashford, December 3, 1993, vol. 477, p. 2.

21. Lamar Lenoir, December 21, 1993, vol. 496, p. 16. Other expressions of this point of view include Henry Kirksey, March 4, 1994; William A. McDougal, December 3, 1994.

22. Ben Fielder, January 27, 1994, interview field notes. Similar stories are told by Ollie Jackson, February 3, 1994, vol. 529, pp. 21–22; Matthew Burks, interviewed by Arvarh Strickland, July 12, 1994 (in author's possession).

23. Dabney Hamner, December 2, 1994; Napoleon Coney, November 28, 1994. See also Charles H. Jones Sr., December 2, 1994.

24. Henry Murphy, October 24, 1994, vol. 595, pp. 26–30.

25. Wilson Evans, interviewed by Orley Caudill, n.d.

26. Sam Jackson, November 20, 1993, vol. 473, p. 8.

27. Dabney Hamner, December 2, 1994. In the 1950s, Hamner became the first black deputy sheriff in Coahoma County since Reconstruction.

28. William Nicholson, interviewed by Roscoe Pickett, November 29, 1993.

29. Nathan Harris, December 2, 1994, vol. 598, pp. 11, 16. Representative POW stories include Emmett J. Stringer, December 3, 1993, vol. 480, pp. 38–39; Alva Temple, December 3, 1993, vol. 481; James B. Jones, December 14, 1993, vol. 499, p. 18.

30. In a typical example, a WAC veteran said of her wartime experiences in the North: "It [racism] was still there, but not in abundance like it was here. In New York, it was just more subtle. They didn't just do you like the people down here [in Mississippi] did you. But I think it was just the same, but you didn't feel it as much" (Brunetta Garner, interviewed by James Nix [December 1993], vol. 495, p. 16). See also Lillian McLaurin, February 17, 1994; Douglas Conner, December 2, 1993, vol. 479, p. 12.

31. Haywood Stephney, December 12, 1994. A retired Scott County garbage collector remembered, "They'd treat you better anywhere than in Mississippi. In Europe and all them places, you was just the same as a white man" (Clell Qualls, March 2, 1994, vol. 609, pp. 14, 22).

32. Eugene Russell Sr., March 3, 1994, vol. 498, pp. 2–3, 9–11, 31, 56–57.

33. Dabney Hamner, December 2, 1994, vol. 606, pp. 16–17, 23. This interviewee reported that he then knocked down his white antagonist! The police, he said, advised him to "overlook those things" in the future but took no other action. The white man actually later apologized. Aware that this series of events had few parallels in the history of contemporary Deep South race relations, the veteran expressed the belief that he could perhaps respond to white insult with such impunity because he was then a letter carrier on leave from military duty, and whites were "afraid to fool with a federal employee." The annals of wartime racial conflict in Mississippi suggest that such immunity was rarely enjoyed by other black federal employees, military or civilian.

34. "The first experience I had reminding me that I was back home," one veteran remembered, came on a bus from Jackson to Hermanville, Mississippi, when he moved the Jim Crow curtain forward so that an overflow of black passengers could sit in seats not used by whites: "The driver reminded me that I'd better not do that again: 'Don't you ever do that on a bus I'm driving. You're back home'" (William J. Heath Sr., September 8, 1994, vol. 596, pp. 14–15). See also Napoleon Harris, December 2, 1994; Sam Jackson, November 20, 1993, vol. 473, pp. 27–29.

35. Roscoe Simmons Pickett, November 5, 1993, vol. 468, p. 7; John C. Berry, November 16, 1993, vol. 471, p. 9; William Nicholson, interviewed by Roscoe S. Pickett, November 29, 1993; Sears Ward, April 19, 1993; Claude Montgomery Jr., December 2, 1994, vol. 597, p.5.

36. Brunetta Garner, interviewed by James Nix [December 1993], vol. 495, p. 2; Douglas Conner, December 2, 1993, vol. 479, pp. 4, 7–8.

37. Clemon F. Jones, September 21, 1994.

38. David W. White, September 30, 1994, vol. 610, p. 11.

39. Wilson Evans II, interviewed by Orley Caudill, n.d.

40. James B. Jones, December 14, 1993, vol. 499, p. 8.

41. Alva Temple, December 3, 1993, vol. 481, pp. 13–14; Benjamin O. Davis Jr., *Benjamin O. Davis: American* (Washington D.C., and London, Smithsonian Institution Press, 1991), 132.

42. One former infantryman, however, a technical sergeant who had been wounded in Italy, did remember the shock and deep humiliation of discovering in 1946 that neither a Purple Heart medal nor an honorable discharge carried any weight at the Forrest County courthouse. Encouraged by his commanding officer, a northern white, to exercise as a civilian

the voting rights he had earned as a soldier, he presented himself, on returning home, to a very angry registrar of voters: "He told me, 'If you don't get out of here, I'll take my foot, kick you plumb out there in the middle of that street,' just like that. Yes sir" (Ollie Jackson, February 3, 1994, vol. 529, p. 15).

43. David W. White, September 30, 1994, vol. 610, p. 12.

44. German Levy, December 21, 1993. vol. 487, pp. 21, 35.

45. William A. McDougal, December 3, 1994. See also Clell Qualls, March 2, 1994, vol. 609, p. 37.

46. Matthew Burks, interviewed by Arvarh Strickland, July 12, 1994, p. 4, Arvark and Strickland to Neil R. McMillen, October 2, 1994; both in author's possession.

47. Ben Fielder, January 29, 1994, vol. 483, pp. 51, 84; Henry Murphy, October 24, 1994, vol. 595, p. 40.

48. John C. Berry, November 16, 1993, vol. 471, pp. 9, 21. But cf. Claude Montgomery Jr., December 2, 1994, vol. 597, p. 18.

49. Charles H. Jones, Sr. December 2, 1994. See also William A. McDougal, December 3, 1994; Dabney Hamner, December 2, 1994; Edward L. Thompkins, April 16, 1994, vol. 592, pp. 24–27; William J. Heath, Sr., September 8, 1994, vol. 596, p. 15.

50. Henry Kirksey, March 4, 1994.

51. Alva Temple, December 2, 1994, vol. 481, p. 29.

52. Henry Murphy, October 24, 1994, vol. 595, p. 33. Other representative examples include Sam Jackson, November 20, 1993, vol. 473, p. 20; Brunetta Garner, interviewed by James Nix [December 1993], vol. 495, 4.

53. Douglas Conner, December 2, 1993, vol. 479, pp. 9, 22, 30; Douglas L. Conner, with John F. Marszalek, *A Black Physician's Story: Bringing Hope in Mississippi* (Jackson, Miss, University Press of Mississippi, 1985), 51.

54. Nathan Harris, December 2, 1994, vol. 598, p. 22.

PART III

✦ ✦ ✦ ✦ ✦ ✦ ✦ ✦ ✦ ✦ ✦

THE NUCLEAR AGE

Myriad Faces of Limited War

CHAPTER 11

✦ ✦ ✦ ✦ ✦ ✦ ✦ ✦ ✦ ✦ ✦

Post-Traumatic Stress Disorder

The Legacy of War

Andrew Wiest, Leslie P. Root, and Raymond M. Scurfield

> Our choices are to be hurtled blindly into the next abyss because we refuse to recognize the living legacy of war, the survivors, or instead to come to terms with the world we have created by understanding the human costs of Vietnam. . . . In the aftermath of any war, it behooves us to recognize the profound price paid by the soldier survivors.[1]

The first portion of this chapter, written by Andrew Wiest, investigates battlefield trauma from a historical perspective.[2]

One of the great philosophical arguments throughout the ages centers on the most basic of human questions: Is mankind inherently good or bad? Are we little more than savages that the world would be better off without, or are we a perfectible species, born of a divine love? If mankind is, in fact, good, then it is quite possible that combat should be considered an inhuman aberration. Indeed, much of history can be seen as an effort to curb the killing instinct of mankind. The rule of law, the beauty of philosophy, and the wonder of religion all point the way to peace. Hammurabi, Lao-Tze, Jesus, and Mohammed, among many others, begged their followers to follow a peaceful path in life. Even so, humanity continues to be violent and often kills in the name of law, philosophy, or religion. Yet for modern mankind, with the exception of a few sociopaths, killing and combat are remarkably difficult.[3] The societal taboo against killing, though not always effective, is powerful. Thus, when one takes part in combat, one must shed the thin veneer of society and take part in the act of killing that is so repugnant to the human experience of nearly four thousand years of history. The result of such action can be devastating and quite traumatic.

The story of trauma and war dates back into the distant past. The earliest Western work on warfare, Homer's *Iliad,* is based on a story of betrayal, death, and revenge that is uncomfortably familiar to many modern veterans. The combat is often one on one and very close range, but the result is the same. Even the mighty Achilles is quite shaken by his role in the carnage. In fact, the *Iliad* can be read as an account of the undoing of the character of Achilles by the ravages of war.[4] Over millennia, the effects of industry and technology have changed the battlefield quite dramatically, but only superficially. Modern, technological war is more extensive and lethal than the ancients could have imagined. Thousands can die in an instant at the hands the frightening array of modern weaponry. Yet, death, maiming, and terror remain at the center of the modern conflict. Modern warriors may meet their enemy only rarely in hand-to-hand combat, as Achilles met Hector, but the rather detached nature of modern battle does not insulate today's soldier from death and killing. Soldiers in present and future wars will still have to face the threat of being killed, the horror of watching a friend die, and the possibility of killing another human being. A Vietnam veteran remembers a kill:

> And I froze, 'cos it was a boy, I would say between the ages of twelve and fourteen. When he turned at me and looked, all of a sudden he turned his whole body and pointed his automatic weapon at me, I just opened up, fired the whole twenty rounds right at the kid, and he just laid there. I dropped my weapon and cried.[5]

In the end, technology has increased the lethality and the trauma associated with battle. This conversion in the nature of war happened quickly at the beginning of the twentieth century. The new, technological battlefield was so traumatic that every army in the

Western world took note of a disturbing new trend, a dramatic rise in the level of psychiatric casualties in battle. Science had changed the battlefield, but combat itself was now so intense that many soldiers broke under the strain. Since the diagnosis of "shell shock" in World War I, psychiatrists and psychologists have grappled with the effects of modern battle on the human psyche. It was and is a vitally important yet formidable struggle. Military powers throughout the world are understandably more interested in devoting the fruits of science and technology to winning battles, not caring for the minds and souls of those who fight those battles. In recent years, however, the scientific healers have begun to gain ground. This chapter thus proposes to identify and illuminate the history of war trauma, post-traumatic stress disorder (PTSD) and its treatment. In so doing, we will examine the two most important wars in twentieth-century American history: the Second World War and Vietnam. We then switch gears, turning to a discussion of post-traumatic stress disorder, its causes, and its treatment.

War, with its risks of being killed or wounded and the added possibility of becoming a killer, has always caused psychological difficulties for combatants. The Greek historian Herodotus, in writing about the Persian War, relates the story of a Spartan soldier who was so shaken by his experience in battle that his comrades referred to him as "the Trembler." The soldier was so shamed by his reaction to conflict that he committed suicide.[6] The reaction of the Spartan is illustrative of how the military world dealt with psychiatric casualties until the mid–twentieth century. A soldier who cracked in battle or exhibited signs of trouble adjusting after the close of the conflict was considered to be "weak." The man who had trouble with the stress of war was somehow psychologically predisposed to mental disorder, and the stress of war merely separated the wheat from the chaff.[7] As military history progressed, the psychological casualties remained in its wake. They were sometimes ridiculed, sometimes killed, sometimes scorned, and often forgotten. Only in very rare occasions were such casualties cared for in any meaningful way.

The first clinical effort to understand the results of stress in combat took place during the United States Civil War. The surgeon general of the Union Army, William Hammond, noticed that a significant number of soldiers in the conflict suffered from varying levels of psychological harm due to the horror of war. He labeled the least virulent form of mental disability "nostalgia." In his study he discovered that 3.3 men per 1,000 suffered from the malady badly enough to warrant removal from battle. In addition, some 26.8 men per 1,000 incurred the more debilitating mental disorders of "paralysis" or "insanity."[8] Though science here had finally noticed the problem of the mental stress of battle, no real study was conducted, and no military leaders learned a lasting lesson. This is apparent in a U.S. government report on illness and treatment in the Civil War, published in 1888. The report stated that a certain type of soldier was prone to nostalgia, "young men of feeble will, highly developed imaginative faculties, and strong sexual desire." Another report, though, noted that the explosion of a shell nearby sometimes caused "compression of the brain," which in turn led to several mental disorders.[9]

World War I, with its heightened lethality and static nature, caused a veritable flood of psychological injuries. As early as December 1914, the British medical journal *The Lancet* began to consider mental illness in battle.[10] The ferocity of the war slowly took its toll, and increasing numbers of men on both sides of the front lines became mental casualties. Their symptoms ranged from paralysis to aimless wandering to loss of bladder and bowel control. As the percentage of mentally ill grew, psychologists and military leaders attempted to discover the nature of the problem. It is easy to see why they were so concerned, for the losses due to mental illness were staggering. The United States sent nearly 2 million men to Europe in World War I. Of this number, 116,516 were killed and another 159,000 were put out of action due to psychiatric problems.[11] The soldiers themselves knew why the number was so high. Wilfred Owen explains in his poem "Mental Cases,"

> These are men whose minds the Dead have ravished.
> Memory fingers in their hair of murders,
> Multitudinous murders they once witnessed.[12]

The common term used by most to describe mental illness in battle was *shell shock*. The prevalent belief was that the majority of psychiatric cases in battle were physical in origin: constant shelling led to chronic concussion and hemorrhages in the cerebellum, causing the victim to appear insane.[13] There remained soldiers, however, who exhibited symptoms without having suffered an extended period of shelling. These men were again considered "weak" and predisposed to mental illness. Lord Moran, in his *Anatomy of Courage*, was a compassionate observer of mental illness in battle. Even he, though, still considered many of war's victims to be weak:

> There were those who were plainly worthless fellows. One without moral sense . . . was sitting there with his head in his hands at the bottom of the trench. . . . He was just a worthless chap, without shame, the worst product of the towns.[14]

The lack of understanding regarding psychological injuries naturally led to a variety of treatments, most of them ineffective. Some soldiers received care, while others were simply ignored. Finally, many soldiers who were plainly suffering from mental trauma were executed as deserters or for dereliction of duty. Austria's methods for dealing with shell shock were perhaps the most disturbing. Threats of imprisonment and execution forced the malingerer back into the fray. When those threats failed, Austrian authorities resorted to electric shock treatment, some so severe that many patients died or committed suicide to avoid further torture. Some did return to battle, much worse for their stint in an Austrian hospital. The Austrian experience, though brutal, led to a major step forward in the treatment of psychological injuries. At the close of the war, Sigmund Freud was asked his opinion of the treatments. Though he did not reject electric shock out of hand, he did state his belief that the injuries suffered by these men were psychological in nature.[15] In the interwar years, Freud's belief that mental illness in battle was due to psychological and not physical damage gained general acceptance. Freud also made an error that would have lasting impact. He posited that once the stress of battle ended, soldiers would return to normal. He saw no reason to believe that war could cause long-lasting psychic damage.

The United States was not prepared, in many ways, for World War II. It was fairly common belief in 1941 that the stress of combat had something to do with mental breakdown in battle. However, there were not even psychiatrists with each U.S. division to deal with such problems until March 1944.[16] The American armed forces did take steps to make certain that men who were predisposed to "weakness" were not allowed into battle. As a result, the military rejected two of every five enlistees and draftees as mentally unfit.[17] Even with this precaution, the numbers of psychiatric casualties were staggering. The United States lost 500,000 dead in World War II; in comparison 1,393,000 soldiers were debilitated due to the traumatic effects of battle.[18] It was plain to the U.S. military that such losses were unacceptable and had to be explained.

In 1944, Dr. William Menninger, neuropsychiatric consultant to the surgeon general, appointed a commission of five psychiatrists to investigate the problem in detail. Their conclusion was that the cumulative stress of battle, from the trauma of killing to the tragic deaths of close friends, slowly wore down the psychic defenses of many soldiers. In the end, then, each man had his breaking point, beyond which lurked "combat exhaustion." At its worst the disorder would result in the complete mental and emotional destruction of the soldier.[19] The military took the study very seriously and began to develop proposed solutions for the problem, including limiting the exposure of soldiers to combat and providing psychiatric intervention before problems became pronounced. Such innovations would ensure that the United States entered future conflicts better prepared to deal with "combat exhaustion."

Though the military was learning how better to deal with the trauma of the battlefield, an attendant problem had gone almost unnoticed. A significant, though unknown, percentage of combat veterans never recovered from the trauma and guilt of war or simply had tremendous trouble readjusting to civilian life, belying the assumptions established by Freud. Psychological damage did not "wear off" but continued to manifest itself. Military and mental health professionals began to look at this persistence in a meaningful way only after World War II. The number of veterans suffering from such continued difficulties seemed to be quite small, aiding in the conclusion that veterans who suffered from such chronic symptoms had "personality disturbances" dating back to their childhood.[20] Some of the "weak" had somehow slipped past the testing process and were now paying the price. There were some studies regarding these troubled veterans,[21] but the number of veterans suffering from continued problems seemed so small that few paid their problems any real heed.

The critical question, in the post-Vietnam world, is why so few World War II veterans suffered the lingering problems associated with the trauma of war. There can be little doubt that the levels of battlefield trauma in World War II and Vietnam were quite high. Warner Hamlett recalled his experience at Omaha Beach:

> Private Gillingham fell beside me white with fear. He seemed to be begging for help with his eyes . . . then I heard a shell coming and dove into the sand facedown. Shrapnel rose over my head and hit all around me. It took Gillingham's chin off, including the bone, except for a small piece of flesh. He tried to hold his chin in place as we ran toward the seawall. He made it to the wall where I gave him a morphine shot. He stayed with me for thirty minutes until he died. The entire time he remained conscious and aware that he was dying.[22]

Nick Uhernick wrote of his experience in Vietnam:

> Simultaneously everyone leveled his weapon at him and fired. "Jesus Christ!" somebody gasped behind me as we watched his body reverse course back toward the trees; chunks of meat and bone flew through the air and stuck to the huge boulders. One of our rounds detonated

> a grenade the soldier carried, and his body smashed to the ground beneath a shower of blood.[23]

It is indeed difficult to discern any difference in the levels of horror exhibited in battlefield experiences in World War II and Vietnam. The circumstances surrounding the horror of the two conflicts were quite different, however.

World War II was the good war. The continental security of the United States had been threatened, indeed violated, by two world powers that Americans could identify as evil. The nation rallied as never before to a wartime cause it knew to be good and right. Young men enlisted or were drafted and trained as a unit. This group of fast friends served together throughout the war. As many veterans attest, the friendships made in training and forged in battle surpassed even the love of a man for his family. As a result, "our boys" went off to play their role in the cathartic struggle with the goodwill of an entire nation behind them and with a wonderfully close support group near at hand. During their time in battle these young men, who averaged twenty-five years of age, indeed witnessed traumatic, soul-wrenching events. They killed together and died together.

When the war was over, it was clear to these men that they had emerged victorious in the service of a just cause. Their road home from the war was torturously slow but therapeutic. Entire units of battle-hardened veterans languished in Europe or the Pacific, then made the long ocean journey home via ship. As a result these men, the closest friends of a lifetime, had a chance to vent, rage, weep, and console one another regarding the most important and frightening events of their lives. Once home, they were met by a nation reveling in the joyous rapture of victory. The nation was just as unified in its support of its returning sons and daughters as it had been in support of the war itself. There were parades, hugs, and prayers to meet the returning heroes of our greatest war. The parades eventually gave way to Veterans of Foreign Wars (VFW) meetings, Memorial Day gatherings, and the massive thank-you of the GI Bill.

All this provided welcome relief to soldiers who had witnessed events often too ghastly to discuss. Our boys had won. They had been able to work through their losses, guilt, and grief with their closest friends, and they were given a national hug and a permanent support group.

The World War II veterans, then, had much aid in overcoming the trauma of war. One additional factor, though much more controversial, worked in their favor. During the war, Brigadier General S. L. A. Marshall conducted after-action interviews with soldiers regarding their combat experiences. What he discovered was shocking. Only 15 to 20 percent of U.S. soldiers who saw combat actually ever fired their weapons.[24] As a result, some 80 percent of U.S. combat soldiers never actually took part in that most traumatic of events, the killing of another human being. This was instrumental in helping these soldiers' readjustment after the close of the conflict. Though Marshall's methods and conclusions still spark controversy in the historical profession, the results of the study sent shock waves through the military establishment. Such a poor rate of fire was simply unacceptable. In the end, it seems that training was to blame. Soldiers in World War II had been trained to fire their weapons rapidly and accurately. They were often expert marksmen but not killers. The training regimen before our next major conflict would address that dilemma—it would train Americans to kill.[25]

The United States military had learned much about psychiatric casualties during World War II and put its newfound knowledge to the test in Korea and Vietnam. During Korea, which came fast on the heels of the Second World War, the rate of psychiatric casualties fell considerably. Techniques used there were perfected in Vietnam, and the number of psychiatric casualties in that conflict fell to an all-time low. In World War II, 10 percent of all battlefield casualties were psychiatric in nature; that number fell to 1 percent in Vietnam.[26] The practices that led to this level of success were varied in nature. The U.S. Army made certain that its soldiers had a high level of amenities in Vietnam. From Budweiser to frequent rest and relaxation, the American soldier in Vietnam had it all. Also, in an effort to limit a soldier's exposure to combat, the military limited the typical combat tour of duty in Vietnam to one year. Finally, the armed forces chose to implement the practice of treating psychiatric injuries quickly near the front lines and reintegrating them into their units as soon as possible.[27] The results of these policies were instant and dramatic. It seemed that the United States Army had mastered the problem of battle trauma.

Our Vietnam experience was not to be so easy, though, as families, loved ones, and veterans themselves began to notice problems among the population of returned Vietnam veterans even before the conflict drew to a close. Symptoms varied but included guilt, social alienation, catastrophic nightmares, intrusive thoughts, sleep disturbance, overly aggressive behavior, and flashbacks.[28] In short, the trauma of war followed the men and women home to haunt them there. These veterans were suffering from varying levels of post-traumatic stress. In its rush to forget Vietnam, the nation also forgot these veterans, who in many ways were never able to return home. For decades, only a small, dedicated number of veterans and mental health professionals worked to save these men. The story of post-traumatic stress, its effects on veterans and others, and its eventual treatment forms the second portion of this chapter. What must concern us here is the historical question of why post-traumatic stress disorder is associated with Vietnam. In Vietnam there were very few cases of psychiatric injury in battle but record numbers of cases of post-traumatic stress. It was World War II in reverse. What made Vietnam different?

Vietnam was not the good war. American soldiers fought for a culture and country that they did not understand, thousands of miles away from home, against an enemy that did not seem to threaten the United States in a direct way. In a confusing turn of events for the fighting man, many of his South Vietnamese allies did not even want him to be in the country. In turn, America entered the Vietnam War united in its support for the conflict but never at the same level of support shown for the great moral crusade of World War II. As the conflict wore on, more and more of the American public began to turn against the war, and even against those sent to fight it. The soldier of the Vietnam era, then, never even had the wholehearted support of his own nation or of the nation for which he was fighting.

The one-year tour of duty and the draft process did even more to destroy the traditional support groups for soldiers in battle. The average age of combat soldiers in Vietnam was nineteen.[29] This is the period when most adolescents establish their stable personality structure. For many young men in Vietnam, the trauma of war interrupted this important process, forever disrupting their lives and leaving them with untold subsequent problems.[30] The one-year tour of duty was designed to protect these young men from an overlong exposure to combat but caused many unexpected difficulties. Most service members entered Vietnam as single replacements for soldiers who had reached the conclusion of their tour. The replacements quickly would become focused on the end of their own tour of duty. The war started the day a soldier entered Vietnam and ended 365 days later. The goal was simply to survive during that period. This simple fact reduced the continuity of the war dramatically. Soldiers felt little relationship between themselves and those who came before or would come after.

The Vietnam War was, then, a very individual and personal thing. Soldiers entered the country singly, without that circle of friends that was so important to their individual well-being. A new soldier was considered to be a "fuckin' new guy," often to be ignored and certainly not to be trusted in battle. Slowly the replacement would integrate into the unit, only to see any newfound friends rotate out or become casualties. Friendships in Vietnam were sporadic, fleeting things, leaving many soldiers in quiet, solitary desperation. In more clinical terms, a high percentage of units in Vietnam lacked that important sense of integrity that acts as a buffer to protect the individual from the overwhelming stresses of combat.[31] Finally, the one-year tour of duty acted as a palliative. Soldiers could suppress their feelings of remorse and guilt, awaiting their return home when all would be well again. Increasing numbers of troops also relied on prophylactic drugs and alcohol. These tactics, though often effective in deadening pain in the short run, only put the reckoning off until later. Young men and women in Vietnam then entered the war and combat alone, had few real friends for support, and lived for the end of their tour, that mythic date when all would be well with the world again. It was a fantasy that would collapse around many veterans with devastating results.

The combat in Vietnam was also different from that in World War II in many ways. Most American forces were trained in conventional warfare. Vietnam, however, was in no way conventional. The enemy was elusive, seemingly everywhere and nowhere, and commonly struck U.S. forces through the use of mines and booby traps. The enemy often did not wear uniforms, and American troops were often forced to kill women and children combatants, causing high levels of emotional trauma. There were no discernible front lines, and victory came to be defined in terms of the "body count." American forces often did not seize and retain control of enemy territory. Instead, U.S. forces swept through an area inflicting and taking losses, only to return to the same area a few weeks later and go through the same process over again. To many, these processes did not make sense, and the only outcome of the war seemed to be the endless production of maimed, wounded, and dead.[32]

Training also had another effect on the soldier in Vietnam. The military had to achieve higher rates of fire than the 15 to 20 percent seen in World War II. As a result, military training began to focus on aspects of desensitization to killing as well as on marksmanship. Such training involved shooting at human-shaped targets that would fall down when struck. In training, every aspect of killing on the battlefield was rehearsed, visualized, and conditioned. One Vietnam veteran recalls:

> The Vietnam era was, of course then at its peak, you know, the kill thing. We'd run PT [physical training] in the morning and every time your left foot hit the deck you'd have to chant "kill, kill, kill, kill." It was drilled into your mind so much that it seemed like when it actually came down to it, it didn't bother you, you know? Of course the

> first one always does, but it seems to get easier—not easier, because it still bothers you with every one that, you know, that you actually kill and you know you've killed.[33]

In the end, the new training was quite effective. S. L. A. Marshall was sent to the Korean War to investigate the results of the training and discovered that 55 percent of the infantry engaged in combat fired their weapons. In a later study in Vietnam, Marshall concluded that an amazing 90 to 95 percent of soldiers involved in combat fired their weapons.[34] The U.S. Army was now a much leaner, more lethal force. Now, however, an astounding 95 percent of soldiers in combat had taken some part in the killing. Thus the levels of trauma and guilt associated with war rose considerably in and after Vietnam.

The war ended for most veterans of Vietnam after their one-year tour of duty on a solitary flight home on the freedom bird. Unlike Second World War veterans, who had months to discuss the horrors of war with beloved comrades, Vietnam veterans were rushed out of the war singly and back into society with very little fuss or time to adapt. Soldiers simply had no time to work through their trauma with caring comrades. The experience of one veteran was all too typical:

> One day I was out in the bush, killing gooks, seeing buddies get killed, covered in mud, trying to sleep at night with the threat of ambush by the VC and two days later I was trying to talk to my family at the dinner table. I couldn't tell them what I had been through. They couldn't have understood it.[35]

In addition to having precious little time to decompress, veterans often returned to ignorance and apathy at best, and hostility at worst, on the part of much of the American public. The Vietnam veteran did not win his war and did not return to find parades and a national welcome home. Many Americans had turned against the war, and others had even turned against the soldiers themselves. One veteran recalls his homecoming:

> On returning from Vietnam, minus my right arm, I was accosted twice . . . by individuals who inquired, "Where did you lose your arm? Vietnam?" I replied, "Yes." The response was, "Good. Serves you right."[36]

Another veteran recalls a less dramatic but possibly more common response of apathy:

> After returning home I took a job with a television station as a still photographer. One day a young girl came around to interview me for the company newsletter as a new employee. She asked me all the usual questions, and then asked, "What is your most recent employment experience?" I puffed up with pride and responded, "I just got back from three tours of duty in Vietnam." She did not bat an eye and went on to ask me where I had gone to school. I was flabbergasted. I had given three years of my life to my country and had seen many friends die. I could have accepted rage or praise on her part, but her utter indifference to my suffering was astounding.[37]

The national support structure that was there to help the World War II veteran deal with guilt and trauma was simply not present for the Vietnam veteran. For many Vietnam-era veterans, this was all the more difficult to take because they could remember taking part in the national welcome home for their World War II veteran fathers—a welcome denied them. In addition, a segment of American society helped make things immeasurably worse for the returning soldier. America had sent its beloved sons to war, had asked them to take part in that most horrible of human events. When they returned, some loved them unconditionally but often could not understand the horrific nature of their conflict or help them. Others treated them with hostility or indifference. There would be no memorial or parade for decades as America stumbled all over itself to sweep the trauma that was Vietnam under the historical carpet. The veterans were painfully aware of this process. They knew that they were unwanted losers in a society full of winners. Many were unable to deal with the pain of their rejection and sank back into themselves, denying their own pain. Lt. Col. Dave Grossman deals with this process eloquently in his book *On Killing*:

> The Vietnam vet . . . is suffering an agony of guilt and torment created by society's condemnation. During and immediately after Vietnam our society judged and condemned millions of returning veterans as accessories to murder. At one level many, even most, of these horrified, confused veterans accepted society's media-driven, kangaroo-court conviction as justice and locked themselves in prisons of the worst kind, prisons in their own mind. A prison whose name is PTSD.[38]

The second part of this chapter, written by Leslie Root and Raymond Scurfield, investigates war-related trauma and its treatment from a clinical psychological perspective.

Post-traumatic stress disorder is defined as psychological wounding that can occur in the aftermath of "events that are so extreme or severe, so powerful, harmful and/or threatening that they demand extraordinary coping efforts,"[39] and that subsequently alter the individual's sense of self, others, and the world. Combat or other war-zone stressors represent one of the most extreme or severe forms of trauma, because there are often a series of repeated, unpredictable human-induced events, as opposed to natural disasters, that occur over a prolonged period. Comprehending this level of trauma may be difficult. One veteran describes it this way:

> Just imagine taking all the horrible and traumatic events one could possibly experience in a lifetime and cram them into one year at the impressionable age of 18 or 19 and have them occur every day.[40]

PTSD is most often associated with the Vietnam War, for good reason. The National Vietnam Veterans Readjustment Study (NVVRS), the largest national mental health epidemiological survey ever conducted on a U.S. veteran population, discovered alarmingly high rates of PTSD even twenty years after the war had ended. Of the 3.14 million veterans who served in Southeast Asia during the Vietnam War, more than 829,000 are suffering from some degree of post-traumatic stress.[41] The NVVRS data revealed that almost 25 percent (15.2 percent male; 8.5 percent female) of theater veterans in 1990 had full-blown PTSD. An additional 18.9 percent (11.1 percent male, 7.8 percent female) of theater veterans suffer some degree of PTSD symptoms or "partial PTSD" but do not meet the full diagnostic criteria.

To link PTSD too closely with Vietnam, however, would be a mistake. A common misconception is that PTSD does not occur among World War II and Korean War veterans. Although the prevalence of PTSD among World War II and Korean War theater veterans appears to be somewhat lower than among Vietnam theater veterans,[42] a significant number of these veterans also suffer from PTSD. Partly due to cultural definitions of manhood and bravery, a very low percentage of World War II veterans sought psychological help in readjusting to their postwar world. Since the close of the Vietnam War and the subsequent acceptance of PTSD as a true psychological disorder, increasing numbers of World War II and Korean War veterans are coming forward who had quietly struggled against the disorder for decades. Recent studies of World War II and Korean War veterans receiving psychiatric treatment in fact indicate that they suffer from levels of PTSD that approximate that of Vietnam veterans.[43] Furthermore, a study assessing the prevalence of mental disorders in prisoner-of-war survivors (POW) found even higher rates of PTSD: 88 percent for Korean War POWs, 76 percent for World War II Pacific theater POWs, and 54 percent for World War II European theater POWs.[44] The trauma of war, then, appears to be universal across time.

Not all veterans exposed to combat or war-zone trauma, however, develop PTSD. Thus exposure to trauma in and of itself does not always result in PTSD. Pre-trauma factors, such as prior life experiences and coping style, may serve as either risk or protective factors in post-traumatic stress reactions. Post-trauma factors including the individual's response to trauma, how the survivor makes sense of the trauma, and the quality of available social supports may also influence the development of longer-term post-traumatic stress reactions.

The most important factors affecting the likelihood of developing

PTSD, however, are the severity, duration, and proximity of an individual's exposure to the traumatic event.[45] The NVVRS findings revealed that the prevalence of PTSD and other postwar adjustment problems is significantly higher among Vietnam veterans with high levels of exposure to combat than among their military peers who did not serve in Vietnam and among Vietnam veterans with lower levels of exposure.[46] Thus, the greater the extent and severity of exposure to war trauma, the greater the risk for developing PTSD.[47] Sutker and Allain concluded:

> As trauma events become more universally brutal; more horrific, gruesome, and prolonged; and more threatening to life, the greater the likelihood that negative psychological sequelae will develop . . . that as trauma experiences become more severe and extreme, eventually all victims succumb to psychological distress.[48]

Those exposed to the mental trauma of combat, then, are the most likely to suffer from PTSD. Other factors of the combat experience, however, can make the threat of PTSD even more likely. Soldiers who are wounded in battle, for instance, suffer elevated rates of PTSD (over 30 percent).

> It was terrible, right after I was hit, I was lying there, all exposed, couldn't move and no one could get to me. I was in such pain, and I was terrified I was going to bleed to death or be hit again.[49]

The wounding itself was compounded by many additional traumatic aspects of the actual medical evacuation process, from the battlefields of Vietnam to field hospitals and eventually back to stateside medical facilities for, in many cases, long-term rehabilitation:

> There were shots all around me as I was being put on the chopper; I was so scared, and when the chopper took off, it got hit and we crashed. I was even more terrified. We then got put on a second chopper, and I was so scared the whole time that we'd be shot down again.[50]

> You know, we learned real quickly about what would happen if we argued with [hospital] staff "too much," or came across as "too demanding." They would raise our meds! So we quickly learned the

ropes—if we wanted more meds to numb out more, then just be disruptive and a pain in the ass—they'd pump us full. If we didn't want more meds then must be quiet and not bother anyone.[51]

Other findings indicate that ethnic minority veterans are at greater risk for developing PTSD. For example, two large-scale national studies of Vietnam veterans found that African American and Hispanic Vietnam veterans had current post-traumatic stress reactions that were significantly higher than those manifested by Caucasian veterans. The Matsunaga Vietnam Veterans Project reported elevated rates of lifetime prevalence of PTSD among African American (35.4 percent) and Hispanic (33.7 percent) Vietnam veterans, in contrast to 19.9 percent for their Caucasian counterparts.[52] The project[53] also reported a similarly higher lifetime prevalence of PTSD among Southwest (45.3 percent) and Northern Plains (57.2 percent) American Indian and Native Hawaiian (38.1 percent) Vietnam veterans.

There has been much speculation about these findings and the experiences of ethnic-minority veterans in Vietnam and afterward. It has been postulated that African American veterans were in a somewhat conflictual, position in which they perceived themselves as being in a "white man's war" against a third-world country and against a people of color who were devalued, maligned, abused, and helpless.[54] The Vietnam War was quite racist in terms of rampant characterization by the U.S. military, both in basic training and in Vietnam, of the Vietnamese as being "gooks," "chinks," "slopes," and "slant-eyes":[55]

How could you kill them and not be affected by it on a deep level? They're just like you in so many ways.[56]

—African American Vietnam veteran

American Indians also seem to have experienced psychological tension and ambivalence in that they were fighting a white man's war while serving a government that historically had betrayed their own people.[57] Other studies[58] indicate that Asian American Vietnam veterans were in the very vulnerable and particularly anguished position of having fought people who were of the same race or who looked very much like their own family members and relatives:

> I had to fight part of who I was.[59]
>
> —Asian American Vietnam veteran

Many ethnic-minority veterans were laying their lives on the line in Vietnam, while back home race riots and race-related relations in the United States were at a heightened state of tension—further conflicting many ethnic-minority veterans while they were in-country and complicating their postwar readjustment after their return to the United States:

> We are taking the young black men who have been crippled by our society and sending them 8,000 miles away to guarantee liberties in Southeast Asia which they had not found in Southwest Georgia and East Harlem. So we have been repeatedly faced with the cruel irony of watching Negro and White boys on TV screens as they kill and die together for a nation that has been unable to seat them together at the same school. So, we watch them in brutal solidarity burning the huts of a poor village, but we realize that they could never live on the same block in Detroit.[60]

This problem was compounded by the fact that both clinicians and ethnic-minority veterans would avoid discussion of race-related issues in the treatment of PTSD:

> Hey, I'm the only black Vietnam veteran in this [therapy] group with all white vets—and you expect me to bring up the racism shit that I had to deal with in Nam!?[61]

Most veterans who suffer from PTSD were unable to receive meaningful treatment for the disorder for some years, for it was only in 1980 that post-traumatic stress disorder was recognized as a psychiatric disorder by the American Psychiatric Association in its *Diagnostic and Statistical Manual for Psychiatric Disorders, 2d ed.* (*DSM-II*).[62] The most recent revision of the official stance is contained in

DSM–IV and recognizes two categories of stress disorder: acute stress disorder and post-traumatic stress disorder. "Acute stress disorder" describes reactions to extreme stressors from *two days to four weeks* after exposure, and PTSD describes reactions *more than four weeks* after exposure. The current understanding and definition of PTSD is of fundamental importance and states:

> A. The person has been *exposed to a traumatic event* in which *both* of the following were present: (1) the person experienced, witnessed, or was confronted with an event or events that involved actual or threatened death or serious injury, or threat to the physical integrity of self or others; and (2) the person's response involved intense fear, helplessness, or horror.
>
> B. The traumatic event is *persistently reexperienced* in one (or more) of the following ways: (1) recurrent and intrusive distressing recollections of the event; including images, thoughts, or perceptions; (2) recurrent distressing dreams of the event; (3) acting or feeling as if the traumatic event were recurring; (4) intense psychological distress at exposure to internal or external cues that symbolize or resemble an aspect of the traumatic event; and (5) physiological reactivity on exposure to internal or external cues that symbolize or resemble an aspect of the traumatic event.
>
> C. Persistent avoidance of stimuli associated with the trauma and numbing of general responsiveness (not present before the trauma), as indicated by three (or more) of the following: (1) efforts to avoid thoughts, feelings, conversations associated with the trauma; (2) efforts to avoid activities, places, or people that arouse recollections of the trauma; (3) inability to recall an important aspect of the trauma; (4) markedly diminished interest or participation in significant activities; (5) feeling of detachment or estrangement from others; (6) restricted range of affect (e.g., unable to have loving feelings); and (7) sense of a foreshortened future (e.g., does not expect to have a career, marriage, children, or a normal lifespan).
>
> D. Persistent symptoms of *increased arousal* (not present before the trauma), as indicated by two (or more) of the following: (1) difficulty falling or staying asleep; (2) irritability or outbursts of anger; (3) dif-

ficulty concentrating; (4) hypervigilance; and (5) exaggerated startle response.

E. Duration of the disturbance (symptoms in Criteria B, C, and D) is more than one month.

F. The disturbance causes clinically significant distress or impairment in social, occupational, or other important areas of functioning.[63]

Reading the clinical criteria does not quite explain the reality of chronic and severe PTSD. War is not a memory for PTSD veterans; it is continually reoccurring through nightmares and what some veterans refer to as "daymares" (intrusive images)—vivid recollections of events and glimpses of horror that they cannot forget. The simplest things can "trigger" a recollection—a baby crying, a helicopter, the smell of diesel fuel, the heat, the rain, an American flag. When they recollect an aspect of their trauma, they also experience intense emotions such as fear, terror, rage, sorrow, and shame. In an effort to avoid such distressing memories and feelings, they learn to go numb and detach themselves from their feelings. They typically avoid people and places that resemble or symbolize their traumatic experiences, which results in their staying at home and being isolated from others. They often do not understand what is happening to them and are frightened, angry, and ashamed of their symptoms. They have great difficulty talking to others about what they are experiencing. Their world begins to shrink—but the nightmares and intrusive thoughts still come. They might drink alcohol or take drugs to anesthetize themselves and escape their memories—if just for a while—or they may get themselves to the point of exhaustion just to be able to get a few hours of sleep. They typically sleep only a few hours at a time, awake in the very early morning hours, and are often unable to return to sleep. They are tired and begin the day knowing they will remember again.

Each time the original trauma or some aspect of it is reenacted, the associated fear is activated to varying degrees. Their experiences and the constant anticipation of reexperiencing them leave them in a persistent heightened state of apprehension. They are always ready and waiting for something to happen. Some know in their minds that their fear is unfounded, but they cannot stop themselves. They engage in rituals to manage their anxiety—nightly routines of checking the "perimeter," or walking around their property and checking doors and windows to make sure they are secure. They may avoid crowds. When sitting, they typically will keep their backs to a wall and "scan" the environment. They are hypersensitive to the slightest noise, and with each unfamiliar sound, the ritual must be reenacted. Returning to sleep is often impossible. Veterans with more severe PTSD may sleep with their clothes on, on couches and floors, with a weapon close by. They cannot help themselves. They may be very ambivalent about their fear—it significantly interferes with everyday functioning—but they feel powerless to overcome it. In contrast, they value highly the protective nature of their fear and hyperarousal.

One consequence of PTSD is the trauma survivor's tendency to attempt to control all aspects of his or her environment in order better to manage fear and anxiety. Survivors' inability totally to control themselves and circumstances around them, however, coupled with sleep deprivation, leads to chronic irritability. Anger outbursts are common in many and are often reactions to feeling stressed or out of control, attempts to keep others from getting too close, or expressions of the hatred they feel for their sense of incompetence. Others will deal with their fears, anxieties, and anger by maintaining a stoic presence, do not express or access their emotions, and seem "emotionally distant" to others.

Their lives become about managing their memories, their emotional turmoil, and their unsuccessful attempts to cope, and soon they are overwhelmed, making life a continuous struggle, or they maintain severe isolative patterns. This often leads to feelings that life is empty and meaningless, and they see no future. They become sad and depressed. Veterans with severe PTSD are often preoccupied with and flirt with suicide:

> Even today, I feel like so much of me died in Vietnam, that at times I wished all of me had died over there. For those who came back, the

> price of living is never easy or cheap. Laughter and happiness is rare. The nightmares, the flashbacks, the pains, waking up soaked in sweat, hyper-alert, and sleeplessness . . . are the norm. The sounds and smells of combat, the smell of sweat and dust, of the damp earth and vegetation, of the hot sun and exhaustion, of ambushes and firefights to full-blown battles, and of blood and death, enter my daily life. The moans of the wounded, some cursing, others calling for their mother, someone screaming for the corpsman or moans of "Oh God, Oh God." Like so many other Vietnam vets, I feel so much rage in me that it exhausts me and isolation is my only sanctuary.[64]

War veterans commonly experience a lasting moral guilt—a violation of values by either what they did or what they did not do.[65] This often results from the veteran's failure to recognize that traumatic situations rarely present "good" choices and "bad" choices; rather, the options are often "bad" choices and "worse" choices (e.g., to kill or be killed). This dilemma is illustrated by the following two examples:

1. A veteran was having nightmares and feeling a tremendous amount of guilt over not stopping other soldiers from torturing Vietnamese prisoners of war. He did not permit himself at the time, and for many years afterward, to realize that his interference might have resulted in his own death.
2. Another veteran was experiencing guilt over his inability to prevent the death of a young Vietnamese girl. Someone yelled "Grenade!" and he responded instantly. He turned to see a girl about three or four years old, walking, carrying a grenade in her hand. He could not bring himself to kill her like as had been trained to do, so he attempted to shoot the grenade out of her hand. Her injuries were severe, and she died. He believed he should have been able to save her. He discounted his well-intentioned attempt to save her due to the adverse outcome of his actions.

For many American military personnel, such choices and exposure to the consequences were fraught with immense individual and collective guilt and anguish—which have been replayed over and over again postwar through intrusive memories, nightmares, and guilt-ridden angst.[66]

One Asian American Vietnam veteran's guilt reflects a conflicted ethnic identity, torn between being an American soldier and being of Asian racial and Asian American cultural heritage:

> One side of me is reaching out with my heart while my hands and arms are stabbing them, abusing them. . . . I was laughing at them, which was a mask for feeling like crying. I had to become a super-gook to survive, to prove I was American. I have tremendous guilt for not shedding a tear over them, those who were my brethren. I lost great respect for myself for what I did.[67]

A complicating factor regarding PTSD is that individuals who suffer from the disorder are at increased risk for developing other mental health problems. Veterans diagnosed with PTSD also reveal high rates of alcoholism, depressive disorders, and anxiety disorders, regardless of the war in question.[68] The relationship of a co-morbid condition of substance abuse with PTSD can be very complex, for initial substance abuse was often directly triggered during the war or began or was initiated after the war at least partly as a form of self-medication to mask haunting memories and feelings.[69]

An individual was engaged in long-range reconnaissance. He and his buddies were regularly provided with combination barbiturate-amphetamine products by one of their commanders prior to going out on patrol. In addition, they purchased amphetamines locally and used both drugs regularly as part of their combat routine. Eliciting this particular history was of considerable importance in evaluating the individual's return to drug use some fourteen years later.[70]

Vietnam veterans with PTSD also have higher rates of marital problems and divorce, more occupational instability, and more legal problems than Vietnam veterans without PTSD.[71] Furthermore, PTSD appears to increase one's risk for health problems.[72] Recent neurobiological research findings among PTSD patients suggest alterations in brain structures and significant neurobiological consequences resulting from failure to cope successfully with traumatic stress.[73] These findings suggest that PTSD can be a persistent mental illness.[74]

Society and the military must work diligently, for the sake of soldiers and civilians alike, to lessen the traumatic impact of war. The most obvious solution is to cut the problem off at its source and end war, the primary source of traumatic injury. This solution seems quite remote, however. If soldiers then *must* fight wars, can we minimize or reduce the psychological damage war inflicts on soldiers? Yes, but only if the current prevailing societal denial, ignorance, and lack of collective responsibility toward the enduring impact of war on our soldiers are resolved. Secondary prevention includes those strategies that either reduce a soldier's vulnerability or increase his or her resiliency toward the development of war-related stress reactions and PTSD. Our current understanding of PTSD suggests that these strategies encompass military factors, societal factors, and psychological factors.

The single most effective strategy the military can use is to *limit* exposure to war, thus limiting exposure to war-zone trauma. It is important to note, however, that exposure to trauma of *any* duration increases the risk of traumatic impact on the person so exposed. While exposure to protracted and repeated trauma is highly associated with the development of PTSD, there are significant "singular impact" traumas that also can be the primary or sole etiological factor to the subsequent development of PTSD. Thus, by definition, serving in a war zone puts participants "at risk" to the subsequent development of PTSD—regardless of predeployment strategies to try to ameliorate or prevent PTSD. The screening for known risk factors for PTSD of recruits designated for combat occupational specialties would identify high-risk individuals and allow for more appropriate assignments.

In his book *Achilles in Vietnam,* Jonathan Shay suggests the military must make key changes if soldiers are to be protected from war trauma. He encourages pragmatic and structural ideas such as rigorous and realistic training, effective leadership, and maintenance of unit cohesion. Unit cohesion versus individual rotation was a key difference between World War II and Vietnam and possibly explains the higher prevalence rate of stress disorders among Vietnam veterans, as well as the profound sense of alienation and associated rage found among this group of veterans. Unit cohesion may function as a buffer by serving as a social support network during the post-trauma adjustment period. Shay also advocates the destigmatization and valuation of grief work and encourages more open and honest expressions and ceremonies to honor lost comrades.

Shay suggests the need for more difficult cultural changes within our military training strategies: a policy of respect for rather than dehumanization of the enemy, discouragement of the valuation of the revenge-driven soldiers, and discouragement of using humiliation and degradation as training tactics to incite rage or "fighting spirit." One veteran who suffered physical abuse in basic training described its impact:

> By the time I got out of Basic, I had so much rage built up in me that I *wanted* to go to Vietnam—so I could *kill* and prove to them that I was a man . . . what I hate the most is that I became abusive towards others just like I was abused.[75]

From a societal perspective, secondary prevention requires an abolishment of the "collusion and sanitization of silence about the true impact of war on its participants and the full extent of the horror of war."[76] Thus, if we are to engage in war, then we, as a society, must accept responsibility for such actions. Commitment of troops must occur only when absolutely necessary, with a clear declaration of war and with clear and obtainable military objectives. We must be willing to support our troops during and, if necessary, long after the military action. We must acknowledge the inevitable longer-term impact on many of our soldiers and commit resources to aid in the amelioration of such residual effects. Finally, we must change our attitude from one of denial and degradation to one of acknowledgment, acceptance, and respect for psychological casualties among our soldiers. Many veterans continue to carry the shame and stigma associated with having war-related problems. They believe they are "weak" and somehow have failed themselves, their comrades, and society. The prevailing societal and military myths that "heroes do not or should not have any problems" and "time heals all wounds"[77] have contributed to the chronicity of PTSD by inhibiting veterans' willingness to access available care.

From a psychological perspective, secondary prevention would include implementation of several strategies. First, aggressive and pervasive education of military personnel[78] and their family members[79] on trauma and its impact would aid in the prevention and early detection of traumatic stress reactions. Second, systematic psychological debriefings would aid in educating survivors about normal reactions to trauma exposure and warning signs for PTSD, facilitating venting of emotions, and providing a supportive environment in which to explore the meaning of trauma. Debriefings also serve to model an adaptive approach to coping with trauma. Finally, debriefings aid in the early detection of acute stress disorder and link high-risk individuals with resource personnel—our best chance of preventing PTSD.

Tertiary prevention encompasses the treatment of PTSD. The treatment of chronic PTSD is as complex as the disorder itself. Generally speaking, the greater the time between the traumatic experience and the initiation of treatment, the poorer the treatment prognosis.

The avoidance behavior associated with the diagnosis of PTSD clearly creates a barrier to treatment for the PTSD veteran. He is likely to struggle continually to forget, deny, and suppress his war trauma in the hopes that it will go away. His self-image of being weak or a failure for not having the self-control to manage his problems hinders his ability to ask for help. Many veterans seek treatment only after a crisis of some sort that essentially taxes their ability to cope, such as marital problems, employment problems, legal problems, financial problems, substance abuse, physical disability, or retirement. They frequently do not connect their postwar life problems with their traumatic war experiences. Often, they literally have become exhausted from their chronically anxious postwar life and discover that characteristic ways of denial, avoidance, detachment, and hyperarousal themselves become problems and may not be working as well as they used to. After resolution of the immediate crisis, they may tend to retreat and avoid treatment in an effort to avoid dealing with the trauma and associated feelings—until the next crisis. As time goes on, the PTSD veteran becomes exhausted and even more fearful of his inability to contain the raw emotions he has buried somewhere deep inside. Consider the following quote:

> I'm getting older now, but what I experienced in Vietnam becomes a kaleidoscope of events and the emotional impact it has on me continues. Today I feel lost, empty, detached, alone and dead. I'm afraid to let go and feel. I'm afraid I might become lost and never return.[80]

Given the complexity and persistence of PTSD in war-trauma survivors and the lack of knowledge about the course of trauma over the life span, treatment models must be both comprehensive and flexible. Treatment of war-related trauma typically occurs in a comprehensive network of PTSD specialty programs in the Department of Veterans Affairs hospitals, offering a broad-based approach that provides a range of services including assessment and intervention in biological, psychological, social, vocational, and spiritual areas of functioning. Flexible and accessible treatment is essential to allow for the veteran to reconnect to treatment during periods of exacerbation and compromised functioning, which may reoccur throughout a lifetime.

Despite the existence of several well-established and successful treatment approaches, no definitive treatment approach has emerged. Nevertheless, there are several treatment principles that guide most therapeutic interventions and programs[81] for chronic war-related PTSD, and they are presented below.[82]

Establishing Safety and Developing a Therapeutic Alliance

> Why should *I* trust *you*? How can you *possibly* understand what I have been through?[83]

The first visit with PTSD veterans is often critical. The veteran has likely contemplated seeking help many times before but has always backed out. He is likely in crisis and at the end of his rope. Creating an atmosphere that is supportive, nonjudgmental, and honest is crucial to enlisting the veteran's support with treatment.

Education on Trauma Reactions and Recovery Process

> An abnormal reaction to an abnormal situation is normal behavior.[84]

It is not uncommon for veterans to enter treatment and say, "I knew there was something wrong with me, but I didn't know what it was." They are often relieved to know that they are not unique in their experiences. Educating the veteran on the impact of trauma can help normalize his reactions to trauma.

Skills Acquisition and Symptom Management

The veteran likely has entrenched maladaptive coping patterns that temporarily result in some relief but that over the long term actually complicate the disorder (e.g., co-occurring substance abuse). The veteran must be aided in identifying his maladaptive patterns and taught more adaptive skills and management strategies (i.e., anger and anxiety management) to reduce distress to tolerable levels; otherwise the veteran will return to entrenched maladaptive coping strategies.

Therapeutic Reexperiencing of the Trauma or Some Aspect of the Trauma

The key ingredient for the potential for successful resolution of trauma is a careful therapeutic reexposure to the trauma or some aspect of the trauma, ideally in a therapeutic context.[85] For many veterans, participation in a therapy group with other veterans provides a safe and understanding milieu.

> When I got back from Nam, the *only* people I could possibly relate with were other Vietnam veterans—and they were the *last* people I wanted to talk to [because I was trying to forget about the war].

With enough exposure trials, the veteran gains the ability to tolerate, explore, and work through the trauma. One creative example of exposure therapy has been the subsequent return of veterans to the war zone. In this therapeutic strategy, the vividly traumatic memories that have become "frozen in time" are challenged by the veteran returning to the war zone years later, long after hostilities have ceased. During this journey, the old, recurring traumatic memories are challenged by a new set of much more recent and nontraumatic memories. Subsequently, the veteran is better able to foster a "more peaceful co-existence" in the here and now.[86]

Reconstruction of the Self and Integrating the Traumatic Experience

Typically, the trauma survivor sees herself as fragmented by the trauma. Vietnam veterans commonly refer to themselves as having two parts or sides, the "good" part and the "bad" part. The task becomes helping the veteran realistically to acknowledge the trauma, both its negative and positive impact,[87] and gain an understanding and acceptance that he is the product of all his life experiences.

Reconnection with Others, Such as Family Members, Members of One's Community, and God/Higher Power

[While in Vietnam] I called out many times to God, and . . . he didn't answer. He left me alone . . . to do what I had to do. . . . How could he have deserted me like that—when I most needed him?[88]

Trauma often has a profound impact on the survivor's ability to find meaning and purpose in life. In his book *Broken Connection,* Robert J. Lifton referred to this as existential malaise. If treatment is to be successful, then the PTSD veteran must reconnect with others, even if it is only with other veterans, in an attempt to reconnect and find renewed meaning in life.

Realistically, successful treatment for chronic war-related PTSD reduces the frequency and intensity of core symptoms of the disorder, but it likely will not eliminate them forever. The trauma of war is literally unforgettable, and thus the goal of treatment is to learn eventually how to *coexist more peacefully* with one's traumatic memories. What successful treatment can hope to accomplish is to reduce the frequency and intensity of the PTSD symptoms, replace maladaptive coping with more adaptive coping, improve interpersonal functioning and in turn improve social support, aid in the acceptance and integration of the trauma into one's life, and aid in the reconnection to others and, one hopes, to a more meaningful present and future.

There is hope, as evidenced by one veteran's description of his treatment experience:

At one time I didn't care about living. I came to treatment in 1991 and it took about three years of sessions and medication before my life started to get better. I'm better with family and friends. I can go to church now. The problems are still there, but at least I can deal with them. It hasn't been until now, after 8 years, that I've noticed a big difference. I'm not well, but at least I have some happiness now.[89]

War has been a constant in the history of the world. Given the present global situation, it would seem quite naive to contend that mankind would become peaceful in the foreseeable future. Warfare and violence will remain a part of the human experience. Continuing advances in technology threaten to make future wars ever more violent and lethal. As a result, the trauma levels of future wars will remain high and quite possibly increase. Such a forecast makes caring for the mental health of the warrior more important than ever before. It is society that asks men and women to place themselves in harm's way for its own benefit. It is society, then, that must take responsibility for their care. Over two thousand years ago, the philosopher Sun Tzu realized the relationship between society and its soldiers. We would do well to remember his admonishment to treat soldiers as our own "beloved children,"[90] for that is exactly who they are.

NOTES

1. S. M. Sonnenberg, A. S. Blank, and J. A. Talbot, eds., *The Trauma of War: Stress and Recovery in Viet Nam Veterans,* (Washington, D.C.: American Psychiatric Association Press, 1985), 34.

2. The views expressed in this chapter are solely the authors' and do not necessarily reflect those of the U.S. Department of Veteran Affairs or the VA Gulf Coast Veterans Health Care System.

3. The best recent study of the psychology of killing and combat is Lt. Col. Dave Grossman, *On Killing* (Boston: Little, Brown and Company, 1995).

4. Jonathan Shay, *Achilles in Vietnam* (New York: Simon and Schuster, 1994).

5. Grossman, *On Killing,* 88.

6. Steve Bentley, "A Short History of PTSD," *Veteran* (January 1991): 13.

7. The idea that only those who were predisposed to mental problems suffered breakdown in war or subsequent problems remained important until the Vietnam War. As in Sonnenberg et al., *Trauma of War,* 16–17, "the etiology of pathology lies not in adult but in childhood traumas." The idea is still current among many laypeople and veterans to this day. Doubt-

less a very high percentage of the U.S. public still view veterans suffering from PTSD to be "weak" or simply "crazy." Such a belief may well be an understandable defense mechanism, i.e., I am not "weak," and if called upon to do my duty I could do it without difficulty.

8. Peter G. Bourne, *Men, Stress and Vietnam* (Boston: Little, Brown and Company, 1970), 9–10.

9. Anthony Babington, *Shell Shock* (London: Leo Cooper, 1997), 15–19. The most valuable recent study of psychiatric disorders in the American Civil War is Eric T. Dean, *Shook Over Hell* (Cambridge: Harvard University Press, 1997).

10. Babington, *Shell Shock,* 45. This work is the most comprehensive regarding psychiatric casualties and World War I.

11. Bentley, "Short History of PTSD," 15.

12. Wilfred Owen, *The Collected Poems of Wilfred Owen,* ed. C. Day Lewis (London: Chatto and Windus, 1963), 69.

13. Bourne, *Men, Stress and Vietnam,* 10–11.

14. Lord Moran, *The Anatomy of Courage* (London: Constable, 1945).

15. Bourne, *Men, Stress and Vietnam,* 12–13.

16. Ibid., 15.

17. Joel Brende and Erwin Parson, *Vietnam Veterans: The Road to Recovery* (New York: Plenum Press, 1985), 67.

18. Bentley, "Short History of PTSD," 15.

19. Brende and Parson, *Vietnam Veterans,* 67–68.

20. Ibid., 68.

21. Dean, *Shook Over Hell,* 39–40.

22. Ronald Drez, *Voices of D-Day* (Baton Rouge: Louisiana State University Press, 1994), 208–9.

23. Grossman, *On Killing,* 265–66.

24. Ibid., 3–4.

25. Ibid., sec. 7.

26. Sonnenberg et al., *Trauma of War,* 20.

27. Ibid.

28. M. Keith Langley, "Post-Traumatic Stress Disorders among Vietnam Combat Veterans," *Journal of Contemporary Social Work* (December 1982): 595–96.

29. William E. Kelly, ed., *Post Traumatic Stress Disorder and the Veteran Patient* (New York: Brunner/Mazel, 1985), 151.

30. Jim Goodwin, "The Etiology of Combat-related Post Traumatic Stress Disorders," in *Readjustment Problems among Vietnam Veterans,* Disabled American Veterans, n.d., 10.

31. Ibid., 8.

32. Ibid., 10.

33. Grossman, *On Killing,* 251.

34. Ibid.

35. Brende and Parson, *Vietnam Veterans,* 72–73.

36. Grossman, *On Killing,* 276.

37. Notes from an interview by Andrew Wiest with John W. Young, who served as a staff sergeant leading a squad in the Mekong Delta in Vietnam. He has since become an invaluable part of my Vietnam War class at the University of Southern Mississippi. He has taught me more about the war than any book ever will and has become a fast friend. It is to him that this chapter is dedicated.

38. Grossman, *On Killing,* 289.

39. D. Meichenbaum, *A Clinical Handbook/Practical Therapist Manual for Assessing and Treating Adults with Post-traumatic Stress Disorder (PTSD)* (Waterloo, Ontario: Institute Press, 1994), 17.

40. Author's oral interview with Vietnam veteran, Root, 1999.

41. R. A. Kulka, W. E. Schlenger, J. A. Fairbank, R. L. Hough, B. K. Jorday, C. R. Marmar, and D. S. Weiss, *Trauma and the Vietnam War Generation: Report of Findings from the National Vietnam Veterans Readjustment Study* (New York: Brunner/Mazel, 1990), v.

42. D. D. Blake, T. M. Keane, P. R. Wine, C. Mora, K. L. Taylor, and J. A. Lyons, "Prevalence of PTSD Symptoms in Combat Veterans Seeking Medical Treatment," *Journal of Traumatic Stress* 3 (1990): 21–22.

43. A. Fontana and R. Rosenheck, "Traumatic War Stressors and Psychiatric Symptoms among World War II, Korean, and Vietnam War Veterans," *Psychology and Aging* 9 (1994): 30.

44. P. B. Sutker and A. N. Allain, "Assessment of PTSD and Other Mental Disorders in World War II and Korean Conflict POW Survivors and Combat Veterans," *Psychological Assessment* 8, 1 (1996): 21.

45. American Psychiatric Association, *Diagnostic and Statistical Manual of Mental Disorders,* 4th ed. (Washington, D.C.: American Psychiatric Association, 1994), 426.

46. Kulka et al., *Trauma,* 84.

47. These articles support the dose-dependent phenomenon, such that the greater the traumatic exposure to war stressors, the greater the likelihood of association between war trauma and increased rates of mental disorders and PTSD: N. Breslau and G. C. Davis, "Posttraumatic Stress Disorder: The Etiologic Specificity of Wartime Stressors," *American Journal of Psychiatry* 144 (1987): 578–83. Sutker and Allain, "Assessment"; P. B. Sutker, M. Uddo-Crane, and A. N. Allain, "Clinical and Research Assessment of Posttraumatic Stress Disorder: A Conceptual Overview," *Psychological Assessment* 3 (1991): 520–30.

48. Sutker and Allain, "Assessment," 22.

49. R. M. Scurfield and S. Tice, "Interventions with Psychiatric and Medical Casualties from Vietnam to the Gulf War and Their Families," *Military Medicine* 157,, 2 (1992): 90.

50. Ibid.

51. Ibid.

52. M. Friedman, "The Matsunaga Vietnam Veterans Project," *PTSD Research Quarterly* 9, 4 (Fall 1998): 7.

53. Ibid.

54. E. R. Parson, "The Intercultural Setting: Encountering Black Vietnam Veterans," in Sonnenberg et al., *Trauma of War,* 369. For an account of African American veterans in Vietnam, see Wallace Terry, *Bloods: An Oral History of the Vietnam War by Black Veterans* (New York: Ballantine, 1984).

55. R. W. Eisehnart, "You Can't Hack It Little Girl: A Discussion of the Covert Psychological Agenda of Modern Combat Training," *Journal of Social Issues* 31, 4 (1975): 18; E. R. Parson, "Ethnicity and Traumatic Stress: The Intersecting Point in Psychotherapy," in C. R. Figley, ed., *Trauma and Its Wake: The Study and Treatment of Posttraumatic Stress Disorder* (New York: Brunner/Mazel, 1985), 318.

56. Parson, "Ethnicity," 318.

57. See T. Holm, "Historical Perspective: Warriors All," in *Report of the Working Group on American Indian Vietnam Era Veterans* (Washington, D.C.: Readjustment Counseling Service, Department of Veterans Affairs, 1992), 8–11; and D. Johnson, "Stress, Depression, Substance Abuse and Racism," in *Report of the Working Group,* 35–38.

58. For further information, see R. Hamada, C. M. Chemtob, R. Sautner, and R. Sato, "Ethnic Identity and Vietnam: A Japanese-American Vietnam Veteran with PTSD," *Hawaii Medical Journal* 47 (1987): 100–109; C. Loo,

"Race-Related Trauma and PTSD: The Asian American Vietnam Veteran," in *Journal of Traumatic Stress* 7 (1994): 1–20; and C. Loo, K. Singh, R. M. Scurfield, and B. Kilauano, "Race-Related Stress among Asian American Veterans: A Model to Enhance Diagnosis and Treatment," *Cultural Diversity and Mental Health* 4, 2 (1998): 75–90.

59. Loo, "Race Related Trauma," 40.

60. A statement made by Dr. Martin Luther King Jr., in H. Bryant, "The Black Veteran," *Stars and Stripes—The National Tribune* 105 (June 1983): 5.

61. Author's oral interview with Vietnam veteran, Scurfield, 1975.

62. See American Psychiatric Association, *Diagnostic and Statistical Manual of Mental Disorders,* 2d ed. (Washington, D.C.: American Psychiatric Association, 1980).

63. American Psychiatric Association, *Diagnostic and Statistical Manual,* 4th ed., 427–29.

64. Author's oral interview with Vietnam veteran, Root, 1999.

65. This information on moral pain comes from R. Marin, "Living in Moral Pain," in *Psychology Today* (November 1981): 68–80.

66. Ibid.

67. C. Loo, "Race-Related Posttraumatic Stress Disorder," in *Readjustment Counseling Services Report on Asian Pacific Islander Veterans,* (Washington, D.C.: Readjustment Counseling Service, U.S. Department of Veterans Affairs, 1998), 88.

68. Sutker and Allain, "Assessment," 21.

69. L. R. Daniels and R. M. Scurfield, "War-related Post-traumatic Stress Disorder, Chemical Addictions and Habituating Behaviors," in J. B. Williams and J. F. Summer, eds., *The Handbook of Post-Traumatic Therapy* (Westport, Conn.: Greenwood Press, 1994), 206.

70. J. Lipkin, R. M. Scurfield, and A. Blank, "Post-Traumatic Stress Disorder in Vietnam Veterans: Assessment in a Forensic Setting," *Behavioral Sciences and the Law* 1, 3 (July 1983): 60.

71. Kulka, *Trauma,* 272–73.

72. This information is derived from M. Friedman and P. Schnurr, "The Relationship between Trauma, Post-traumatic Stress Disorder, and Physical Health," in M. J. Friedman, D. S. Charney, and A. Y. Deutch, eds., *Neurobiological and Clinical Consequences of Stress: From Normal Adaptation to PTSD* (Philadelphia: Lippincott-Raven, 1995), 507–24.

73. M. Friedman, "Neurobiological Research on PTSD," *PTSD Research Quarterly* 6, 4 (Fall 1995): 1. For a comprehensive review of neurobiological consequences of stress and trauma, see Friedman, Charney, and Deutch, eds., *Neurobiological and Clinical Consequences of Stress.*

74. M. J. Friedman, "PTSD as a Severe Mental Illness," *National Center for Post-traumatic Stress Disorder Clinical Quarterly* 8, 4 (1999): 63.

75. Author's oral interview with Vietnam veteran, Scurfield, 1996

76. R. M. Scurfield, "Treatment of War-Related Trauma: An Integrative Experiential, Cognitive, and Spiritual Approach," in Williams and Sommer, eds., *Handbook of Posttraumatic Therapy,* 179–203. For in-depth discussion of the collusion of silence about and sanitization of war, see R. M. Scurfield, "The Collusion of Sanitization and Silence about War: An Aftermath of Operation Desert Storm," *Journal of Traumatic Stress* 5, 3 (1992): 505–12.

77. Ibid.

78. T. Oei, B. Lim, and B. Hennessy, "Psychological Dysfunction in Battle: Combat Stress Reactions and Posttraumatic Stress Disorder," *Clinical Psychology Review* 10 (1990): 355.

79. M. J. Peebles-Kleiger and J. H. Kleiger, "Re-Integration Stress for Desert Storm Families: Wartime Deployments and Family Trauma," *Journal of Traumatic Stress* 7, 2 (1994): 189–99.

80. Author's oral interview with Vietnam veteran, Root, 1999.

81. R. M. Scurfield, "Post-Trauma Stress Assessment and Treatment: Overview and Formulations," in C. R. Figley, ed., *Trauma and Its Wake* (New York: Brunner/Mazel, 1985), 241–47; and Scurfield, "Treatment of War Related Trauma," 186–91.

82. A major multisite study of group treatment of PTSD incorporating most of these principles has recently been completed (D. Foy, personal communication with the author, July 1999).

83. Author's oral interview with Vietnam veteran, Scurfield, 1976.

84. V. Frankl, *Man's Search for Meaning: An Introduction to Logotherapy* (Boston: Beacon Press, 1959), 38.

85. J. A. Fairbank and R. A. Nicholson, "Theoretical and Empirical Issues in the Treatment of Post-traumatic Stress Disorder in Vietnam Veterans," *Journal of Clinical Psychology* 43, 1 (1987): 50.

86. A. Gerlock, "Vietnam: Returning to the Scene of the Trauma," *Journal of Psychosocial Nursing* 29, 2 (1991): 7; and R. M. Scurfield, "Vietnam Revisited: A Journey of Healing," *Western Express* 1, 2 (1989): 5.

87. Scurfield, "Post-Trauma Stress Assessment," 246–47; Scurfield, "Treatment of War Related Trauma," 200.

88. Author's oral interview with Vietnam veteran, Root, 1988.

89. Author's oral interview with Vietnam veteran, Root, 1999.

90. Sun Tzu, *The Art of War,* trans. by Ralph Sawyer (New York: Westview, 1994), 215.

CHAPTER 12

Iraq

A Third-World Superpower?

Sean McKnight

In the post–Cold War world, the challenges facing the armed forces of Western nations have changed dramatically. With the possibility of large-scale conventional war all but gone after the fall of the Soviet Union, military challenges in the developing world have taken center stage. The diverse nature of these current flash points has forced Western militaries to be more flexible and has placed a heavy emphasis on intelligence. In many of the conflicts of the future, technologically superior and overwhelming force might not be the answer as it was in the Gulf War. The example of Iraq in fact demonstrates the military strengths and weaknesses of the developing world and how they factor into modern conflict.

Since the establishment of the Ba'thist regime in 1968, Western perceptions of the Iraqi armed forces have oscillated wildly. During the last two years of the Iran-Iraq War (1980–88), the Iraqi military appeared formidable compared to most others in the Middle East. In 1990—in the aftermath of the Iraqi invasion of Kuwait—Saddam Hussein's boasting of Iraqi military might was taken seriously. At other times, however, the Iraqi forces have been regarded as militarily inept; so inept that in the 1970s and 1991 they were unable to prevent what appeared to be the imminent dissolution of the Iraqi state.

This inconsistency in the military assessments of Iraq is disturbing, for while representing an excessive regard for the military strength of a potential adversary inhibits policy, by contrast, seeing such a state as a military "pushover" could be disastrous. The United States—and due to its preponderance, its allies are implicated—has manifested these contradictory tendencies in its perceptions of Iraq over the last quarter of a century. In the aftermath of defeat in Vietnam, America's fear of military entanglements contributed to the world's strongest state "punching beneath its weight." Since victory in the Gulf, however, in dealing with Saddam, Somalia, and arguably Kosovo, the problems of military intervention have been underestimated.

Understanding the complex factors that generate military power is thus a matter of some importance, especially since military intervention has become more likely since the end of the cold war. Iraq is a particularly pertinent case study as unlike most third-world states, it has many of the more obvious ingredients that make for military power, and it seems unlikely that the events of 1991 have curbed the regime's willingness to use force.

It is commonly believed that little was known about the Iraqi military at the time of the Gulf War,[1] and General R. H. Scales (U.S. Army) speaks for many with his view that "the historical insights available from the Iran-Iraq War were meager to say the least."[2] This is a rather odd view. Iraq has been involved in numerous conflicts since its foundation, and the Iran-Iraq War (1980–88) gives us an accurate picture of its strengths and weaknesses. The meagerness of the insights drawn reflects on the lack of serious study of the war rather than the secrecy with which it was waged. Indeed, given the brevity of the 1991 Gulf War, any attempt to understand the Iraqi armed forces must address the eight-year conflict with Iran.

The failure to study the Iraqi military helps explain the wild swings in assessments of their capabilities. This chapter attempts to avoid such oscillations, recognizing that asking why Iraq's armed forces fail to realize their potential strength is a genuine question. Viewing the Iraqi armed forces just through the prism of their 1991 debacle produces as distorted a picture as one derived from a focus on the events of 1988 or the 1991 "blitzkrieg" invasion of Kuwait.

Indeed, to understand the weaknesses and strengths of the Iraqi armed forces, it is necessary to trace their development since the establishment of the modern Iraqi state.

The modern Iraqi state—which started life as a British mandate in 1920—inherited little in the way of an organized military tradition. The armed tribes of Iraq did not produce good natural soldiers; not only were tribesmen very reluctant to serve the new state, but the record of tribal forces in modern war is poor.[3] For example, the Ottoman force raised from Iraq in 1914—in theory thirty-two thousand strong upon mobilization—had performed poorly in the First World War, and in any event, over 40 percent were Kurdish or Turcoman, and most of the 1,350 officers were Turkish.[4]

Recruiting for the new Iraqi army proved difficult, and the lack of volunteers reflected a lack of enthusiasm for the new state. Indeed, the Iraqi government passed a conscription act in 1927 but did not enact it for fear of tribal hostility. Few Iraqis wished to serve as soldiers, and even when conscription was introduced in 1934, the better-off paid the *al-badal* to avoid service. Drawn mainly from the ranks of the Iraqi urban poor, soldiers were poorly motivated and uneducated—as late as the 1970s, many were still illiterate.

The Iraqi officer corps fared better, initially attracting over six hundred ex-Ottoman officers. Subsequently, however, a military career failed to attract the Iraqi elite, and many officers came from the literate lower middle class—the class most receptive to radical ideologies. Indeed, General Abd al-Karim Qasim, who headed the military government from 1958 to 1963, worked as an elementary school teacher.[5] From its inception the Iraqi officer corps was highly politicized, and Pan-Arab nationalism was only the strongest of the various ideologies attracting officer adherents. Drawn from relatively humble social origins, many of the Iraqi officer corps became at best half-hearted defenders of the status quo.

In 1933, however, the army defeated an Assyrian "rising," inspiring celebrations across Iraq, and in 1936 they suppressed a mostly Arab Shi'a tribal rising. The prestige of these "victories" established the armed forces as the strongest of the weak institutions of a weak state. In October 1936, a military coup (organized by General Bakr Sidqi) inaugurated five years in which governments required military support.

Arab nationalists in the army resented British tutelage, and in 1941 they backed Rashid Ali's anti-British government. Britain resolved this dispute with military intervention, though on paper the Iraqi armed forces could have made it difficult for the invading British. In reality, the Iraqi military performance was inept—it took the British only a month to occupy Iraq and install a regime more to their taste. Far from humiliating the Iraqi armed forces, however, this crushing military defeat was seen in Iraq as a heroic event of the nationalist struggle, and the military's political status was further enhanced.

Governments from 1941 to 1958 were dominated by men from the wealthy landed classes, most of whom owed their wealth to the political influence they enjoyed with the Iraqi monarchy. These governments were seen as pro-British and lukewarm to both Arab nationalism and modernization. The monarchy made itself even more unpopular by removing nationalists from the armed forces—by 1943, nearly three-quarters of the 1941 army had been purged.[6] These purges were drastic, but many officers continued to espouse revolutionary causes. When the Free Officers movement overthrew the monarchy in 1958, many of those who engineered the revolution were from relatively humble backgrounds.[7]

The Free Officers movement had a radical agenda and quickly passed a redistributionist agrarian reform law. The military was divided, however; some advocated a radical program of domestic reform, others a Pan-Arab nationalist agenda, and, among the nationalists, traditionalists vied with those who blended nationalism with social and economic radicalism. Headed by General Qasim, the government initially inclined toward domestic reform, but his overthrow in February 1963 ushered in a series of more nationalist regimes.

Iraq's rising oil revenues and military government ensured the expansion of the armed forces—between 1958 and 1966, the army doubled its budget.[8] In contrast to the interwar years, the Iraqi military could now no longer be seriously threatened by Arab tribal revolts. Yet the Iraqi army could not defeat the antigovernment Kurds—despite the fact that many Kurds fought for the government[9]—and in May 1966 they suffered a humiliating defeat in the Battle of Handrin. Military weakness was further emphasized by the failure to make good threats to incorporate Kuwait into Iraq. The most damaging Iraqi "defeat" of the 1960s, however, was the failure to make any significant contribution to the common Arab cause in the 1967 Six-Day War.

Waiting to take advantage of any weakness in the military regime was the B'ath Party. In 1963, the Iraqi B'ath Party only had 830 full members,[10] but its blend of Arab nationalism and radical domestic reform appealed to the urban poor as well as the educated lower middle class. In 1963, B'athist militia forces helped overthrow Qasim and the party formed a government, but distracted by vicious internal feuding, they were overthrown by their military allies within months. By 1968, however, the military regime had been weakened by military failure, and factions within the military sought B'athist support to overthrow the government. The principal military plotters—'Abd al-Razzaq al-Nayif, the deputy director of military intelligence, and Ibrahim al-Da'ud, commander of the presidential guard—expected to be the senior partners in any new regime.

The history of the military in modern Iraq and the experience of being in government in 1963 made the B'athists determined to deny the military the opportunity to return to power. The party planned to dispense with its military allies as soon as possible, ensuring that military formations loyal to the B'ath moved into Baghdad.[11] Within weeks, al-Nayif and al-Da'ud were forced into exile, and the B'ath determined "to consolidate the Party's leadership of the army by purging it of suspect elements, re-education and immunization against deviation and insuring that it would always be completely identified with the popular movement and the leadership of the Party."[12] The regime's pursuit of security from its own military—which from an outsider's perspective seems to be a policy of gratuitously enfeebling the military—is one of the main reasons for the longevity of the regime.

Since 1968 in Iraq, there has been a continuous process of purging senior military officers. In the first ten years of the regime, fifteen senior officers were purged, four were exiled, six were imprisoned, four were executed or murdered and one was dismissed.[13] Being an active member of the B'ath Party was no guarantee of safety—General Hardan al-Takriti (deputy premier and a B'athi since 1961) was exiled in 1970 and assassinated the next year. Even members of Saddam's family were not immune: his cousin, the minister of defense Adnan Khairallah, died in a helicopter "accident" in 1988. Some of those purged had certainly been plotting to overthrow the regime. Other victims were executed for military failings—for instance, the executions following the Iraqi loss of Khorramshahr in 1981[14] and similar punishments meted out when Iraq's air defenses failed in 1991. Many of those purged were innocent, however; being too successful as a commander, particularly if you were popular and independent-minded, made you a likely candidate for the wrath of the regime.

The regime also strove to B'athize the armed forces—aiming to create an *al-Jaish al-'Aqa'idi* (ideological army). The B'ath Party Military Bureau and the Department of Political Orientation were tasked in 1968 to propagandize the military, supplemented by B'athists in the military—in particular, the political guidance officers introduced in 1971. B'athist activists were fast-tracked for entry into the officer corps,[15] and senior posts often went to long-term political associates of Saddam, fellow Takritis, and members of his extended family. According to some sources all officers, NCOs, and applicants to military colleges were required to join the party.[16] Only B'athist political activity was tolerated in the armed forces; in particular, the regime has been concerned weed out communist activists.

The main job of watching for any hint of military dissent belongs to the various intelligence organizations serving the regime. For most of the period since 1968, the Mukhabarat has been the most feared, but the Estikhbarat (military intelligence) also has a role, and in recent years the Amn al-Khas (special security division) has become increasingly important.

The regime's close scrutiny of the armed forces and the harsh penalties it exacted for any perceived disloyalty had an impact on the command culture of the Iraqi armed forces. The command culture that the B'athist regime inherited was already "top down," with most decisions referred to very senior officers, but this was accentuated with even quite trivial matters having to be decided by the Revolutionary Command Council (RCC).[17]

So intense has been the process of B'athization of the military, it is tempting to conclude that Tariq Aziz (Iraq's deputy premier) is right to be confident of their loyalty. According to Aziz:

> The second lieutenant who was in the army in 1968 is now a general. When we speak of about the army, they are all our own men. For every officer there is a relative, . . . in the Ministry of the Interior, the Ministry of Information, the Ministry of Foreign Affairs and the Ministry of Communication. So how is this man going to organize a coup against his own people?[18]

Yet, while some of the military "plots" unearthed by the regime were undoubtedly fabrications, there have been serious attempts from within the military to overthrow the government. To counter the very real danger of a military coup, the regime has created parallel military institutions—a policy that emphasizes the importance the regime places on its own survival even at the cost of gross military inefficiency.

Three armed institutions have particular importance to the regime as insurance against its own military. The largest but, since the early 1970s, least effective is the Popular Army. Much more important because it is militarily more potent is the Republican Guard. Less is known about the Amn al-Khas—which has grown in importance in the last fifteen years. In addition to these, there are heavily armed police formations, of which the Border Guard and Mobile Police Strike Force are the most formidable. With the exception of the Popular Army, these are all privileged formations, enjoying higher rates of pay and, in the case of the Republican Guard and Amn al-Khas the most modern military equipment. Even more than the regular military, these formations are officered by party loyalists, Takritis, and members of Saddam Hussein's extended family.

The Iraqi armed forces grew between 1968 and 1980, in terms of numbers of men under arms and equipment. The army expanded its tank park from eight hundred in 1971 to seventeen hundred by 1978 and acquired a further 1,820 armored infantry vehicles.[19] This expansion was funded by the dramatic rise in Iraqi oil revenues, from $476 million in 1968 to $26 billion in 1980;[20] a result of the massive rise in world oil prices, expanding Iraqi output, and Saddam Hussein's bold decision to nationalize Iraq's oil industry in 1972. The regime believed it had built "one of the best fighting structures in the world."[21] Though now quite large, the Iraqi armed forces were no more competent than under the previous regimes. Assessed purely on their military performance, the B'athised Iraqi armed forces performed very poorly in the 1970s.

Iraq sent two of its better armored divisions to the Golan Front during the 1973 Ramadan War. On 12 October, the Iraqis attacked the open flank of the Israeli force advancing toward Damascus. In terms of intention the Iraqis cannot be faulted from a military perspective, but they moved so slowly that the Israelis were able to create an inviting four-and-a-half-mile "box"; the Iraqis dutifully advanced into this trap—the Iraqi Twelfth Armored Brigade alone lost eighty tanks and failed to hit a single Israeli tank.[22] A second major Iraqi attack was launched on 18 October, but all-arms cooperation between different elements of the Iraqi force was close to nonexistent. Iraqi tanks and infantry operated as separate arms; artillery fire was inaccurate and totally uncoordinated with the attacking forces.

Though the failures against the Israelis were serious, more worrying for the regime was their initial failure to defeat the forces of the Kurdish Democratic Party (KDP). Fighting broke out in northern Iraq in April 1974, and, supported by Iran and the United States, the KDP at first got the better of the Iraqi military. The regime faced the defacto breakaway of much of northern Iraq and was forced to make concessions to Iran. Then, finding themselves abandoned by their allies, roughly 70 percent of the KDP *Peshmargas* guerrillas accepted an Iraqi amnesty; the rest retreated into the mountains and across the international frontier.[23] Just as in the 1920s, the Iraqi military had failed, forcing Iraq to seek external assistance to shore up the power of the central government.

Iraq's biggest "defeat" in the 1970s resulted from its decision not to intervene in the 1971 Jordanian Civil War—a decision that seemed like a repudiation of the Arab cause. In contrast, the military disaster that befell Iraqi forces on the Golan in 1973 was seen in Iraq and the wider Arab world as a great victory. Despite overwhelming evidence to the contrary, Baghdad's claim to have saved Damascus was given credence—a "victory" confirming Iraq as a leading Arab state.

From 1920 to 1980, the Iraqi military played an important symbolic role as embodying progress and Arab nationalist strength. They were, however, lacking in fundamental skills, incapable of waging all-arms warfare, and their involvement in politics detracted from their military effectiveness. Although ideological ferment within the military was suppressed by the B'ath after 1968, the traditional divisions of tribe, family, and ethnic or religious group—which the ideologies had often concealed—persisted. The major positive difference between the armed forces of the monarchy and that of the 1970s was that the B'athist armed forces were more numerous and better equipped. The measures the B'ath took against its own armed forces, however, were more rigorous than those of earlier regimes, and these measures had a major impact on the ability of the military to perform well in battle.

But, with the exception of the struggle against the Kurds, the interests of the regime were actually well served by the armed forces.

What seem to be fundamental military problems were actually of little concern for the B'athist regime. The armed forces fulfilled the symbolic requirements of the regime and earned Iraq a place at the "top table" of the Arab world; and the B'ath had neutered them as a political threat. The Iran-Iraq War (1980–88) was to change that relationship and see the military competence of the Iraqi armed forces become entwined with the very survival of the regime.

Iraq attacked Iran on 22 September, halting its advance after five days of fighting. Making it clear his objectives were limited, Saddam hoped Iran would end the war, giving him and his regime a cheap, prestigious victory. Iran was in revolutionary chaos, but Iraq's "pin prick" of an attack left it with an enraged enemy who failed to recognize Iraq's restraint. Iran's refusal to respond to the "subtle" nuances of Iraq's limited-war strategy plunged Iraq into a life-or-death struggle.[24]

The shortcomings of the Iraqi armed forces were illustrated on the first day of the campaign. An attempt to knock out the Iranian airforce on the ground was an embarrassing failure, and within days the Iraqi airforce was largely inactive, despite an operational superiority of at least three to one.[25]

On the ground, Iraq launched three separate shallow thrusts into Iran, the southern thrust into Khuzistan being the most important. In the south, any Iraqi success would place them in control of the important Shatt al Arab waterway and the cities of Khorramshahr and Abadan. Three Iraqi armored divisions and two mechanized divisions were used in this southern thrust—at least two-thirds of Iraq's most effective armored brigades.[26]

Iraq's advance into Khuzistan was inhibited by poor intelligence of the enemy and was excessively cautious. Undertaking virtually no of-road movement, Iraq forfeited the opportunity to wage maneuver warfare. As in 1973, the Iraqis failed to coordinate their different arms; tanks, infantry, artillery, and aircraft all operated separately from one another. Initial gains were not consolidated because of inadequate infantry support. As an example of this shortcoming

Iraqi armor entered Susangerd on 28 September, but in the absence of infantry, Iranian Revolutionary Guards were able to reoccupy the area quickly. Where infantry were available, Iraqi troops were reluctant to close with the enemy, preferring to trust to (very inaccurate) artillery bombardment.

Nonetheless, despite Iraqi shortcomings, they held the advantage over the disorganized and underequipped Iranian forces—an advantage Iraq held until September 1981. Having retrained selected infantry formations to fight in urban areas, they successfully—if at great cost in lives—took Khorramshahr and most of Abadan in the autumn of 1980. In January 1981, the Iraqis demonstrated they had learned from the Israeli Defense Force, enticing an Iranian armored division into a "box" and destroying over two hundred tanks. Although Saddam's strategy had not delivered a short, victorious war, Iraqi forces appeared to have consolidated their modest gains of 1980, and he clearly believed that Iran would soon sue for peace.

Saddam's complacency was shattered by a series of spectacular Iranian victories between September 1981 and May 1982. Despite the material advantages of the Iraqi military, enthusiastic Iranian troops captured over forty thousand Iraqi prisoners, retaking Abadan and Khorramshahr—in Khorramshahr alone, an estimated thirty thousand to fifty thousand Iraqi soldiers lost their lives.[27] Iran then proceeded to carry the war to Iraq, and, in contrast to the limited aims of Saddam in 1980, it aimed to destroy the B'athist regime. For the first time in the brief history of the Iraqi state, the survival of the state was directly linked to its armed forces' ability to perform competently in battle.

It was not possible to represent these military disasters as a "victory." Military problems provoked a domestic political crisis, but Saddam dealt decisively with murmurs of discontent before they became outright revolt. In late June there was a purge affecting the top level of B'athist leadership. Initially, the only execution was Minister of Health Riyadh Ibrahim Hussein—rumor has it that Saddam shot him during a cabinet meeting.[28] Later in the summer, however, some three hundred high-ranking officers were executed, followed by roughly ninety antiregime Shi'a activists in the spring of 1983.[29]

For much of 1981–86 the Iraqi defenses held Iran at bay, but the stability of the front was interrupted by several threatening Iranian advances. For instance, in 1984 the Iranians temporarily severed the Baghdad–Basra highway (Operation Badr). It is tempting to see Saddam as militarily naive, but this is to misunderstand the dilemma into which the war had placed the regime. In fact, fundamental military reforms to encourage a bolder operational style would have created a direct threat to the regime. Making the situation more perilous, the Iraqi people—in contrast to the Iranians—were lukewarm toward the war. Such forces pulled the regime in different directions, and it is unsurprising that they hoped the occasional periods of stability presaged the end of the conflict. Fortunately for Saddam, changes were coming that might stave off defeat falling short of fundamental military reform.

Throughout the war, Iraq enjoyed an immense advantage in materiel, purchasing huge quantities of military equipment—often on favorable commercial terms—from sources as diverse as the Soviet Union, France, Argentina, and China. External support, which funded much of Iraq's weapons procurement, enabled Iraq to mobilize a large percentage of its population for the war effort without forcing an overly austere war economy on the Iraqi people.

Iraq constructed hardened defenses, often in a triangular shape for all-around defense, and thousands of miles of sand berms. Into these defenses they dug large numbers of artillery pieces, emplaced older model tanks, laid millions of mines, and, in the south, made imaginative use of water barriers. Although these defenses appeared formidable, however, they were constructed slowly and lacked depth. Consequently, the Iraqis needed to develop a mobile counterattacking force and an infrastructure of north-south roads to facilitate its deployment. Once Iranian forces penetrated Iraqi defenses, they lacked the capacity to either consolidate or exploit, and given the opportunity, Iraqi armor, artillery, and aerial assets launched effective—if rather unsubtle—counterattacks. On the rare occasions when advancing Iranian forces could not be driven back, as on the Faw Peninsula in 1986, the Iranians paid for their operations with far higher casualties than their Iraqi enemies.

Iraq also developed the military skills of some of its armed forces. This was especially important for the Iraqi airforce, whose pilots had been so outclassed in 1980. Trusted pilots were trained in other Arab states, the Soviet Union, and elsewhere (most notably France). The Iraqi army sought to improve the skills of its troops, although the effort concentrated on a minority of formations, but the main focus was on specifics rather than more fundamental operational matters. The role of political guidance officers altered, reducing their influence on operational decisions, and some of the more incompetent B'ath Party appointees were removed from important military posts,[30] but the security services remained intrusive and ruthless. Most important, the rigid top-down command system of the Iraqi armed forces underwent no serious reform. It was still the case that military decisions of any note required, at the very least, assent from the political leadership.

Despite horrendous losses, Iran continued to launch offensives, pushing Iraq into attempting to end the conflict in other ways. For most of the war, Iraq made rather spasmodic attacks on Iran's war-waging capacity, particularly their oil industry. This strategy, however, which extended the war into the Gulf, was also designed to enrage the Iranians into lashing out indiscriminately. If Iranian attacks forced other powers—especially the United States—into entering the war, then Iraq's problems were over. Iraq also hoped to force Iran to sue for peace by attacking their will to continue the war. Initially, attempts to coerce Iran revolved around city bombing, but as the war developed, the use of medium-range surface-to-surface (S-S) missiles became as important. The most frightening aspect of Iraq's coercive strategy, however, was the chemical weapons (CW) program, the primary motive for which was to intimidate Iran into ending the war. Given this strategy, it is not surprising that Iraq's initial use of CW on the battlefield was very tentative. The increasingly frequent resort to chemical warfare in the last third of the war, though, underlines the CW program's strategic failure. Not only did Iran prove to be harder to provoke than Iraq anticipated, but until late in the war, it was Iraq that proved more malleable than Iran when faced with threats.

In 1986 the Iranians, once again, shattered Iraqi complacency by launching a surprise attack—Wal-Fajr 8—that seized and held the Faw Peninsula. Given Iraqi material advantages, this Iranian offensive was an impressive achievement, but in purely military terms it was less successful than those of 1981–82. Faw had a dramatic effect on Iraq, however, and unlike the defeats of 1982, Wal-Fajr 8 did lead to important changes in nature of the Iraqi armed forces.

The fighting on the Faw Peninsula exposed continuing Iraqi deficiencies: their excessive dependence on armored vehicles, the lack of effective dismounted infantry, a severe degradation in their combat capabilities at night, very inadequate battlefield intelligence, and the continued "stodginess" of Iraqi command and control (C2), stemming from a highly centralized "top-down" structure. The effectiveness of Iraqi C2 was also compromised by the continued politicization of military decision making. In the case of Faw, Saddam intervened to cancel any more counterattacks because he believed the expected high casualties would breed defeatism. Instead, ignoring his military advisers, Saddam ordered an attack on Mehran—which was initially captured but retaken by the Iranians in June. These defeats seem to have convinced the B'athist regime that its 1982–86 recipe for avoiding defeat would be insufficient, and more fundamental military reforms might well be required.

In the aftermath of these 1986, defeats Iraq continued developing the all-arms armored element of their forces. The most visible sign of this was the rapid expansion of the Republican Guard from only seven brigades at the start of 1986 to at least twenty-eight Brigades—more than one hundred thousand men—in 1988.[31] The guard, hitherto deployed only in emergencies, was used as a matter of course in the remaining fourteen months of the war for both counterattacks and offensives. The expanded guard and some army armored formations trained to wage a simple style of all-arms warfare, and for the first time in the war, the Iraqis developed a degree of proficiency in coordinating different arms. This was particularly the case with close air and artillery support, which became much more effective due to the use of properly trained forward observers. Supplementing this, the Iraqis established a Chemical Weapons Corps, integrating CW into their battlefield tactics, and naval infantry brigades trained to fight as dismounted assault infantry for operations in the marshlands of the south. The Iraqis also made a considerable effort to ensure they could redeploy these forces, purchasing over one thousand tank transporters and constructing new roads—including an extra north-south six-lane highway.[32] These forces were encouraged to take tactical initiatives and act flexibly. This developing mobile elite was also trained in a more demanding manner than had hitherto been the case.

This process was underpinned by the development of an Iraqi military-industrial complex that could manufacture smaller weapons, upgrade tanks,[33] build missiles, and extend the range of the Soviet-manufactured Scuds. Indeed, the expansion of Iraq's capacity to manufacture its own conventional weaponry is often overlooked because of understandable concerns about the Iraqi program to develop "weapons of mass destruction."

There were also changes to the Iraqi management of the war. Senior military men were drafted into the National Defense Council, which took over much of the running of the war from the RCC. Although the political elite remained involved in military decision making, they were much less "hands on" than they had been up to 1986. Officers were still purged, disappeared, or the victims of "accidents," but for the first time in the war, Saddam tolerated very limited public criticism of his decisions.[34]

The Iraqi armed forces were becoming more effective in 1987; at the same time, their Iranian opponents—still imbued with a greater fighting spirit—were finding the material balance of the war increasingly unfavorable. At the start of 1987, Iran was outnumbered five hundred to sixty-eight in combat planes, four thousand to one thousand in tanks, 4,000 to 1,360 in other armored vehicles and fifty-five hundred to six hundred in tube artillery.[35] Unlike Iraq, most of Iran's military equipment was technologically basic and its imports were costly. Finally, by 1987 the Iranian people were tiring of the war. The balance of advantage in the war, so long in Iran's favor, was starting to turn.

These changes were not immediately apparent. In the opening days of Iran's 1987 Karbala 5 offensive, Iranian forces surprised the Iraqis, and by 26 January they were within nine miles of Basra. Iraq's initial counterattacks were ineffective because the marshy terrain made it hard to deploy armor. The Iranians sought to press home their advantage but became increasingly vulnerable to Iraqi firepower and counterattacks. The battle continued into late February until it was clear Iranian forces were unable to take Basra and, with more than twenty thousand dead, this offensive was clearly a disaster for Iran. The defeat did not lead to an immediate collapse in the Iranian people's support for the war, but it did mark the beginning of widespread antiwar feeling.

In 1988, Iraq regained the initiative. Using their elite forces, they launched four successful offensives. These attacks were characterized by an effective coordination of different arms, the use of armored forces to strike the enemy to their depth, and an immense volume of accurately delivered firepower—both conventional and chemical. It is important, though, not to overstate the boldness of these operations. Iraqi attacks were elaborately prerehearsed, and the armored thrusts never penetrated beyond twenty miles. Nonetheless, it is clear that these attacks inflicted significant losses on Iran, and although Iraqi claims to have captured 1,298 tanks since March 1988 may be exaggerated, Baghdad was able to invite Western journalists to inspect an impressive array of captured equipment.[36]

Typical of the Iraqi successes, in just eight hours an offensive near Basra in late May regained all the ground lost in 1987. A short intense bombardment, including chemical munitions, preceded the assaulting naval infantry and army formations. The Iranians resisted fiercely for five hours, but the Iraqis then committed Republican Guard armored formations. These broke through to the drier ground, shattering the coherence of the Iranian defenses. The Iraqi advance of just over fifteen miles to the border town of Salamcheh was not spectacular, but they captured 155 artillery pieces and pushed the Iranians beyond artillery range of Basra.

Iraqi success on the ground helped change the attitude of Iran's leadership to the war. It was not just Iraqi battlefield success, however, that forced the Iran to sue for peace. Belatedly, some of Iraq's stratagems for ending the war began to have an impact. The United States entered the war in the Gulf, sinking a large percentage of Iran's navy. Iraq's aerial attacks on Iran, particularly its oil industry, also began to have an effect. Finally, the constant threat of Scud missile attacks and the fear these missiles might carry a chemical payload persuaded over a million inhabitants of Teheran to flee into the countryside. On 8 August, after some prevarication by Iraq, the war finally came to an end—the two belligerents were virtually back where they had started.

The Iran-Iraq War transformed Iraq's armed forces, and in 1988 they could claim, with some accuracy, to be the strongest in the Middle East. Despite growing financial problems—problems that made the option of invading Kuwait attractive—the changes initiated in the Iran-Iraq War continued until 1991. Indeed, the Iraqi armed forces in 1991 appeared so formidable that only a minority of commentators expected that the liberation of Kuwait would turn out to be such a one-sided affair.

The most constant strand in Iraq's drive to strengthen its armed forces was the attempt to acquire more and better weaponry. A significant minority of these weapons were *relatively* modern, and an unspecified number of older pieces of equipment had been upgraded by Iraq. Unfortunately, in the months prior to the liberation of Kuwait, most journalists and military analysts ignored the fact that much of Iraq's military equipment was very old and even their newer equipment was technologically less advanced than that available to the main United Nations Coalition powers. Iraq's T-72 tanks, regularly described in the western media as "state of the art," were older than and inferior to the U.S. M60A1, not to mention the most modern coalition main battle tanks, the British Challenger and U.S. M1A1 Abrams. Similarly, in the air, neither the Iraq's MiG-29 (some of which lacked a "look down–shoot down" capability)[37] nor the French manufactured Mirage F1s (which were preferred by Iraqi pilots) were as advanced as the best coalition combat aircraft. It is important, however, to appreciate that the Iraqi arsenal was technologically second-rate only in comparison with Western military powers and the Soviet Union.

An important change in the Iraqi armed forces was the improvement in Iraqi individual military skills. For instance, Iraqi pilots in 1985 launched accurate bombing raids on Iran's Kharg Island oil terminal.[38] By 1988 a significant element of the Iraqi armed forces had improved their skills—although the nonmechanized elements of the army and the Popular Army still lagged behind. Improvements were the result of experience, realistic training, and, for a few Iraqi pilots, overseas experience. Even after 1986, however, improving the performance of its armed forces was not the only priority for the regime—especially when it involved contact with seditious foreign influences. The sheer quantity of weaponry available allowed for its unskilled, profligate use resulting in stunning inaccuracy. For instance, in the war with Iran, only one in eight Milan and fewer than one in twenty AT-3 Sagger (antitank missiles) had actually hit their targets.[39]

A belated but more important change was the development after 1986 of Iraqi elite forces. Indeed, their very definition of elite forces had undergone a fundamental change. Whereas "elite" had previously been a label denoting political reliability, by 1988 it also implied improved military skills. The elite were encouraged to take tactical initiatives, and the regime moderated its micromanagement of military decisions. The extent to which the elite improved must not be exaggerated, however, a caution that is underlined by the manner in which Iraq attempted through training to "choreograph" precisely its 1988 offensive operations. It is also important to appreciate that not all the Republican Guard of 1991 were this good—the four guards divisions formed (at most) a few months before the 1991 ground war were of a lower caliber. Despite these reservations, the elite constituted a quantum leap in the nature of Iraq's armed forces.

The Iran-Iraq War accelerated the development of Iraq's array of S-S missiles. In August 1987, the Iraqi variant of the Scud-B, the al-Hussein which has a range of six hundred kilometers, was bought into service. The next year, the 750-kilometer range al-Abbas was tested.[40]. These missiles gave Iraq the capability to target Israel, but research to build longer-range weapons with partners such as Egypt, Brazil, and Argentina continued. In 1989, Iraq tested the al-Abid missile, which succeeded in reaching a height of twelve thousand meters before its second and third stages failed.[41] It is important, though, to be skeptical about Iraqi claims concerning their long-range missile program, and in 1991, only the al-Hussein was operational.

The Iraqis' development of CW was the most feared consequence of their war with Iran. During 1985, Iraq created an infrastructure capable of producing large quantities of mustard gas and the nerve agents Sarin and Tabun. By 1989, Iraq produced approximately thirty-five hundred tons of mustard gas per annum, and two thousand tons of nerve gas—a twentyfold expansion since 1985.[42] Iraqi offensives in 1988 were all supported by CW, the mix of persistent with nonpersistent agents demonstrating a developing tactical sophistication. Even in 1988, however, chemical casualties were far less significant than those inflicted by traditional conventional weapons. As late as April, the Iranians claimed to have suffered in the war so far some 260 fatal chemical casualties.[43] Despite relatively primitive Iranian countermeasures, the main impact of Iraqi CW was to force Iranian forces to fight less efficiently and erode their fighting spirit.

For Saddam, CW were "terror" weapons, originally developed to intimidate Iran into ending the war, but it was only in attacks on undefended *Iraqi* civilians, particularly the Kurdish town of Halabja, that CW claimed thousands of fatalities. In 1988, fear of Iraqi CW attack did affect Teheran—pictures of Halabja on Iranian TV and the frequency of al-Hussein strikes created this panic—but this was not the major reason Iran sued for peace. Using a terror weapon devalues its strategic role, and the relatively modest impact of CW on the battlefield undermined the attempt to frighten Iran into ending the war before 1988. The Iranians—unlike many observers in the outside world—did not see CW as the "poor man's atomic bomb."

After the 1991 Gulf War, the inspectors of the United Nations Special Commission (UNSCOM) were able to get a clearer idea of the extent of Iraq's CW program, and Iraq admitted to UNSCOM that it had produced 127,941 filled and unfilled "special" CW munitions.[44] Most of these were designed to be delivered by artillery or from the air, but there were a small number of experimental warheads designed for the al-Hussein missile. It was by no means clear, however, that these warheads would have worked or even if they could have been delivered by the al-Hussein, that they would have been effective. Many Iraqi chemical munitions were of a rather crude design, and leakage made them dangerous to their handlers. The Iraqis do not appear to have succeeded in manufacturing "dusty" mustard (a more dangerous variant) and had clear quality problems with their nerve agents. There is no doubt that Iraq developed an extensive chemical arsenal, but its potential falls well short of that claimed by Saddam or feared by the outside world.

Harder to assess is Iraq's drive to acquire nuclear and biological weapons. The Iraqi nuclear program explored several avenues to develop weapons, supporting this effort with an elaborate clandestine procurement effort. The program was not at the weaponization stage in 1991, and to make it so, Iraq still needed to acquire a large number of finished components from abroad.[45] In 1995, Iraq admitted it had weaponized biological agents and that it had conducted field tests with biological agents since 1987, and in 1990 it had filled twenty-five "special" al-Hussein warheads.[46] It seems these warheads were not specifically designed for a biological payload and that they had not been tested—indications that in 1990 Iraq did not yet have a real biological warfare capability.[47] But in 1990, Iraq initiated a program to produce biological toxins on an industrial scale.[48] Without the interruption of the 1991 Gulf War, Iraq might have developed a small number of nuclear devices and would have acquired a real biological warfare capability.

The importance to the Iraqi regime of their unconventional weapons is that they give a means of making military gestures without incurring the costs of a conventional conflict. Saddam's bellicose language serves to conceal his fears of the consequences of Iraq suffering high military or civilian casualties. Throughout the Iran-Iraq War, Saddam's decisions were often influenced by this fear, and until 1988 it was always Iraq that "blinked first" in the several "wars of the cities." Unconventional weapons fit well into the pre-1980 Iraqi tradition of military gestures, and in many ways, their increasing importance underlines some of Iraq's enduring military weaknesses.

The fundamentals of the relationship between the Iraqi military and the B'athist regime were established in the 1970s, and the Iran-Iraq War did not alter them to any important extent. The armed forces continued to be watched by the various security agencies; there were regular purges of the military; promotion normally went to the politically reliable rather than the competent; and the regime still fostered large armed institutions outside the control of the regular military authorities.

In 1990 the regime was still well insured against its own military. The regime's insistence on creating loyalist military institutions is underlined by important changes in the status of the Republican Guard. The rapid expansion of the guard and the increasing recourse to it in battle against the Iranians devalued it as the regime's "insurance policy." This has led, in effect, to a less-trusted "outer" guard, an "inner" guard consisting of the Baghdadi Division, and, according to some sources, several Special Republican Guard brigades.[49] In addition, armed formations outside the Republican Guard, such as the Amn-al-Khas, and the Haras al Ra'is al-Khas (Saddam's bodyguard), have increased importance. The Amn al-Khas by 1991 controlled the equivalent of a heavy armored division. Since then, Saddam has underlined its importance by making Qusay (his youngest son) its commander.

The important development of the Iraqi elite should not blind us to the military defects of most of the Iraqi army. Ordinary Iraqi formations fought the Iran-Iraq War with little enthusiasm, and it was fear of the regime that ensured they "went through the motions." Iranian attacks normally saw the rapid collapse of ordinary Iraqi formations, and, once freed from the fear of their own secret police, Iraqi soldiers surrendered with alacrity. Most soldiers in Iraqi infantry formations were Shi'a Arabs who, in the main, did not feel so strongly for their coreligionists from Iran that they would rise in revolt against Iraq, but equally, the regime in Baghdad could not inspire a strong fighting spirit in these soldiers.

Since its foundation, the Iraqi state has been unable to command the loyalty of a large proportion of its people. Tribal loyalties, religious divisions, and non-Arab minorities—of which the Kurds are by far the most numerous—all contribute to the lukewarm attitude many Iraqis have toward their state. A minority of Iraqi regimes, notably Qasim's from 1958 to 1963, have stressed "Iraqiness," focusing on social and economic reform. The popularity of the Iraqi Communist Party (ICP) in the early 1960s and the thousands of Iraqis who unsuccessfully fought to defend the Qasim regime demonstrate that their agenda appealed to many Iraqis. But both Qasim and the ICP lost to rivals who placed Arab nationalism at the heart of their ideology. On the whole, Iraqi regimes have seen Arab nationalism as the most suitable ideology for forging an identity as a modern nation-state. Iraq's political elite, however, is mainly drawn from its Sunni Arab citizens, and in Iraq, Arab nationalism has become entwined with Sunni preeminence, diluting its appeal for Iraq's Shi'a Arabs. Understandably, Iraq's Kurds are indifferent or outright hostile to the Arab nationalist ideology. At best, whichever ideology an Iraqi regime has espoused evokes strong support only from a significant minority but inevitably alienates other Iraqis.

Many commentators on Iraqi affairs believed that the B'athist regime had succeeded in imposing a measure of unity on the peoples of Iraq. Up to 1980, the B'athist regime dramatically improved the lot of the ordinary Iraqi. To dynamic modernization and Arab nationalism the Iran-Iraq War seemed to join a new pride in being Iraqi—Shi'a Arabs and many Kurds fighting loyally for their country, culminating in both a victory over Iran and an affirmation of the maturity of the Iraqi state. With the benefit of hindsight, it can be seen that this view was almost totally incorrect.

The Iran-Iraq War was claimed by the B'athist regime as a victory, and just as in previous wars, this "triumph" owed little to events on the battlefield. The majority of Arab states, and the Western world—perhaps because of its relief that the "fundamentalist threat" had diminished—uncritically accepted Saddam's claims. In reality, the war had won Iraq nothing of any value, and unlike earlier Iraqi "triumphs," the costs of surviving the war with Iran were staggering.

In assessing the enthusiasm with which Iraqis supported the war—as with so much of the analysis of the Iran-Iraq War—received wisdom went from one extreme to the other. That Shi'a Arabs did not rise in rebellion in support of Iran does not mean they loyally embraced the Iraqi state. Several thousand Iraqi Shi'a continued their armed opposition to the state during the war, joined by thousands of Iraqi prisoners of war. Even more deserted, finding a temporary refuge in the southern marshlands. If Iran had been able to advance farther into southern Iraq, it seems the best the B'athist regime could hope for from the majority of its Shi'a citizens was acquiescence to (rather than support for) Iranian occupation. Equally, that most Iraqi soldiers did fight does not mean they fought wholeheartedly; in fact, "going through the motions" more accurately describes how ordinary Iraqi soldiers fought the war.

Since 1991, the main divisions within Iraq—Kurds, Sunni Arabs, and Shi'a Arabs—are reasonably well known, but reality is more complex. To take just the Kurds: in the northern "safe haven" are two major armed factions, as well as the Turkish Kurdish Workers Party (PKK); to the south are Kurdish forces loyal to Baghdad. Even these factions, however, understate the fragmentation of loyalty in Kurdish Iraq, and it is to their *Agas* (tribal chiefs) that most of the Kurds in northern Iraq give their allegiance. The KDP, which is the most formidable and best known of the armed opponents of the B'athist regime, illustrates the problems of application of the Western terminology of nationalism and ideology to Iraq. At first glance, the KDP espouses a combination of Kurdish nationalism and Marxism, but it is also the militia of a powerful Kurdish *Aga*. Not only has the KDP fought the Iraqi army, but surprisingly for a supposedly nationalist group, it has fought Iranian Kurds, the PKK, and the Iraqi-based Patriotic Union of Kurdistan (PUK). Indeed, it has been argued that there is no such thing as a Kurdish movement,[50] and it is clear that Iraq's Kurds cannot be seen as a cohesive block. Similarly, the Arabs of Iraq, both Sunni and Shi'a, are divided into smaller groups, whose loyalties are as much parochial and traditional as inspired by larger ideologies that bind states together. These divisions weaken challenges to the regime but also make it impossible to create a large military animated by common purpose and loyalty to Baghdad.

With its financial resources, ruthless totalitarian institutions, and massive external aid, Saddam's regime imposed sufficient control over Iraqis to survive the Iran-Iraq War. The evidence of this war, reinforced by the uprising of 1991 and Baghdad's subsequent vilification of Shi'a Iraqis,[51] is that a new Iraqi nationalism had not been created. The hope of Iraq's first king (Faisal I) that the army would become "the spinal column for forming a nation"[52] has yet to be realized; instead, the Iraqi armed forces reflect the divisions of Iraq. The failure of the Iraqi state to foster a comprehensive Iraqi nationalism is an important, and often ignored, factor holding back the Iraqi armed forces from realizing their potential. By a mixture of patronage, B'athist ideology, and traditional ties of tribe and family, however, the regime created a significant minority—mainly but not exclusively Sunni Arab—who were prepared to fight for the Iraqi B'athist state. The Iran-Iraq War had a dramatic impact on the size and military capabilities of this loyalist minority, and the Iraqi mobile elite became a much more formidable force in terms of commitment to the regime, as well as in its skills and equipment.

Discussion of the 1991 Gulf War can polarize into a debate on whether its outcome is a comment on U.S. military superiority or Iraqi military flaws. Proponents of the first argument focus on the advantages enjoyed by the United States; for instance, America's official accounts do not dwell on Iraqi weaknesses and support the idea that the U.S. armed forced have experienced a recent technology-based "revolution in military affairs." The view that the war was won by the United States, rather than lost by the Iraqis, is reinforced by positive accounts of the Iraqi mobile elite's performance as it opposed the U.S. Seventh Corps' outflanking movement.[53] Despite its merits, this is a dangerously convenient view, validating the massive military expenditure of the United States, enabling analysts to ignore their embarrassing overestimates of the Iraqi military, and playing well to Middle Eastern audiences reluctant to confront another Arab military defeat. Conversely, it can be argued that Iraq's inability to make the best of its military assets was central to the outcome of the 1991 Gulf War. Focusing on Iraq's failings supports a more skeptical view of American military achievements, and it is understandable that several Russian military analysts have taken this view.[54] Stressing Iraqi weaknesses, however, falls into the pattern of oscillating to the opposite extreme in assessing the Iraqi military and risks ignoring significant improvements in their capabilities since 1980.

Iraq did score some symbolic successes, such as their Scud attacks and the Khafji Operation. That these Iraqi "successes" were, in military terms, dismal failures does not deprive them of their political utility. Even an event as revealing of Iraqi weakness as their airforce flying to Iran, could, by wrong-footing the coalition, reap political dividends for the Iraqi regime. The Iraqi war effort aimed for more than symbolic victories, however, and there is little doubt that in 1990–91, the Iraqis displayed capabilities they did not possess in 1980.

The Iraqi plan in 1991 reveals their growing military sophistication. Their strategy was in part mobile: close behind Iraqi fortifications were six divisions for immediate counterattack; a further six divisions were held deeper in Kuwait (one mechanized and five armored divisions); and behind these were Republican Guard armored formations as the theater reserve, capable of deploying anywhere in the Kuwaiti theater of operations within twenty-four hours. These deployments were supported by some well-constructed fortifications, and a system of land lines and microwave relays—a communications system that the coalition airpower was unable to prevent functioning.

The performance of the mobile elite against the coalition demonstrates capabilities developed during the Iran-Iraq War. Three Republican Guard divisions redeployed to face the U.S. Seventh Corps in southern Iraq: the Tawakalna (a mechanized division), Medina (an armored division), and Adnan (essentially a motorized infantry division). Assorted Iraqi army formations, of which the most important was the Twelfth Armored Division, also moved to face the Seventh Corps. It took considerable operational mobility to get these forces to the "right" position, especially in view of coalition aerial supremacy. The Iraqis moved quickly enough to place a coherent force in front of the American advance; indeed, for several hours the U.S. Second Cavalry Regiment had the distinction of being "the only American unit obviously outnumbered and outgunned during the campaign."[55] These Iraqi forces demonstrated a commendable fighting spirit, and individual Iraqis displayed tactical skill. For instance, a few tank commanders kept motionless and "cold," enabling them to engage U.S. armored vehicles from the rear.[56] Finally, in several of these clashes the Iraqis attempted, albeit unsuccessfully, to wrest the initiative from the advancing coalition forces.

Iraqi clashes with Seventh Corps, however, were very one-sided. In the main, coalition armor engaged Iraqi vehicles at ranges considerably beyond Iraqi capabilities. The U.S. Bradley infantry fighting vehicle faced Iraqi T-72s on favorable terms, not to mention the much more formidable M1A1 Abrams tank. Domination of the airspace above the battlefield gave the coalition forces immense advantages in terms of intelligence and firepower support. The AH-64 Apache antiarmor helicopter proved highly effective—for instance, it was Apache attacks that broke the resistance of the Iraqi Tenth Armored Division, losing just one helicopter to enemy action.[57] It is difficult to believe Iraqi military skills could ever compensate for these technological weaknesses.

The battles in southern Iraq suggest that the 1991 Gulf War was more a question of U.S. might than Iraqi weaknesses. Even within this scenario of American strength, however, some revealing Iraqi weaknesses came to light. The Iraqis fought fiercely but ineffectively, partly because they assumed enemy armor would remain stationary, expecting set-piece battles rather than fluid, high-tempo armored engagements. A good example of this was the attempt by two brigades of the Medina to halt the U.S. First Armored Division. The Medina planned to engage from strong defensive positions in rocky terrain, drive the Americans back into a *wadi,* and then pummel them with artillery. Not only was the Medina unable to slow the U.S. advance, however, but, very revealingly, the Iraqi artillery stuck with the plan and plastered the empty wadi.[58]

In any event, the fighting in southern Iraq involved only a minority of Iraqi forces. In contrast to the efforts of the mobile elite, most of the Iraqi forces fought with little enthusiasm. Just as in the Iran-Iraq War, ordinary Iraqi troops surrendered quickly once cut off from their rear—so rapid was this collapse that it is virtually impossible to assess anything other than the fighting spirit of the soldiers. So poor was the quality of resistance of most of the Iraqi armed forces that the coalition plan was unbalanced by the speed with which Kuwait was liberated. Arguably, the mobile elite were militarily competent, but in the case of the bulk of the Iraqi armed forces, it seems likely they could have been defeated by an enemy much less formidable than the United States.

The changes in the Iraqi armed forces since 1980 are of importance. The Iraqi mobile elite developed a style of waging war that required a fundamental change in its nature. Although the bulk of the Iraqi armed forces have not changed in nature from 1980, they are much better armed and still, in the main, firmly under the control of the regime. As a result, the Iraqi armed forces had become capable of winning more than symbolic victories against their neighbors. The decisive speed of their invasion of Kuwait stands in stark contrast to their fumbling, inept performance invading Iran in 1980. The rapidity with which forces loyal to Baghdad put down the post–Gulf War revolt—a revolt that was, by far, the largest uprising ever faced by an Iraqi regime—underlines the significance of these improvements.

There is little doubt that the 1991 Gulf War, and the subsequent rising against Saddam, caused a massive reduction in Iraq's military strength on paper. Immediately after the war the United States Central Command (CENTCOM) estimated Iraq had lost 3,847 tanks and 2,917 artillery pieces.[59] Five years after the Gulf War, Iraq probably had less than half the military equipment and military manpower it had possessed in 1990.[60] Iraq's problems with its military equipment are understated by these figures, because since 1988, Iraq has ceased to be given favorable access to the world arms market. Iraq's isolation has deprived it of sources of sophisticated weaponry; its Arab brothers are no longer bankrolling the Iraqi war machine; and Iraq's state revenues have plummeted. Periodic scares concerning Iraqi weapons of mass destruction make exciting headlines, but UN-imposed inspection regimes have dramatically curtailed Iraq's unconventional weapons programs. Only two of Iraq's Scuds (modified and unmodified) are unaccounted for,[61] and over forty thousand chemical munitions have been destroyed.[62] Even after all of the wartime losses and peacetime restrictions, however, Iraq still possesses one of the largest military establishments of the Middle East.

Important elements of the Iraqi military elite survived the Gulf

War, and their survival proved decisive in the defeat of the subsequent rising against the regime. The most dramatic change in the post-1991 reorganization of the Iraqi armed forces has been in the sharp reduction in infantry divisions and the concentration on rebuilding the Republican Guard. It was the emergence of a military elite, not the well-armed but reluctant mass of the Iraqi armed forces, that made the Iraqi military of 1988 so different from that of 1980. This elite, however, has reverted to its previous role as the guarantor of the regime against internal rather than external enemies, and it has ceased to train for large-scale operations.[63] Indeed, whereas, at the time, attacking Iran in 1980 seemed to offer reasonable prospects of success, in 1999 it is almost impossible to imagine a second such war being contemplated.

In many ways, it was the Iran-Iraq War that was the aberration in Iraqi military history—an aberration that led to the overmighty Iraq of 1990. In contrast, the 1991 Gulf War joins the long list of Iraqi conflicts in the twentieth century hailed by ruling regimes as a triumph—whatever the actual results of the fighting. The current regime cannot address the fundamental weakness that afflicts most of its armed forces—a weakness that stems from both the nature of the state and the B'athist regime. More than any previous Iraqi regime, however, Saddam's is able to attract the dedicated support of an important minority within his armed forces. Iraq might not have been a regional superpower, but in 1990 it possessed a military power its Middle Eastern neighbors feared. If Iraq escapes its current isolation before its military elite atrophies from their excessive internal focus, it would once again be a feared regional military power.

NOTES

1. As can be seen in *Conduct of the Persian Gulf War: Final Report to Congress* (Washington, D.C.: U.S. Department of Defense, 1992), xviii.

2. Brig. Gen. R. H. Scales, *Certain Victory: The US Army in the Gulf* (London: Brassey's, 1994) 112.

3. For instance, near Ahwaz on 3 March 1915, approximately five

thousand tribal cavalry (Bani Turuf and Bani Lam tribes) surprised a British column of under eight hundred men. The Indian army force lost sixty-four dead, the tribal force lost over two hundred—competent conventional forces in such circumstances would have destroyed the column.

4. Scales, *Certain Victory,* 73.

5. P. Marr, *The Modern History of Iraq* (Boulder, Colo.: Westview Press, 1985), 176.

6. Committe Against Repression and for Democratic Rights in Iraq (CARDRI) *Saddam's Iraq* (London: Zed Books, 1986) 208.

7. Majid Khadduri, *Republican Iraq* (London: Oxford University Press, 1969) 18.

8. Samir al-Kabir (Kanan Makiya), *Republic of Fear* (London: Hutchinson Radius, 1989) 22.

9. Marr, *Modern History of Iraq,* 179.

10. Al-Kabir, *Republic of Fear,* 228.

11. Ibid, 42–44.

12. *Revolutionary Iraq 1968–1973: The Political Report Adopted by the Eighth Regional Congress of the Arab B'ath Socialist Party–Iraq* (Baghdad: The Party, 1974) 169.

13. Al-Kabir, *Republic of Fear,* 292–95.

14. Ibid., 28.

15. Ibid., 26

16. CARDRI, *Saddam's Iraq,* 216. In 1976 full B'ath Party membership was ten thousand Iraqis. Clearly, this was a requirement to establish oneself as a supporter of the party.

17. During the 1970s, this was effectively the "cabinet" of the B'athist regime.

18. R. F. Oxenstierna, *Saddam Hussein in the Post-Gulf War: The Phoenix of Iraq* (London: Gulf Centre for Strategic Studies, 1992).

19. J. Pimlott and S. D. Badsey, *The Gulf War Assessed* (London: Arms and Armour, 1992), 24.

20. Marr, *Modern History of Iraq,* 242 and 336.

21. *Revolutionary Iraq 1968–73,* 171.

22. C. Herzog, *The War of Atonement* (London: Weidefield and Nicolson, 1975), 138–39.

23. Marr, *Modern History of Iraq,* 234.

24. These matters are well explored by E Karsh in his 1987 Adelphi

Paper *The Iran Iraq War: A Military Analysis* (London: International Institute for Strategic Studies, 1981). Karsh also argues that the Iraqi limited-war strategy might have succeeded if they had pressed home their initial advantage with greater vigor.

25. A. H. Cordesman and A. B. Wagner, *The Lessons of Modern Warfare: The Iran Iraq War* (Boulder, Colo.: Westview Press, 1990), 81 to 84.

26. Ibid., 87.

27. Ibid., 140.

28. E. Karsh and I. Rautsi, *Saddam Hussein* (London: Brassey's, 1991) 166.

29. Ibid., 167.

30. S. Chubin and C. Tripp *Iran and Iraq at War* (London: Tauris, 1988) 117.

31. Cordesman and Wagner, *Lessons of Modern Warfare,* 354–55.

32. Ibid., 356.

33. See Captain J. M. Warnford, "The Tanks of Baghdad," in *Armor* (November–December 1990).

34. Karsh and Rautsi, *Saddam Hussein,* 192.

35. *The Military Balance 1986/87* (London: International Institute of Strategic Studies, 1986), 96–98.

36. Cordesman and Wagner, *Lessons of Modern Warfare,* 395–96.

37. Mike Gaines, "Paper Tigers," *Flight International* (9–15 January 1991): 20.

38. Cordesman and Wagner, *Lessons of Modern Warfare,* 211–12

39. Ibid., 443.

40. *Conduct of the Persian Gulf War,* 13.

41. A. H. Cordesman, *Iran and Iraq* (Boulder, Colo.: Westview Press, 1994) 237.

42. Ibid., 249–50.

43. Cordesman and Wagner, *Lessons of Modern Warfare,* 516.

44. UN Document S/1999/94 (29 January 1999), 48.

45. International Atomic Energy Agency, *Annual Report* (Vienna: IAEA, 1992).

46. UN Document S/1999/94, 77.

47. Ibid., 81.

48. Ibid., 116.

49. Cordesman, *Iran and Iraq,* 200.

50. S. C. Pelletiere, *The Kurds and Their Agas* Strategic Studies Institute U.S. Army War College, (Carlisle, Penn. 1991), 17.

51. Kanan Makiya (Samir Al-Khalil), *Cruelty and Silence* (London: Penguin Books, 1994) 101–2.

52. CARDRI, *Saddam's Iraq,* 203.

53. See S. Bourque's article "Correcting Myths about the Persian Gulf War," *The Middle East Journal* 51, 4 (Autumn 1997) 566–83.

54. See Stephen J. Blank, *The Soviet Military Views of Operation Desert Storm* (Carlisle, Penn.: Strategic Studies Institute, U.S. Army War College, 1991).

55. *Conduct of the Persian Gulf War,* 279.

56. T. Carhart, *Iron Soldiers* (New York: Pocket Books, 1994) 247–8.

57. *Conduct of the Persian Gulf War,* 670.

58. Carhart, *Iron Soldiers,* 294.

59. *Conduct of the Persian Gulf War,* 294.

60. A. Rathmell, *The Changing Military Balance in the Gulf* (London: Royal United Services Institute Whitehall Paper Series, 1996), 24.

61. UN Document S/1999/94, 14.

62. Ibid., 48.

63. Cordesman, *Iran and Iraq,* 200.

CHAPTER 13

Russia and Warfare in the Postindustrial Age

Michael Orr

The Soviet army was built for warfare in the industrial age, and it was undoubtedly among the most successful of the world's armies during that period. It is no coincidence that Engels and Lenin, the founders of Marxist military theory, were writing as the Industrial Revolution came to its peak. Marxism was a political doctrine of and for the nineteenth century, and it, in turn, ensured that the military doctrine of the Soviet Union remained deeply and uniquely embedded in the assumptions of industrial society. After the debacle of the First World War, the Soviet state strove to adapt the Russian military machine to the conditions of industrial warfare. From the 1920s onward, it became almost impossible to separate the development of the Soviet economy from that of the armed forces; nor was it easy to say where the priority lay. The Soviets believed that, despite the glittering distractions of individual battles and even campaigns of maneuver and strategic brilliance, victory in war was fundamentally determined by attrition between industrial economies. Even the introduction of nuclear weapons did not fundamentally alter this approach, since Soviet nuclear strategy remained in its essence a war-fighting and not a deterrence strategy.[1] While other states built their armed forces to protect their society and economy, during most of its history the Soviet Union developed its society and the economy to support the armed forces. As Chris Donnelly writes,

"The USSR was not merely a state with a military machine, the state was a military machine." In fact, most students of Soviet history concur that the Soviet economy was basically a war economy, a distortion that was probably the major reason for the system's collapse.[2]

Since that collapse, Russia has had to confront a post–cold war world characterized by what can be termed *postindustrial* conflicts. This chapter looks at the two wars in Chechnya as case studies in this confrontation. It takes as its starting point an understanding that the Soviet armed forces were highly successful in waging the industrial warfare for which they were intended. It also maintains that Russia has experienced almost complete continuity in military affairs over the last decade—regardless of the many political changes that have occurred. This continuity is particularly striking in the field of military thought, where there is no perceptible break between the Soviet and modern Russian periods. Some changes have, of course, occurred in recent years, but they have been few in number and may well have been natural developments in Soviet military thinking that would have arisen even if the Soviet Union had survived.[3]

One of the most important features of Soviet-style systems that persists in Russia is the ruling party's direct control of military affairs through the concept of "military doctrine." In Western states there is an effective distinction between defense policy, which is set by the government of the day, and the military's guiding principles, or doctrine, whose development is part of the professional competence of the nation's armed forces. In the Soviet Union, however, doctrine encompassed much more. Indeed, it was defined as nothing less than "the nation's officially accepted system of scientifically founded views on the nature of modern wars and the use of armed forces in them, and also on the requirements arising from these views regarding the country and its armed forces being made ready for war."[4] Significantly, the latest Russian document on military doctrine—now approved by President Vladimir Putin—has a similarly broad scope and depth.[5]

In many respects, the Soviet approach to military doctrine was a source of strength, enabling the state to match the West in military power from a much weaker economic base. Soviet military doctrine produced a vast array of missiles, tanks, guns, and other weapons, thus making the Soviet Union a superpower. By the playing the most significant role in defeating Nazi Germany, the Red Army validated its doctrine and proved the competence of its commanders, and it remained a formidable force for long thereafter. Hence Russia's more recent, apparent decline in military power does not indicate that the Soviet military machine came to resemble a hollow shell, crumbling within. It did not become less powerful but rather, like the battleship in the Second World War, simply less relevant.

The dominance of Soviet military theory came under threat in the late 1970s from two sources. The first was identified at a very early stage as a so-called revolution in military affairs.[6] Although Soviet military theorists were quick to identify how the revolution might change warfare, it soon became obvious the Soviet Union was even more ill prepared for this transformation than Czarist Russia had been for the first industrial revolution. This time the revolution was based on the microchip, and the Soviet Union fell further behind its likely opponents in the field of information technology with every year that passed. A rigidly centralized economy ruled by five-year plans would not cope with a technology that updated and reinvented itself as rapidly as the computer did. Traditional Soviet standards of quality control did not favor the production of the new technology either. After all, tolerances that were acceptable in a T-34 tank would be fatal for a cruise missile. But even more important than the ability to produce the technology was the ability to use it. How could the information revolution be applied in a society in which information was a state monopoly? Potential soldiers in the West would be growing up in societies in which the computer was a household item. Future Soviet recruits lived in a country where even a typewriter had to be registered with the KGB.

This obvious disparity did not go unnoticed in the Soviet Union. In fact, Soviet leaders, who already held a basically "industrial" view of war, now focused too strongly on the technological weaknesses brought them by the information revolution, thus paying scant attention to the human element in war. They also neglected a significant change in the nature of warfare itself. Military operations in the postindustrial age are not, after all, destined to be mere repetitions of previous wars, updated only by the use of "smarter" and "sexier" weapons. In fact, they might not be wars in the traditional sense at all, involving conventional battles and clearly defined victories. Although we should not repeat the errors of those who, a century ago, argued that wars between developed states had become so terrible in their consequences that they effectively abolished themselves, it may be safe to assume that the era of mass mobilization for war is indeed passing.

In any case, the Russian armed forces clearly remain fatally handicapped by the doctrine they have inherited and have tried to preserve, regardless of whether facing "high-tech" wars against developed states or "low-tech" counterinsurgencies and peace-keeping operations. Their senior officers are like their counterparts in Russian mills and factories, surrounded by rusting capital plant and poorly trained workforces, convinced that their problems would be solved if only the state could once again give them the funds for new equipment and a bigger workforce. They have not noticed that there is no longer a market for their product.

This problem becomes clear when one examines the Russian military leadership's attitude to manpower difficulties, which have surfaced with prominence in the Chechen wars. No one doubts that manning has been one of the new Russian armed forces greatest problems, but the problem has been compounded because of an attempt to retain a Soviet system. In 1990, the Soviet Union had a population of over 280 million and maintained armed forces of over 5 million men, of whom at least 3 million were conscripts. The ground forces, the most manpower dependent of the armed services, had an active strength of nearly 2 million men and an order of battle of two hundred–plus divisions. Only about a fifth of these approached full strength; another fifth were at something over half strength; and the rest were basically cadre formations. As in the model developed by the nineteenth-century Prussian army, the aim of conscription was to provide not just a large active army but rather an even more significant mobilization capacity. Today, the Russian population is about 140 million, but because of demographic problems, budget deficiencies, and widespread evasion of military service, the armed forces struggle to maintain a force of 1.2 million men. (In 1999, only 13.8 percent of the relevant age group were actually available for conscription.)[7] The ground forces, with an establishment of three hundred thousand men, until recently had an order of battle of over sixty divisions, but almost all of them were only cadres. As the number of conscripts actually serving has fallen, so the pool of trained reservists is also declining, restricting the possibility of mobilizing cadre formations in emergency.

At the same time, the quality of the conscript intake has declined dramatically since the Soviet era. It is easy for the better educated to avoid military service legally now, and many regions of the Russian Federation will not enforce the conscription laws. Those who do serve are often physically unfit (about a third of those who reach the military commissariats prove to be medically unfit to serve) and poorly educated. But soldiering in the twenty-first century is not about marching and firing a rifle, as it was for most of the world's armies a hundred years ago. Physical fitness is still essential, but mental fitness to operate complicated weapons systems is even more crucial. Increasingly, the armies of developed states are becoming all-professional forces. In 1996, President Boris Yeltsin, seeking reelection at a time when the public's opposition to conscription was growing during the unpopular Chechen war, decreed that the Russian armed forces would be all volunteer by 2000. That commitment was quietly dropped after the election, and there is no sign now that conscription will be abandoned any time in the foreseeable future.

There are two main reasons why the Russian army is unable to change its manning policy. First, its attempts to recruit "contract servicemen" or professional soldiers have not been successful. When contract service was introduced in the mid-1990s, military leaders hoped that most volunteers would come from the ranks of conscripts who would want to stay on after their period of service.

This would provide an intake that already possessed an up-to-date training and who could develop their skills as professionals. In fact, the experience of conscription in the modern Russian army has actually deterred volunteers from reengaging as contract servicemen. Instead, contract servicemen largely fall into two categories. About half are women, generally the wives of serving officers or *praporshchiki* (warrant officers) who cannot find alternative employment in remote garrison areas. They usually serve in medical, communications, or administrative posts. Most of the others are men who return to the armed forces after several years in civilian life. They tend to join because they are unemployed or unemployable. About 20 percent of contract servicemen leave within their first year of service, either because they are rejected as unsuitable or because they choose to serve no longer. Those who remain seek static and safe administrative postings, so that combat units probably have a lower proportion of contract servicemen than the average, even though their need is the greatest.

The second reason for the absence of all-regular manning is that the general staff and the rest of the Russian high command remain committed to maintaining a total war mobilization capability. But while such a stance saved the Soviet Union in the Great Patriotic War, it now appears a source of weakness, not strength. The roots of the traditional Russian mind-set that opposes reform, however, are deep seated. There is the belief that a state with such vast land frontiers can be secure only with a mass army. Although this belief has been questionable since the introduction of nuclear weapons, its advocates justify it with dubious comparisons of manpower-to-frontier ratios in other states. Russian history, moreover, is seen as a sequence of foreign invasions, and a feeling of vulnerability to encircling hostile powers remains. To make matters worse, every addition to NATO's order of battle is assumed to be directed against Russia. Last, possession of large armed forces has always brought Russia international respect and influence, although it is arguable that it has necessarily brought security.

Thus the need for a mass army remains embedded in Russian

doctrine, challenged only from outside the system by what are seen as attempts to destroy the system, not to improve it. Consequently, by the mid-1990s the Russian army had an overlarge, undermanned order of battle and almost no deployable formations. Although the number of formations has been drastically cut since 1997, overall manpower remains too high in comparison with the country's resources. Semi-official studies have suggested that the Russian armed forces need to be halved in size to be supportable. But such suggestions remain anathema to the dinosaurs of the general staff organization-mobilization department, whose head, Lieutenant General V. Putilin, recently dismissed a question about all-regular armed forces on the grounds that contract servicemen cost on average 2.4 times as much as a conscript and such an increase in personnel costs could not be afforded. The possibility of reducing numbers, as other armies have done when moving to an all-volunteer basis, clearly never occurred to him.[8]

As the Chechen experience demonstrates, conscript service on the Russian system actually detracts from combat effectiveness. Most recruits are trained in frontline units, so that training programs become a tedious cycle of repetitive individual and small-unit training, and at any time a fair proportion of soldiers are not fully trained. Force structures are determined by the cadre-mobilization system and by the experience of industrial wars. Because of the manning problem, combined with the structural weakness, force deployments, as in Kosovo in 1999 and in both Chechen wars, have involved improvised or temporary structures. And the new president, Putin, does not appear to be moving in the right direction: among his first decrees on military matters have been attempts to reinstate premilitary training for teenagers and to call up twenty thousand reservists for training. Although the Russian military has frequently complained about the collapse of the premilitary and reserve training programs since the end of the Soviet Union, Putin would have at least offered a glimmer of hope if he had challenged the very need to restore such unpopular elements of the Soviet system in the first place.

The full extent of the inadequacy of traditional Russian doctrine and consequent manning policy to provide for the wars Russia actually has to fight, as opposed to those studied in the military academies, has become vividly clear during the two Chechen wars (1994–96 and 1999 onward). Russian performance in those wars can thus serve as a test of the Russian armed forces' capabilities in postindustrial warfare.

Are the Chechen wars really examples of "postindustrial warfare"? They are certainly not wars between two developed industrial societies, so they are not classically defined industrial wars. All through the industrial age, however, the armies of developed states have at various times fought counterinsurgencies and wars of colonial conquest, either category of which the Chechen operations might fall under. Indeed, the British army in the nineteenth century developed a doctrine of "small wars" because they were such a common feature of its operations. In the standard textbook, Colonel C. E. Callwell's *Small Wars; Their Principles and Their Practice,* they are defined as

> all campaigns other than those where both the opposing sides consist of regular troops. It comprises the expeditions against savages and semi-civilized races by disciplined soldiers, it comprises campaigns undertaken to suppress rebellions and guerrilla warfare in all parts of the world where organized armies are struggling against opponents who will not meet them in the open field.[9]

In the nineteenth century, the Imperial Russian Army had its own experience of such wars, including half a century's campaigning in the Caucasus. Now, once again, "organized armies struggling against opponents who will not meet them in the open field" will, in fact, be a characteristic feature of postindustrial warfare, especially as the technological superiority of leading armies becomes more pronounced.

The Chechen wars are typical of this new type of small war because the rebels, although unable to match the Russians overall in technology, still obtained significant successes by exploiting individual items of modern technology. Some of their weapons have been as modern as the Russians' (and often acquired from Russian sources), and civilian technology has also been pressed into service to gain an advantage. For example, in the first war, rebel communications based on mobile telephones were more reliable than the Russian army's military radios. In the second war, the Russians shut down the North Caucasus cellular phone network at the start of the campaign, but the Chechens continued to use satellite communications. The Chechens also proved more capable in the media war in the first campaign, because they had appreciated its importance, while the Russians were still thinking in terms of a Soviet-style controlled media. This was a major reason for the Russian defeat; having lost the "information war," the army lost the support of the Russian people. Thus the Chechen wars show elements of the "asymmetrical warfare" that is now becoming a popular piece of jargon in Western staff colleges and that is likely to be a feature of postindustrial warfare.

The first failure of the Russian military system in the Chechen wars lay in its lack of doctrine for such operations. The historical experience of Russian insurgency campaigns had been neglected, although it forms a considerable database. The nineteenth-century campaigns in the Caucasus and Central Asia, operations against the Basmatchis in Central Asia in the 1920s, and the pacification of western Ukraine and the Baltic States after 1945 were considered irrelevant to a war against NATO and therefore not studied. They are not even mentioned in the military history textbooks of the military academies. Since 1995, articles on these subjects have been appearing in the Russian military press and may have influenced planning for the second Chechen war, but it is hard to detect a fully-grown doctrine in the literature.

The emergence of such a doctrine has also been hampered by a neglect, even a contempt, for outside experience. Western armies were involved in a series of counterinsurgency campaigns during the decolonization that followed the Second World War. This ensured that their doctrines were never totally focused on the NATO–Warsaw Pact confrontation. Although a few articles considering

Western counterinsurgency campaigning have appeared in the press in recent years, they have almost all been written by civilians in civilian journals. The most glaring example of the Russian military's refusal to broaden its focus is, of course, the neglect of its own experience in the Afghan War of 1979–87. It is not surprising that the Soviet army was not prepared for the war in terms of operational and tactical doctrine, but it is remarkable that it was so slow to adapt to Afghan conditions and so quick to revert to "real soldiering" once the war finished. The Frunze Military Academy did publish one tactical study of the war, based on the experiences of its students, but the lessons learned generally refer to conventional operations as much as insurgency.[10]

This failure of vision was reflected in the Russian army's unpreparedness for the first Chechen war. Because the army considered internal operations to be outside its sphere, the staff had not studied the campaign area to assess its influence on tactics. Even a brief map study would have suggested that mountain warfare and operations in built-up areas were likely to feature. The Soviet army had passed on a doctrine for "operations in special conditions," but it generally acknowledged that those sections of the combat regulations were rarely practiced. The doctrine itself was reasonably sound, if old-fashioned; it emphasized that operations in special conditions required a much more combined arms approach at the lowest levels of command and greater initiative by junior commanders. This was something that never featured in platoon, company, and even battalion training programs. Training for major wars assumed that flexibility and skill in operational art, the Soviet army's main contribution to military science, could compensate for unsophisticated tactics. But in Chechnya, the Russians had to learn again that counterinsurgency and fighting in built up areas are the domain of platoon commanders.

Besides this weakness in tactical doctrine, by not studying their own and other experience of this sort of war, the Russian army failed to understand the importance of political factors in counterinsurgency. Operational and tactical plans were developed without consideration of their political impact, nor was there any realistic political guidance to influence planning. To give two obvious examples, Russian tactics in the battles in Grozny probably caused more casualties to the ethnic Russian population than the Chechen rebels, and the ceasefires announced in Moscow were not implemented on the ground, compromising any hope of negotiations.

Once the Russians began to deploy forces for the first war, the faults in their force structures and their manning system were immediately obvious. Because too many formations had been retained on a limited manpower base, there were almost no deployable formations or units. Even in the North Caucasus Military District, Eighth Army Corps, formerly Eighth Guards Army based in East Germany, could scarcely scrape together four thousand men for a war on its own doorstep. As a result, the Russian army resorted to so-called composite units. The term was used in the Great Patriotic War, when it was necessary to amalgamate units that had suffered such heavy casualties that they were no longer combat effective. By starting the war with such units, the Russian army was admitting that it was not combat effective even in peacetime. The composite regiments and battalions were composed of men from a variety of units, thrown together without time to train together. Men often served in completely different specializations to those they had been trained for; radar operators were expected to become snipers overnight.

It is not surprising that the first Chechen campaign has become a byword for incompetence at every level. In a sense, it hardly mattered that the Russian army had no doctrine for the operation, because they had no time to train their soldiers in any doctrine at all. The supposedly elite units such as the airborne troops may have been better than the average but still met their fair share of setbacks in battle. There is little point in reciting the catalog of disasters; the Russian army failed because it could not produce competent soldiers organized into effective units and led by professionally educated commanders. As one of the first of the nonindustrial wars, it demonstrated that the military's professional standards must be as high or higher in these new "small wars" than in major wars of the past.

The first Chechen war was followed by a round of recrimination and calls for reform. There were personnel changes and the order of battle was drastically cut. Looking back, we can now see that from at least 1997, military leaders preparing for a new Chechen war that was to restore Russia's military reputation. The opportunity to launch the war came in August 1999, and although by March 2000 the generals were claiming victory, fighting continued and the "victorious" army suffered a number of setbacks, with embarrassing casualty rates thereafter.

The move for military reform has been driven by a group of generals associated with the North Caucasus Military District (NCMD) and with the first war. Its leader is undoubtedly general Anatoliy Kvashnin, promoted in 1997 from command of the NCMD to be chief of the general staff, and its diagnosis of the Russian army's illness seem to run on the following lines: In 1996 the army was stabbed in the back by the politicians when it was on the brink of winning the war. The generals were also let down by their troops, but this was largely the fault of the politicians for starving the army of men and funds for many years, making it impossible to form an effective force. There were some failings in operational command, but they were the fault of generals such as then Minister of Defense, General Grachev, who was a Yeltsin crony, not a star of the general staff.

Kvashnin's group planned for change, but with a realistic assessment of what was or was not practical. In fact, it was possible to reduce the number of formations so that those that remained could be properly manned. As many units as possible were nominated as permanent readiness forces, with full wartime equipment sets and at least 80 percent of the wartime manpower establishment. These forces were also to be trained for armed conflicts on Russia's borders. In the event, however, it was only possible to raise ten permanently ready formations, three airborne divisions, and three divisions and four brigades in the ground forces. There was a heavy concentration of these forces in the NCMD. In addition, other military districts managed to create permanently ready regiments within divisions that would otherwise require up to thirty days to mobilize. Thus the basis of a deployable force had been created by 1999.[11]

The training of these units emphasized cooperation with troops from the other "force ministries," such as the internal troops or border guards, an area that was identified as a major weakness in 1994–96. There was an attempt to shift away from the old Soviet-style rigid linear tactics at the lowest levels, with new references to the use of fire and movement in the infantry section. There was also a pause in October, between the first operation in Dagestan and the main campaign in Chechnya, which was devoted to shakedown training at every level of command. The leadership, however, seems to have accepted that, given the poor material of the basic soldier and the inadequacies of junior officer training, it was impossible to approach the standards required in fully professional armies. But these limitations could be by-passed by modifying tactics. If the Russian infantry were not going to perform well in direct combat with Chechen guerrillas, whether in towns or mountains, then the Russians were to avoid that sort of battle and fight with the weapons the rebels could not match, artillery and airpower.

The scale of the Russian application of firepower to the Chechen war can be measured by the admission by Marshal Sergeyev, the minister of defense, in March 2000 that strategic reserves had been reduced by over 30 percent. These were war stocks that had been built up over nearly fifty years for a world war. The tactics were obvious from the start of the advance into Chechnya. Russian forces would surround a target village and attempt to persuade the local leaders to force the rebel garrison to leave. If this failed, the Russians would bombard the village from the air and by artillery until it was safe for Russian occupation. The same policy was applied on a larger scale to Grozny. Commanders on the ground were not subject to artificial deadlines, as had been the practice in the first war. They were allowed to minimize their own casualties even if that meant loss of tempo. Observers have talked about the "Russian steamroller" since 1914, but never with more justification than today.

At the heart of such tactics is a refusal to adapt to the special circumstances of counterinsurgency. The methods might be called "Procrustean tactics," meaning that if the tactics do not fit the situation, one adjusts the situation—not the tactics. In the short term, these methods have been successful, as a flood of refugees has removed much of the guerrillas' local support, thereby draining the water from the fish. But when the refugees return to their ruined homes, will they settle down under Russian rule or nurse their grievances for the next round of conflict? Given Chechnya's history, it is certainly possible that Russia's victory will be not last long.

The operational planning of the second Chechen war clearly reflects an industrial and attritional approach to warfare. The Russian army has effectively reverted to the tactics of middle period of the First World War, which were embodied by the slogan "Artillery conquers, infantry occupy." The occupation phase will require different skills and put the infantry back in the leading role. The Russians reached this stage of the war in 1996, when most of Chechnya was occupied and the war seemed almost over. In reality, however, Russian occupation failed to extend beyond the ground immediately around their blockhouses. The rest of the country belonged to the guerrillas, particularly at night. In August 1996 the rebels struck back and reoccupied Grozny and other major towns. Although the generals claimed they could have counterattacked successfully if the politicians had not insisted on making peace, the fact is that Russia had lost the war because the Russian people would no longer support it.

As the second war reaches this phase of occupation, any weaknesses of morale and leadership will be exposed. What may also be become apparent to the generals is that in modern wars operations cannot be isolated from politics. When the first exchanges between Soviet and Western military academies or staff colleges took place, Soviet officers were astonished that their Western counterparts studied politics and society for much of their time, visiting factories or listening to lectures from politicians. The Soviets, by contrast, devoted two or three years to a detailed and purely technical study of operations. They became very proficient within that enclosed world but never saw it in context.

The first question a Western staff officer asks nowadays when planning a campaign is "What is the end state?" In other words, what is the political situation we hope to achieve at the end of the operation? If that question was asked in the headquarters of the NCMD or in the general staff operations directorate in Moscow in the summer of 1999, the answer seems to have been "The Russian flag flying all over Chechnya and Mr. Putin elected as president." But this answer does not make up a true end state, since it does not include a political settlement in Chechnya. Short of genocide or deportation, the Russian army is going to have to live with the Chechen population and its political leadership. By refusing to deal with Aslan Maskhadov between 1996 and 1999, Moscow undermined the only Chechen leader who could broker a settlement acceptable to both sides. At every turn in the campaign since August, the political consequences of military actions have not been assessed, nor have military actions really been subordinated to political objectives.

In the long term, therefore, the Russian armed forces will have to face the consequences of their refusal to accept that the world has changed and that warfare has inevitably changed also. These changes are partly technological, the aspects of which are studied within the Russian system. But although this may be an age of technology, human factors remain decisive. Only by opening its thinking to this broader range of influences can the Russian military possibly cope with the demands of warfare in the twenty-first century.

NOTES

1. The Soviets believed nuclear weapons might deter war altogether, or they might deter an opponent from using strategic nuclear weapons during a prolonged conventional war. In general, Soviet exercise scenarios seem to have assumed that theater and tactical nuclear weapons would be used in a

war between NATO and the Warsaw Pact. William E. Odom *The Collapse of the Soviet Military* (New Haven: Yale University Press, 1998), 66–71.

2. For the best account of the Soviet military machine, see C. N. Donnelly *Red Banner—The Soviet Military System in Peace and War* (Coulsdon, Surrey: Janes's Information Group, 1988). A valuable account of the machine in the last years of the Soviet Union is found in Odom, *Collapse.*

3. For this reason, the author will spare the reader tedious phrases such as "Soviet-Russian"; when referring specifically to the Soviet or Russian period, the appropriate term will be used. Where, however, there is a continuity between the practices of the Soviet army and the modern Russian army, *Russian* will cover both.

4. *Dictionary of Basic Military Terms* (Moscow, 1965), trans. United States Air Force (Washington, D.C.: U.S. Government Printing Office, 1977), 37.

5. The document describes doctrine as "the aggregate of basic official views on the prevention of wars and armed conflicts, their character and means of conducting them and the organization of the activities of the state, society and the citizen to secure the military security of the Russian Federation and its allies." See "Voyennaya Doktrina Rossiyskoy Federatsii Proyekt," in *Krasnaya Zvezda,* Russia's military newspaper *Red Star* (Moscow, 9 October 1999), 3.

6. On the idea of a revolution in military affairs, which has generated a tremendous amount of discussion in professional military journals over the last decade, see also the above chapters by Chin and McKnight in this book.

7. V. Putilin, "Kto Vstanet Pod Boyevye Znamena," in *Krasnaya Zvezda* (31 March 2000—Internet version).

8. *Krasnaya Zvezda* (16 February 2000), 1.

9. C. E. Callwell, *Small Wars: Their Principles and Their Practice* (London: H. M. Stationery Office, 1906), 21.

10. L. W. Grau, ed. *The Bear Went over the Mountain: Soviet Combat Tactics in Afghanistan* (Washington, D.C.: National Defense University Press, 1996).

11. For a discussion of Russian force structuring and tactics in the second Chechen war, see M. J. Orr, "Russia's Chechen War," *Jane's Defense Weekly* (8 March 2000), 32.

CONTRIBUTORS

✦ ✦ ✦ ✦ ✦ ✦ ✦ ✦ ✦ ✦ ✦ ✦

STEPHEN BADSEY is a senior lecturer in the Department of War Studies of the Royal Military Academy Sandhurst. In the past he has worked for the Imperial War Museum and for the British Broadcasting Corporation. He presently holds a senior research fellowship at De Montfort University and has been a guest lecturer and speaker at numerous universities and other institutions. He has published extensively on the military-media relationship and is a fellow of the Royal Historical Society.

NIALL J. A. BARR is a senior lecturer in the Department of War Studies, Royal Military Academy Sandhurst. He studied for his degree and doctorate at the University of St. Andrews. He is an expert in the development and history of the British veterans' movement and has a deep interest in the military history of both world wars. Having recently worked with J. P. Harris on a collaborative study, *Amiens to the Armistice: The B.E.F. in the Hundred Days Campaign 8 August–11 November 1918* (London: Brassey's, 1998), he is currently researching the Alamein campaign of 1942.

ERIC BOBO is completing his Ph.D. at the University of Southern Mississippi on the development of the jet and radar during World War II. He also works at the Mississippi Department of Archives and History and is adjunct instructor at Copiah-Lincoln Community College.

BRIAN BOND is professor of military history at King's College, London. His publications include *The Pursuit of Victory: From Napoleon to Saddam Hussein* (Oxford: Oxford University Press,

1996) and *War and Society in Europe, 1870–1970* (Oxford: Oxford University Press, 1986).

WARREN CHIN is a senior lecturer in the Defense Studies Department at the British Joint Services Command and Staff College and was formerly a lecturer in the War Studies Department of the Royal Military Academy Sandhurst. He has written articles on technology and war and is currently working on a contemporary history of the British weapons acquisition process.

PADDY GRIFFITH's recent books include *Forward into Battle* (Novato, Calif.: Presidio, 1990); *Battle Tactics on the Western Front 1916–18* (New Haven: Yale University Press 1994); and *The Art of War of Revolutionary France* (London: Greenhill Books, 1998). Now an independent historian, he was a senior lecturer at the Royal Military Academy Sandhurst.

An assistant professor of history at the University of Southern Mississippi, GEOFFREY JENSEN is a former visiting senior lecturer in the Department of War Studies, Royal Military Academy Sandhurst. He has published articles on the military in Spain, and he is author of *Irrational Triumph: Cultural Despair, Military Nationalism, and the Ideological Origins of Franco's Spain,* a book that grew out of his 1995 Yale University Ph.D. dissertation.

CHRIS MCCARTHY works at the Imperial War Museum in London. His publications include books on the World War I battles of Passchendaele and the Somme.

SEAN MCKNIGHT is deputy head of the Department of War Studies at the Royal Military Academy Sandhurst. He has published articles on the Second World War and the Iran-Iraq War, the latter interest arising from his having lived and worked in the Middle East.

ROBERT MCLAIN is currently finishing his Ph.D. at the University of Illinois at Urbana-Champaign. His current research interests include transnational aspects of Indian nationalism and the Indian Army in the Great War.

NEIL R. MCMILLEN, professor of history at the University of Southern Mississippi, is the author of *The Citizens' Councils: Organized Resistance to the Second Reconstruction* and *Dark Journey: Black Mississippians in the Age of Jim Crow,* for which he received the Bancroft Prize. He also edited the recent book *Remaking Dixie: The Impact of World War II on the American South.*

MICHAEL ORR joined the Royal Military Academy Sandhurst to teach military history and war studies in 1969, after reading modern history at Balliol College, Oxford. In 1984 he transferred to the Soviet Studies Research Centre (now Conflict Studies Research Centre, or CSRC) as an analyst of Soviet army tactics. He has lectured widely to British and NATO military audiences and to academic and other civilian institutions. Within CSRC, his research is now concentrated on structural change in the Russian ground forces and Russian peace-support operations. His publications include a study of the Battle of Dettingen and a number of articles on military history, Soviet army tactics, Russian peace-support operations, and Russian military reform.

LESLIE P. ROOT is a clinical psychologist specializing in the assessment and treatment of war-related trauma and post-traumatic stress disorder (PTSD). She has been the director of the Post-Traumatic Stress Disorder Clinical Team at the Veterans Administration Gulf Coast Veteran's Health Care System since 1993. She is active in training mental health professionals in the assessment and treatment of PTSD.

A Vietnam veteran, RAYMOND M. SCURFIELD, D.S.W., is an internationally recognized expert in war-related PTSD. He had a distinguished career in clinical leadership and research with the United States Veterans Administration, has made over 220 presentations, and has authored or edited forty publications, including *Ethnocultural Aspects of PTSD.* He is now an assistant professor in social work at the University of Southern Mississippi–Gulf Coast.

G. D. SHEFFIELD is senior lecturer in War Studies, King's College London, based at Britain's Joint Services Command and Staff College, where he is Land Warfare Historian on the Higher Command and Staff Course. He is also adjunct professor in the Department of History, University of Southern Mississippi.

Andrew Wiest is associate professor of history at the University of Southern Mississippi. His publications include *Passchendaele and the Royal Navy,* several articles on World War I, and *The Illustrated History of the Vietnam War.* A former senior lecturer at the Royal Military Academy Sandhurst, he codirects (with Geoffrey Jensen) the Center for the Study of War and Society at the University of Southern Mississippi.

INDEX